Silver Nano/Microparticles

Silver Nano/Microparticles: Modification and Applications

Special Issue Editors

Bong-Hyun Jun
Won-Yeop Rho

MDPI • Basel • Beijing • Wuhan • Barcelona • Belgrade

MDPI

Special Issue Editors
Bong-Hyun Jun
Konkuk University,
Korea

Won-Yeop Rho
Chonbuk National University,
Korea

Editorial Office
MDPI
St. Alban-Anlage 66
4052 Basel, Switzerland

This is a reprint of articles from the Special Issue published online in the open access journal *International Journal of Molecular Sciences* (ISSN 1422-0067) from 2018 to 2019 (available at: https://www.mdpi.com/journal/ijms/special_issues/silver_nanoparticles)

For citation purposes, cite each article independently as indicated on the article page online and as indicated below:

LastName, A.A.; LastName, B.B.; LastName, C.C. Article Title. *Journal Name* **Year**, *Article Number*, Page Range.

ISBN 978-3-03921-177-7 (Pbk)
ISBN 978-3-03921-178-4 (PDF)

Contents

About the Special Issue Editors

Bong-Hyun Jun received his M.S. and Ph.D. degrees from Seoul National University, School of Chemical and Biological Engineering (2009). He worked at the Seoul National University (2009–2010) and at the University of California, Berkeley (2011–2012). He is now a professor at the Department of Bioscience and Biotechnology, Konkuk University (2013–current). He has been serving as a member of the board of directors of the Korean Society of Industrial and Engineering Chemistry (2015–current) and of the Korean Peptide and Protein Society (2013–current). Prof. Jun's work at Konkuk has been mainly focused on metal- or semiconductor-based optical nanoparticles and their applications.

Won-Yeop Rho received his B.S. in Chemical Engineering from Chonbuk National University and his M.S. and Ph.D. in Chemical Engineering and Chemistry from Seoul National University in 2006 and 2013, respectively. After his postdoctoral experiences at Seoul National University, Chonbuk National University, and Konkuk University, he joined Chonbuk National University in 2018 as a professor at the School of International Engineering and Science. His research focuses on organic/inorganic nanomaterials for solar energy, solar water splitting, and photocatalysis.

International Journal of
Molecular Sciences

|MDPI|

Editorial

Silver Nano/Microparticles: Modification and Applications

Bong-Hyun Jun

Department of Bioscience and Biotechnology, Konkuk University, Seoul 143-701, Korea; bjun@konkuk.ac.kr

Received: 24 May 2019; Accepted: 24 May 2019; Published: 28 May 2019

Nano/micro-size particles are widely applied in various fields. Among the various particles that have been developed, silver particles are among the most important because of their favorable physical, chemical, and biological characteristics [1]. Thus, numerous studies have been conducted to evaluate their properties and utilize them in various applications, such as diagnostics, antibacterial and anticancer therapeutics, and optoelectronics [2–8]. The properties of silver particles are strongly influenced by their size, morphological shape, and surface characteristics, which can be modified by diverse synthetic methods, reducing agents, and stabilizers [9].

This Special Issue provides a range of original contributions detailing the synthesis, modification, properties, and applications of silver materials, particularly in nanomedicine. Nine outstanding papers describing examples of the most recent advances in silver nano/microparticles are included.

Lee et al. comprehensively described the synthesis of silver nanoparticles by various physio-chemical and biological methods and elucidate their unique properties which are useful for applications such as for developing antimicrobial agents, biomedical device coatings, drug delivery carriers, imaging probes, and diagnostic and optoelectronic platforms [10]. The underlying intricate molecular mechanisms behind the plasmonic properties of silver nanoparticles on their structures, potential cytotoxicity, and optoelectronic properties were also discussed.

Several innovative silver-based nanomaterials have been introduced in bio-applications. Kang et al. reported a functionalized β-cyclodextrin (β-CD)-immobilized silver structure as a drug carrier [11]. Synthesized β-CD derivatives, which have beneficial characteristics for drug delivery including hydrophobic interior surfaces, were immobilized on the surface of silver-embedded silica nanoparticle to load doxorubicin (DOX). DOX release and its effects on cancer cell viability were studied. Liu et al. reported polydopamine (PDA)-assisted silver nanoparticle self-assembly on a sericin (SS)/agar film with potential wound dressing applications [12]. They prepared an SS/agar composite film, and then coated PDA on the surface of the film to prepare an antibacterial silver nanoparticle-PDA-SS/agar film, which exhibited excellent and long-lasting antibacterial effects. Radtke et al. studied silver ion release processes and the mechanical properties of surface-modified titanium alloy implants [13]. Dispersed silver nanoparticles on the surface of titanium alloy (Ti_6Al_4V) and titanium alloy modified with a titania nanotube layer (Ti6Al4V/TNT) as substrates were prepared using a novel precursor with the formula $[Ag_5(O_2CC_2F_5)_5(H_2O)_3]$ and may be suitable for constructing implants with long-term antimicrobial activity. The properties of silver nanoparticles have been widely studied, including by surface-enhanced Raman scattering (SERS). Pham et al. reported the control of the silver coating on Raman label-incorporated gold nanoparticles assembled as silica nanoparticles for developing a strong and reliable SERS probe for bio-applications [14]. A SERS-active core Raman labeling compound shell material based on Au–Ag nanoparticles and assembled on silica nanoparticles can be used to solve signal reproducibility issues in SERS.

Humans and the environment are becoming increasingly exposed to silver nanoparticles, raising concerns about their safety. Liao et al. focused on the bactericidal and cytotoxic properties of silver nanoparticles [15]. Silver nanoparticles have been reported to be toxic to several human cell lines. In their paper, the state-of-the-art of applications in antimicrobial textile fabrics, food packaging films, and wound dressings of silver nanoparticles in addition to the bactericidal activity and cytotoxic effect in mammalian cells are presented. Fehaid et al. conducted an in-depth study of the toxicity of the size-dependent effect of silver nanoparticles [16]. Since tumor necrosis factor α (TNFα) is a major cytokine that is highly expressed in many diseased conditions, the size-dependent effect of silver nanoparticles on the TNFα-induced DNA damage response was studied. Yan et al. focused on the impacts of silver nanoparticles on plants [17]. They summarized the uptake, translocation, and accumulation of silver nanoparticles in plants and described the phytotoxicity of silver nanoparticles towards plants at the morphological, physiological, cellular, and molecular levels. The current understanding of the phytotoxicity mechanisms of silver nanoparticles were also discussed.

Silver particles can also be used as ink. Mo et al. summarized silver nanoparticle-based ink with moderate sintering in flexible and printed electronics [18]. They developed methods and mechanisms for preparing silver nanoparticle-based inks that are highly conductive under moderate sintering conditions and applied the ink to a transparent conductive film, thin film transistor, biosensor, radio frequency identification antenna, and stretchable electronics. The authors summarized their perspectives on flexible and printed electronics.

Silver nano/microparticles are emerging for use in next-generation applications in numerous fields including nanomedicine. The potential benefits of using silver as a prominent nanomaterial in the biomedical and industrial sectors have been widely acknowledged. This Special Issue highlights outstanding advances in the development of silver nano/microparticles as well as their modification and applications.

Acknowledgments: This work was supported by Konkuk University in 2018.

References

1. Sun, Y.G.; Xia, Y.N. Shape-controlled synthesis of gold and silver nanoparticles. *Science* **2002**, *298*, 2176–2179. [CrossRef]
2. Campion, A.; Kambhampati, P. Surface-enhanced Raman scattering. *Chem. Soc. Rev.* **1998**, *27*, 241–250. [CrossRef]
3. Hahm, E.; Cha, M.G.; Kang, E.J.; Pham, X.H.; Lee, S.H.; Kim, H.M.; Kim, D.-E.; Lee, Y.-S.; Jeong, D.-H.; Jun, B.-H. Multilayer Ag-Embedded Silica Nanostructure as a Surface-Enhanced Raman Scattering-Based Chemical Sensor with Dual-Function Internal Standards. *Acs Appli. Mater. Interfaces* **2018**, *10*, 40748–40755. [CrossRef] [PubMed]
4. Kim, H.M.; Kim, D.M.; Jeong, C.; Park, S.Y.; Cha, M.G.; Ha, Y.; Jang, D.; Kyeong, S.; Pham, X.H.; Hahm, E.; et al. Assembly of Plasmonic and Magnetic Nanoparticles with Fluorescent Silica Shell Layer for Tri-functional SERS-Magnetic-Fluorescence Probes and Its Bioapplications. *Sci. Rep.* **2018**, *8*, 10. [CrossRef] [PubMed]
5. Pham, X.H.; Hahm, E.; Kim, T.H.; Kim, H.M.; Lee, S.H.; Lee, Y.S.; Jeong, D.H.; Jun, B.H. Enzyme-catalyzed Ag Growth on Au Nanoparticle-assembled Structure for Highly Sensitive Colorimetric Immunoassay. *Sci. Rep.* **2018**, *8*, 7. [CrossRef] [PubMed]
6. Rho, W.Y.; Kim, H.S.; Chung, W.J.; Suh, J.S.; Jun, B.H.; Hahn, Y.B. Enhancement of power conversion efficiency with TiO^2 nanoparticles/nanotubes-silver nanoparticles composites in dye-sensitized solar cells. *Appl. Surf. Sci.* **2018**, *429*, 23–28. [CrossRef]
7. Cha, M.G.; Kim, H.M.; Kang, Y.L.; Lee, M.; Kang, H.; Kim, J.; Pham, X.H.; Kim, T.H.; Hahm, E.; Lee, Y.S.; et al. Thin silica shell coated Ag assembled nanostructures for expanding generality of SERS analytes. *PloS ONE* **2017**, *12*, 13. [CrossRef] [PubMed]
8. Pham, X.H.; Shim, S.; Kim, T.H.; Hahm, E.; Kim, H.M.; Rho, W.Y.; Jeong, D.H.; Lee, Y.-S.; Jun, B.H. Glucose Detection Using 4-mercaptophenyl Boronic Acid-incorporated Silver Nanoparticles-embedded Silica-coated Graphene Oxide as a SERS Substrate. *Biochip J.* **2017**, *11*, 46–56. [CrossRef]
9. Tao, A.R.; Habas, S.; Yang, P.D. Shape control of colloidal metal nanocrystals. *Small* **2008**, *4*, 310–325. [CrossRef]
10. Lee, S.H.; Jun, B.-H. Silver Nanoparticles: Synthesis and Application for Nanomedicine. *Int. J. Mol. Sci.* **2019**, *20*, 865. [CrossRef] [PubMed]

11. Kang, E.J.; Baek, Y.M.; Hahm, E.; Lee, S.H.; Pham, X.H.; Noh, M.S.; Kim, D.E.; Jun, B.H. Functionalized β-Cyclodextrin Immobilized on Ag-Embedded Silica Nanoparticles as a Drug Carrier. *Int. J. Mol. Sci.* **2019**, *20*, 315. [CrossRef] [PubMed]
12. Liu, L.; Cai, R.; Wang, Y.; Tao, G.; Ai, L.; Wang, P.; Yang, M.; Zuo, H.; Zhao, P.; He, H. Polydopamine-Assisted Silver Nanoparticle Self-Assembly on Sericin/Agar Film for Potential Wound Dressing Application. *Int. J. Mol. Sci.* **2018**, *19*, 2875. [CrossRef] [PubMed]
13. Radtke, A.; Grodzicka, M.; Ehlert, M.; Muzioł, T.; Szkodo, M.; Bartmański, M.; Piszczek, P. Studies on Silver Ions Releasing Processes and Mechanical Properties of Surface-Modified Titanium Alloy. Implants. *Int. J. Mol. Sci.* **2018**, *19*, 3962. [CrossRef] [PubMed]
14. Pham, X.-H.; Hahm, E.; Kang, E.; Son, B.S.; Ha, Y.; Kim, H.M.; Jeong, D.H.; Jun, B.H. Control. of Silver Coating on Raman Label. Incorporated Gold Nanoparticles Assembled Silica Nanoparticles. *Int. J. Mol. Sci.* **2019**, *20*, 1258. [CrossRef] [PubMed]
15. Liao, C.; Li, Y.; Tjong, S.C. Bactericidal and Cytotoxic Properties of Silver Nanoparticles. *Int. J. Mol. Sci.* **2019**, *20*, 449. [CrossRef] [PubMed]
16. Fehaid, A.; Taniguchi, A. Size-Dependent Effect of Silver Nanoparticles on the Tumor Necrosis Factor α-Induced DNA Damage Response. *Int. J. Mol. Sci.* **2019**, *20*, 1038. [CrossRef] [PubMed]
17. Yan, A.; Chen, Z. Impacts of Silver Nanoparticles on Plants: A Focus on the Phytotoxicity and Underlying Mechanism. *Int. J. Mol. Sci.* **2019**, *20*, 1003. [CrossRef] [PubMed]
18. Mo, L.; Guo, Z.; Yang, L.; Zhang, Q.; Fang, Y.; Xin, Z.; Chen, Z.; Hu, K.; Han, L.; Li, L. Silver Nanoparticles Based Ink with Moderate Sintering in Flexible and Printed Electronics. *Int. J. Mol. Sci.* **2019**, *20*, 2124. [CrossRef] [PubMed]

International Journal of
Molecular Sciences

MDPI

Review

Silver Nanoparticles: Synthesis and Application for Nanomedicine

Sang Hun Lee [1] and Bong-Hyun Jun [2,*]

[1]   Department of Bioengineering, University of California Berkeley, Berkeley, CA 94720, USA;
     shlee.ucb@gmail.com
[2]   Department of Bioscience and Biotechnology, Konkuk University, 1 Hwayang-dong, Gwanjin-gu,
     Seoul 143-701, Korea
*   Correspondence: bjun@konkuk.ac.kr; Tel.: +82-2-450-0521

Received: 30 January 2019; Accepted: 15 February 2019; Published: 17 February 2019

Abstract: Over the past few decades, metal nanoparticles less than 100 nm in diameter have made a substantial impact across diverse biomedical applications, such as diagnostic and medical devices, for personalized healthcare practice. In particular, silver nanoparticles (AgNPs) have great potential in a broad range of applications as antimicrobial agents, biomedical device coatings, drug-delivery carriers, imaging probes, and diagnostic and optoelectronic platforms, since they have discrete physical and optical properties and biochemical functionality tailored by diverse size- and shape-controlled AgNPs. In this review, we aimed to present major routes of synthesis of AgNPs, including physical, chemical, and biological synthesis processes, along with discrete physiochemical characteristics of AgNPs. We also discuss the underlying intricate molecular mechanisms behind their plasmonic properties on mono/bimetallic structures, potential cellular/microbial cytotoxicity, and optoelectronic property. Lastly, we conclude this review with a summary of current applications of AgNPs in nanoscience and nanomedicine and discuss their future perspectives in these areas.

Keywords: silver nanomaterial; synthesis; characterization; mechanism; cytotoxicity; nanomedicine; diagnostics; optoelectronics

---

1. Introduction

Metal nanoparticles have been used in a wide-ranging application in various fields. Specifically, as shapes, sizes, and compositions of metallic nanomaterials are significantly linked to their physical, chemical, and optical properties, technologies based on nanoscale materials have been exploited in a variety of fields from chemistry to medicine [1–3]. Recently, silver nanoparticles (AgNPs) have been investigated extensively due to their superior physical, chemical, and biological characteristics, and their superiority stems mainly from the size, shape, composition, crystallinity, and structure of AgNPs compared to their bulk forms [4–8]. Efforts have been made to explore their attractive properties and utilize them in practical applications, such as anti-bacterial and anti-cancer therapeutics [9], diagnostics and optoelectronics [10–12], water disinfection [13], and other clinical/pharmaceutical applications [14]. Silver has fascinating material properties and is a low-cost and abundant natural resource, yet the use of silver-based nanomaterials has been limited due to their instability, such as the oxidation in an oxygen-containing fluid [15]. AgNPs, therefore, have an unrealized potential compared to relatively stable gold nanoparticles (AuNPs) [6]. Previous discoveries have shown that the physical, optical, and catalytic properties of AgNPs are strongly influenced by their size, distribution, morphological shape, and surface properties which can be modified by diverse synthetic methods, reducing agents and stabilizers [8,16]. The size of AgNPs can be adjusted according to a specific application—e.g., AgNPs prepared for drug delivery are mostly greater than 100 nm to accommodate for the quantity of drug to be delivered. With different surface properties, AgNPs can also be formed into various shapes, including rod, triangle, round, octahedral, polyhedral, etc [17]. Moreover, AgNPs are used in antimicrobial applications with proven antimicrobial characteristics of Ag^+ ions. These exceptional properties of AgNPs have enabled their use in the fields of nanomedicine, pharmacy, biosensing, and biomedical engineering.

In this review, we present a comprehensive and contemporaneous view of the synthesis of AgNPs by various physio-chemical and biological methods, as well as the mechanism of action based on their unique properties. In addition, the review focuses on the characteristics of the optical and physio-chemical properties of AgNPs. Further, insights into understanding how various factors affect these distinct characteristics are discussed. Finally, promising applications of AgNPs in the biomedical field from nanomedicine to optoelectronics, including their anti-cancer or anti-bacterial activity, are presented.

2. Synthesis and Characterization of AgNPs

2.1. Synthesis of AgNPs via Top-Down and Bottom-Up Methods

As mentioned above, numerous types of silver nanostructures with distinctive properties have been used in various biomedical fields [18]. In particular, silver nanomaterials of varying sizes and shapes have been utilized in a broad range of applications and medical equipment, such as electronic devices, paints, coatings, soaps, detergents, bandages, etc [19]. Specific physical, optical, and chemical properties of silver nanomaterials are, therefore, crucial factors in optimizing their use in these applications. In this regard, the following details of the materials are important to consider in their synthesis: surface property, size distribution, apparent morphology, particle composition, dissolution rate (i.e., reactivity in solution and efficiency of ion release), and types of reducing and capping agents used. The synthesis methods of metal NPs are mainly divided into top-down and bottom-up approaches as shown in Figure 1A. The top-down approach disincorporates bulk materials to generate the required nanostructures, while the bottom-up method assembles single atoms and molecules into larger nanostructures to generate nano-sized materials [20]. Nowadays the synthetic approaches are categorized into physical, chemical, and biological green syntheses. The physical and chemical syntheses tend to be more labor-intensive and hazardous, compared to the biological synthesis of AgNPs which exhibits attractive properties, such as high yield, solubility, and stability [14]. The following sections discuss diverse synthesis methods in detail, from the synthesis of spherical AgNPs to shape-controlled Ag colloids, as well as how size-controlled AgNPs are synthesized. The sections also aim to introduce various routes of synthesis and their mechanisms, elucidating how shape- and size-controlled synthesis of AgNPs can be achieved through appropriate selection of energy source, precursor chemicals, reducing and capping agent, as well as concentration and molar ratio of chemicals.

2.2. Physical Method

The physical synthesis of AgNP includes the evaporation–condensation approach and the laser ablation technique [21,22]. Both approaches are able to synthesize large quantities of AgNPs with high purity without the use of chemicals that release toxic substances and jeopardize human health and environment. However, agglomeration is often a great challenge because capping agents are not used. In addition, both approaches consume greater power and require relatively longer duration of synthesis and complex equipment, all of which increase their operating cost.

The evaporation–condensation technique typically uses a gas phase route that utilizes a tube furnace to synthesize nanospheres at atmospheric pressure. Various nanospheres, using numerous materials, such as Au, Ag, and PbS, have been synthesized by this technique [23]. The center of the tube furnace contains a vessel carrying a base metal source which is evaporated into the carrier gas, allowing the final synthesis of NPs. The size, shape, and yield of the NPs can be controlled by changing the design of reaction facilities. Nevertheless, the synthesis of AgNPs by evaporation–condensation through the tube furnace has numerous drawbacks. The tube furnace occupies a large space, consumes high energy elevating the surrounding temperature of the metal source, and requiresa a longer duration to maintain its thermal stability. To overcome these disadvantages, Jung et al. demonstrated that a ceramic heater can be utilized efficiently in the synthesis of AgNPs with high concentration [24].

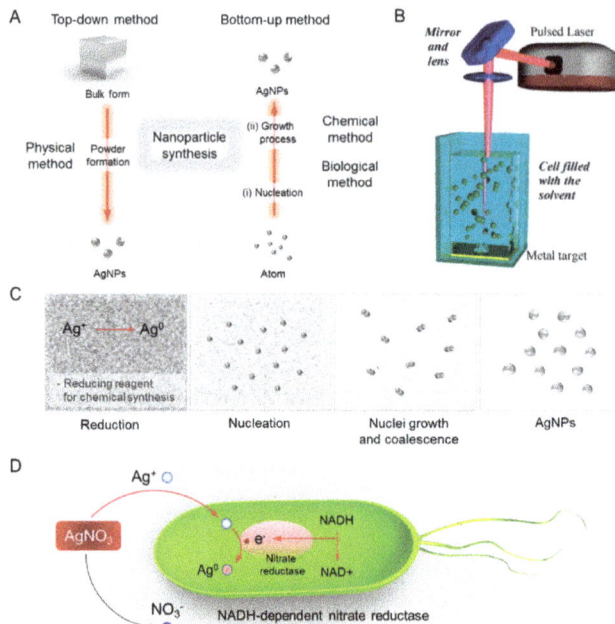

Figure 1. Diverse synthesis routes of silver nanoparticles (AgNPs). (**A**) Top-down and bottom-up methods. (**B**) Physical synthesis method. Reprinted with permission from [21]. Copyright 2009 Royal Chemical Society. (**C**) Chemical synthesis method. (**D**) Plausible synthesis mechanisms of green chemistry. The bioreduction is initiated by the electron transfer through nicotinamide adenine dinucleotide (NADH)-dependent reductase as an electron carrier to form NAD$^+$. The resulting electrons are obtained by Ag$^+$ ions which are reduced to elemental AgNPs.

Another approach in physical synthesis is through laser ablation. The AgNPs can be synthesized by laser ablation of a bulk metal source placed in a liquid environment as shown in Figure 1B. After irradiating with a pulsed laser, the liquid environment only contains the AgNPs of the base metal source, cleared from other ions, compounds or reducing agents [25]. Various parameters, such as laser power, duration of irradiation, type of base metal source, and property of liquid media, influence the characteristics of the metal NPs formed. Unlike chemical synthesis, the synthesis of NPs by laser ablation is pure and uncontaminated, as this method uses mild surfactants in the solvent without involving any other chemical reagents [20].

2.3. Chemical/Photochemical Methods

Chemical synthesis methods have been commonly applied in the synthesis of metallic NPs as a colloidal dispersion in aqueous solution or organic solvent by reducing their metal salts. Various metallic salts are used to fabricate corresponding metal nanospheres, such as gold, silver, iron, zinc oxide, copper, palladium, platinum, etc. [26]. In addition, reducing and capping agents can easily be changed or modified to achieve desired characteristics of AgNPs in terms of size distribution, shape, and dispersion rate [27]. The AgNPs are chemically synthesized mainly through the Brust–Schiffrin synthesis (BSS) or the Turkevich method [20,28–30]. The strength and type of reducing agents and stabilizers should be taken into consideration in synthesizing metal NPs of a specific shape, size, and with various optical properties. More importantly, as stabilizing agents are typically used to avoid aggregation of these NPs, the following factors need to be considered for the safety and effectiveness of the method: choice of solvent medium; use of environment-friendly reducing agent; and selection of relatively non-toxic substances.

Nucleation and growth of NPs are governed by various reaction parameters, including reaction temperature, pH, concentration, type of precursor, reducing and stabilizing agents, and molar ratio of surfactant/stabilizer and precursor [31]. The chemical reduction of these metal salts can be accomplished by various chemical reductants, including glucose ($C_6H_{12}O_6$), hydrazine (N_2H_4), hydrazine hydrate, ascorbate ($C_6H_7NaO_6$), ethylene glycol ($C_2H_6O_2$), N-dimethylformamide (DMF), hydrogen, dextrose, ascorbate, citrate (Turkevich method), and sodium borohydride (BSS method) [32,33]. Brust and co-workers have invented the most widely used synthesis method in producing thiol-stabilized AuNPs and AgNPs [30]. As shown in Figure 1C, silver ion (Ag^+) is reduced in aqueous solution, receiving an electron from a reducing agent to switch from a positive valence into a zero-valent state (Ag^0), followed by nucleation and growth. This leads to coarse agglomeration into oligomeric clusters to yield colloidal AgNPs. Previous studies using a strong reductant (i.e., borohydride) have demonstrated the synthesis of small monodispersed colloids, but it was found to be difficult to control the generation of larger-sized AgNPs. Utilizing a weaker reductant, such as citrate, resulted in a slower reduction rate, which was more conducive to controlling the shape and size distribution of NPs [34].

Stabilizing dispersive NPs during a course of AgNP synthesis is critical. The most common strategy is to use stabilizing agents that can be absorbed onto the surface of AgNPs, avoiding their agglomeration [35]. To stabilize and to avoid agglomeration and oxidation of NPs, capping agents/surfactants can be used, such as chitosan, oleylamine gluconic acid, cellulose or polymers, such as poly N-vinyl-2-pyrrolidone (PVP), polyethylene glycol (PEG), polymethacrylic acid (PMAA) and polymethylmethacrylate (PMMA) [27]. Stabilization via capping agents can be achieved either through electrostatic or steric repulsion. For instance, electrostatic stabilization is usually achieved through anionic species, such as citrate, halides, carboxylates or polyoxoanions that adsorb or interact with AgNPs to impart a negative charge on the surface of AgNPs. Therefore, the surface charge of AgNPs can be controlled by coating the particles with citrate ions to provide a strong negative charge. Compared to using citrate ions, using branched polyethyleneimine (PEI) creates an amine-functionalized surface with a highly positive charge. Other capping agents also provide additional functionality. Polyethylene glycol (PEG)-coated nanoparticles exhibit good stability in highly concentrated salt solutions, while lipoic acid-coated particles with carboxyl groups can be used for bioconjugation.

On the other hand, steric stabilization can be achieved by the interaction of NPs with bulky groups, such as organic polymers and alkylammonium cation that prevent aggregation through steric repulsion. For instance, Oliveira et al. described a Brust synthesis-modified procedure for dodecanethiol-capped AgNPs, wherein dodecanethiol could bind onto the surface of nanoparticles and exhibited high solubility without their aggregation in aqueous solution [36]. A phase transfer of a Au^{3+} complex can be carried out from aqueous to organic solution in a two-phase liquid–liquid system, then the complex can be reduced with sodium borohydride ($NaBH_4$) along with dodecanethiol as a stabilization agent. The authors demonstrated that small alterations in parameters can lead to dramatic modifications in the structure, average size, and size distribution of the nanoparticles as well as their stability and self-assembly patterns [16].

Next, the surface of AgNPs conjugated with biomolecules, such as DNA probes, peptides or antibodies, can be used as a target for specific cells or cellular components. Attaching biomolecules to AgNPs can be achieved, for instance, by physisorption onto the surface of NPs or through covalent coupling by ethyl(dimethylaminopropyl) carbodiimide (EDC) to link free amines on antibodies to carboxyl groups. The photochemical synthesis method also offers a reasonable potential for the synthesis of shape- and size-controlled AgNPs although multiple synthesis steps may be required.

Ag nanoprisms can be synthesized by irradiating Ag seed solution with a light at a selected wavelength. Commonly, the synthesis of bipyramids, nanodiscs, nanorods, and nano-decahedron involves a two-step process. Ag seeds prepared in the first step are subsequently grown in the second step by using an appropriate growth solution, by selecting a specific wavelength of light for irradiation, or by adjusting the duration of microwave irradiation. To synthesize distinctively shaped AgNPs, selective adsorption of surfactants/stabilizers to specific crystal facets needs to be controlled, since surfactants/stabilizers can guide growth along a specific crystal axis, generating varied shapes of AgNPs. The absorbance spectra of AgNPs have been reported to reflect changes in the shape of AgNPs. Such changes in UV–Vis–NIR spectra were illustrated during the photochemical synthesis of Ag nanoprisms grown by illuminating small silver NP seeds (λ_{max} of 397 nm) with low intensity LED [32]. As the seeds were converted to nanoprisms, the peak wavelength at 397 nm decreased over time, and new peaks appeared at 1330 nm and 890 nm, representing a localized surface plasmon resonance (LSPR) of the nanoprisms. For instance, the Mirkin group have investigated photo-induced conversion of spherical AgNPs to triangular prisms. Spontaneous oxidative dissolution of small Ag particles enabled the production of Ag^+ ions that could subsequently be reduced on the surface of Ag particles by citrate under visible light irradiation [37].

2.4. Green Chemistry

Recently, the biogenic (green chemistry) metal NP synthesis method that employs biological entities, such as microorganisms and plant extracts, has been suggested as a valuable alternative to other synthesis routes as illustrated in Figure 1D [5,38,39]. It is known that microorganisms, such as bacteria and fungi, play a vital role in remediation of toxic materials by reducing metal ions [40,41]. Quite a few bacteria have shown the potential to synthesize AgNPs intracellularly, wherein intracellular components serve as both reducing and stabilizing agents [42]. The green synthesis of AgNPs with naturally occurring reducing agents could be a promising method to replace more complex physiochemical syntheses since the green synthesis is free from toxic chemicals and hazardous byproducts and instead involves natural capping agents for the stabilization of AgNPs [16].

A plausible mechanism of AgNP formation by the green synthesis was explored in the biological system of a fungus, *Verticillium species* [43,44]. The main hypothesis was that AgNPs are formed underneath the surface of the cell wall, not in the aqueous solution. Ag^+ ions are trapped on the surface of the fugal cells due to the electrostatic interaction between Ag^+ ions and negatively-charged carboxylate groups of the enzyme. Then, as intracellular reduction of Ag^+ ions occurs in the cell wall, Ag nuclei are formed, which subsequently expand by further reduction of Ag^+ ions. The result of transmission electron microscopy (TEM) analysis indicated that AgNPs were formed in cytoplasmic space due to the bioreduction of the Ag^+ ions [45], yielding a particle size of 25 $\pm$ 12 nm in diameter. Interestingly, the fungal cells continued to proliferate after the biosynthesis of AgNPs. Bacteria commonly use nitrate as a major source of nitrogen, whereby nitrate is converted to nitrite by nitrate reductase, utilizing the reducing power of a reduced form of nicotinamide adenine dinucleotide (NADH). Bacterial metabolic processes of utilizing nitrate, namely reducing nitrate to nitrile and ammonium, could be exploited in bioreduction of Ag^+ ions by an intracellular electron donor [46]. In fact, the utilization of nitrate reductase as a reducing agent is found to play a key role in the bioreduction of Ag^+ ions [47]. For instance, Kumar and colleagues have demonstrated a rationale of an in vitro enzymatic strategy for the synthesis of AgNPs, based on α-NADPH-dependent nitrate reductase and phytochelatin [48]. Nitrate reductase purified from a fungus, *Fusarium oxysporum*, was used in vitro in the presence of a co-factor, α-NADPH. The process of AgNPs formation required the reduction of α-NADPH to α-$NADP^+$. Hydroxyquinoline probably acted as an electron shuttle, transferring electrons generated during the reduction of nitrate to allow conversion of Ag^{2+} ions to Ag. As the Ag^+ ions were reduced in the presence of nitrate reductase, a stable silver hydrosol (10–25 nm) was formed and subsequently stabilized by capping peptide. Similarly, AgNPs have been synthesized in various shapes using naturally occurring reducing agents (i.e., supernatants) in *Bacillus* species [49]. In *Bacillus licheniformis*, it was demonstrated that electrons released from NADH were able to drive the reduction of Ag$^+$ ions to Ag^0 and led to the formation of AgNPs. Li et al. also showed the synthesis of AgNPs by reductase enzymes secreted from a fungus, *Aspergillus terreus*, based on a similar NADH-mediated mechanism [50]. The synthesized AgNPs were polydispersed nanospheres ranging from 1 to 20 nm in diameter and exhibited antimicrobial potential to various pathogenic bacteria and fungi. In another example, *Pseudomonas stuzeri* isolated from a silver mine was used for the synthesis of AgNPs in aqueous AgNO$_3$ [51]. The synthesized AgNPs exhibited a well-defined size and distinct morphology within the periplasmic space of the bacteria.

3. Characterization and Property of AgNPs

3.1. Plasmonic Properties

In many applications, surface chemistry, morphology, and optical properties associated with each NP variant require a careful selection to acquire the desired functionality of nanomaterials. In particular, corresponding reaction conditions during the synthesis of silver nanomaterials can be tuned to produce colloidal AgNPs with various morphologies, including monodisperse nanospheres, triangular nanoprisms, nanoplates, nanocubes, nanowires, and nanorods (Figure 2). Nowadays, since the most commonly used Ag and Au nanospheres are isotropic, they are widely utilized nanostructures for nanoantenna, capitalizing the LSPR phenomena caused by the collective oscillation of electrons in a specific vibrational mode at the conduction band near the particle surface in response to light. The optical properties can be varied by changing the composition, size, and shape of NPs which can affect the collective oscillation of free electrons in metallic NPs at their LSPR wavelengths when irradiated with resonant light over most visible and near-infrared regions [52,53]. Endowed with the tunable optical response, the NPs can be utilized as highly bright reporter molecules, efficient thermal absorbers, and nanoscale antenna, all through amplifying the strength of a local electromagnetic field to detect changes in the environment. The shape of silver nanoprisms has a specific peak wavelength that ranges from 400 to 850 nm as a surface plasmon resonance (SPR) band as shown in Figure 3A [54,55]. The SPR band or absorption spectra for nanoprisms can be measured by the UV–VIS spectroscopy, whereby the λ_{max} reflects an alteration in the size, shape, and the scattering color of AgNPs (Figure 3B) [56]. The optical properties of AgNPs have been of particular interest due to the strong coupling of AgNPs to specific wavelengths of incident light. Ag nanospheres are known to have rather short LSPR wavelengths in the violet and blue regions of the visible spectrum.

AgNPs can be utilized in bio-sensing by single nanoparticle spectroscopy, such as dark-field microscopy. Alivisatos and his co-worker described 'plasmon rulers' to monitor distances between two distinct nanoparticles [57]. The distance can be determined by plasmonic coupling of two nanospheres modified at two ends of a single-stranded DNA (ssDNA) probe with biotin on one end and streptavidin on the other end. The authors demonstrated the plasmonic coupling between single pairs of silver and gold nanoparticles to measure the DNA length and tracked the hybridization kinetics over 3000 s. The plasmonic coupling between two distinct nanoparticles led to more pronounced spectral changes based on the dimerization of single nanoparticles. For example, DNA hybridization was responsible for the observed blue-shift in the spectra via an increase in steric repulsion or for the observed drastic red-shift via aggregation, such as DNA wrapping around DNA-binding dendrimers. In addition, the distance between the AuNPs was adjusted by controlling the length of ssDNA and by changing the ionic strength of the buffer. The maximum plasmon resonance (LSPR) shifted to the red region at high salt concentrations (0.1 M NaCl), indicating a decreased distance between the two AuNPs due to the reduced electrostatic repulsion of the particles at high ionic strength environments. Conversely, low salt concentrations (0.005 M NaCl) increased electrostatic repulsions and led to the blue shift of the maximum LSPR. Along with these results, the hybridization of complementary DNA also resulted in a significant blue shift, which is expected considering that the structural property of double-stranded DNA (dsDNA) shows greater stiffness than ssDNA, and hence allowing it to repulse two AuNPs.

Figure 2. Representative images of electron microscopy of synthesized Ag nanostructures, demonstrating that diverse sizes and morphologies are made possible by controlling the reaction chemistry. (**A**) Silver nanosphere [58], (**B**) Silver necklaces [59], (**C**) Silver nanobars [60], (**D**) Silver nanocubes [7], (**E**) Silver nanoprism [61], (**F**) Silver bipyramids [62], (**G**) Silver nanostar [63], (**H**) Silver nanowire [58], (I) Silver nanoparticle embedded silica particle [64]. All figures were reprinted with permission from the publisher of each article.

Several reports have demonstrated that AgNPs absorb electromagnetic radiation in the visible range from 380 to 450 nm, which is known as the excitation of LSPR. The optical properties of AgNPs of different sizes by gallic acid using biological synthesis methods were characterized by Park and colleagues [65]. The authors demonstrated that spherical AgNPs of 7 nm have SPR at 410 nm, while those of 29 nm have 425 nm. In addition, 89 nm-sized AgNPs exhibited a wider band with a maximum resonance at 490 nm. It was noticed that the width of the SPR band was related to the size distributions of NPs. For instance, Lee et al. investigated the dependence in the sensitivity of SPR responses (frequency and bandwidth) that enabled NPs to recognize the changes in their surrounding environment. They also demonstrated how the optical scattering of Au or Ag nanorods with diverse sizes and shapes can affect total extinction [66]. Greater enhancement in the magnitude and sharpness of the plasmon resonance band was observed in nanorods with higher Ag concentration, which could contribute to superior sensing resolution even with a similar plasmon response. As such, Ag nanorods have an additional advantage as better scatterers when compared to Au nanorods.

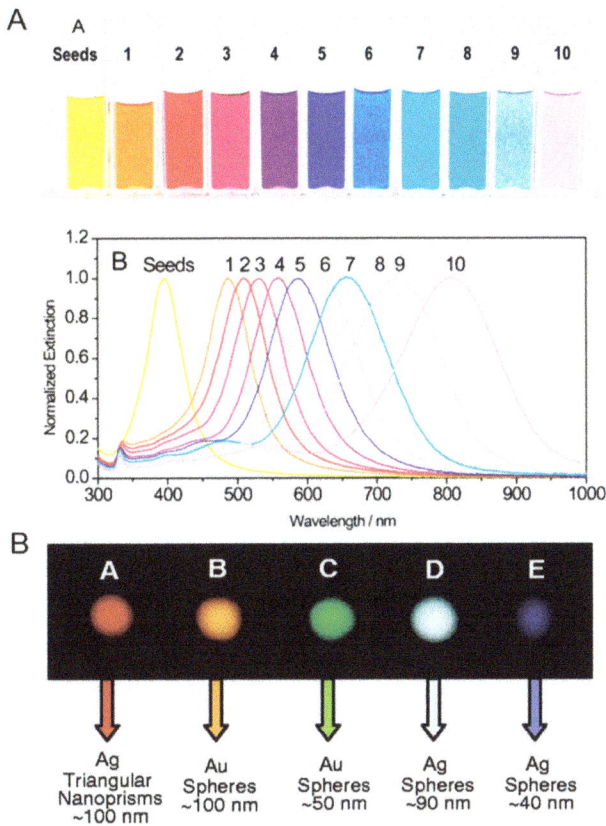

Figure 3. (**A**) Photograph of silver nanoprisms (top) and corresponding optical spectra changes of nanoprisms (bottom). Control on the edge-length of nanoprisms allows the plasmon resonance to be tuned across the visible and near-infrared portions of the spectrum. Reprinted with permission from [55]. Copyright 2008 Wiley-VCH. (**B**) Dark field microscopy images of (left to right) 100 nm diameter silver triangular nanoprism, 90 nm diameter silver nanosphere, and 40 nm diameter silver nanosphere, illustrating the ability to tune the scattering color of silver nanoparticle labels based on size and shape. Reprinted with permission from [56]. Copyright 2001 American Association for the Advancement of Science.

3.2. Chemical Cytotoxicity

One of the current issues in AgNP-based nanomedicine involves nanotoxicity and environmental impact of AgNPs on a nanometer scale. To predict the potential cytotoxic effect of AgNPs, it is necessary to investigate chemical transformation that occurs with AgNPs travelling through the intracellular environment [15]. The use of AgNPs based on their chemical cytotoxic property has received much attention as potent anticancer or antibacterial agents. Despite various hypotheses available, the mechanisms of the antibacterial properties of AgNPs so far have not been established clearly. Based current literature, the proposed cytotoxic mechanisms can be summarized as follows: (i) adhesion of AgNPs onto the membrane surface of microbial cells, modifying the lipid bilayer or increasing the membrane permeability; (ii) intracellular penetration of AgNPs; (iii) AgNP-induced cellular toxicity triggered by the generation of reactive oxygen species (ROS) and free radicals, damaging the intracellular micro-organelles (i.e., mitochondria, ribosomes, and vacuoles) and biomolecules including DNA, protein, and lipids; and (iv) modulation of intracellular signal transduction pathways towards apoptosis. Critical parameters, such as ion release, surface area, surface charge, concentration and colloidal state, can all influence the cytotoxic properties of AgNPs.

The main mechanism of AgNPs regarding their antimicrobial activity can be simplified to their high surface area in releasing silver ions. AgNPs in an aqueous environment are oxidized in the presence of oxygen and protons, releasing Ag^+ ions as the particle surface dissolves. The release rate of the Ag^+ ions depends on a number of factors including the size and shape of NPs, capping agent, and colloidal state. For example, it is well known that antibacterial activity is enhanced with the release of Ag^+ ions from AgNPs onto the bacterial cells [11]. In particular, smaller or anisotropic AgNPs with a larger surface area showed more toxicity and exhibited a faster ion release rate due to high surface energy originating from highly curved or strained shapes of NPs [67]. The small-sized AgNPs also exhibited a superior release rate of silver ion particularly into the Gram-negative bacteria. The shape and higher temperature of AgNPs equally caused a greater degree of toxicity and accelerated the rate of ion release by more effective dissolution [68,69]. Furthermore, the cytotoxic effect of AgNPs arises in a similar concentration range for both bacteria and human cells [70]. Therefore, a higher Ag ion concentration, a faster release rate of the Ag ions, and a larger surface area of AgNP should be considered for the enhanced antimicrobial treatment in clinical medicine [71]. Moreover, the presence of chlorine, thiols, sulfur, and oxygen was shown to strongly impact the rate of silver ion release [72]. Silver ions can interact with thiol groups in critical bacterial enzymes and proteins, and subsequently damage cellular respiration, resulting in cell death. The generation of ROS and free-radicals is another mechanism of AgNPs causing a cell-death process as illustrated in Figure 4. The potent cytotoxic activity of AgNPs, such as antibacterial, antifungal, and antiviral action, is mainly due to their ability to produce ROS and free radical species, such as superoxide anion ($O_2{}^-$), hydrogen peroxide (H_2O_2), hydroxyl radical (OH), hypochlorous acid (HOCl), and singlet oxygen [73]. When in contact with bacteria, the free radicals have the ability to generate pores on the cell wall, which can ultimately lead to cell death [10]. AgNPs can also anchor to the surface of the bacterial cell wall and penetrate it to cause structural changes to the membrane or increase its permeability, all of which trigger cells to die.

Interestingly, the strength of the antibacterial property of AgNPs is correlated with different types of bacterial species, such as Gram-positive and -negative bacteria. This is because these species differ in the architecture, thickness, and composition of their cell wall [74]. It is known that *Escherichia coli* (*E. coli*) which is Gram-negative bacteria is more susceptible to Ag^+ ions than Gram-positive *Staphylococcus aureus* (*S. aureus*). The reason for different susceptibility lies on the peptidoglycan which is a key component of the bacterial cell membrane. The cell wall in Gram-positive bacteria is composed of a negatively-charged peptidoglycan layer with approximately 30 nm in thickness, whereas Gram-negative bacteria have a peptidoglycan layer of only 3 to 4 nm [31,75]. These structural differences, including the thickness and composition of the cell wall, explain why Gram-positive *S aureus* is less sensitive to AgNPs, and Gram-negative *E.coli* displays substantial inhibition even at a low concentration of AgNPs. Loo et al. investigated silver and curcumin NPs against both Gram-positive and Gram-negative bacteria, and the NPs in 100 µg/mL concentration were able to distort matured bacterial biofilms. The sustained anti-bacterial effects of this formulation can be utilized in antimicrobial treatment [76]. Petr Pařil et al. investigated the anti-fungal effects of AgNPs and copper nanospheres against wood-rotting fungi [77]. The AgNP treatment required a very low mass of NPs and exhibited high efficiency against *Tinea versicolor* (*T. versicolor*) fungi in comparison to *Poria placenta* (*P. placenta*) fungi, showing differing anti-fungal effects of AgNPs against white and brown-rot fungi, respectively. The details of antimicrobial properties are beyond the scope of this review and are reviewed elsewhere [9,14,72,78].

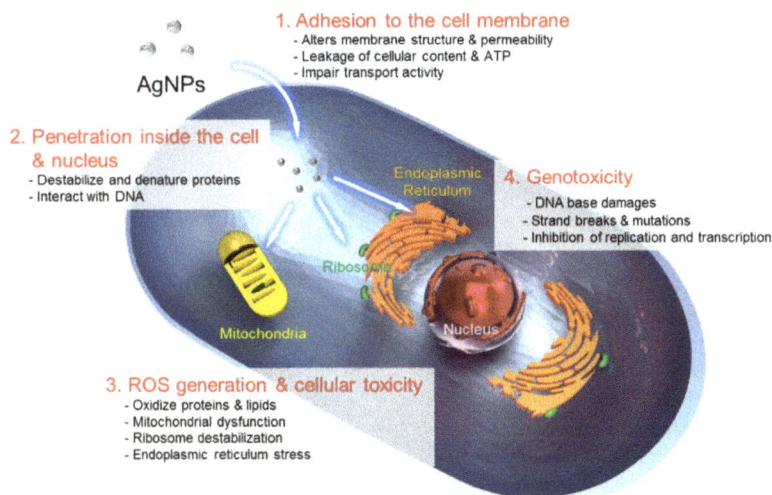

Figure 4. The four main routes of cytotoxic mechanism of AgNPs. 1, AgNPs adhere to the surface of a cell, damaging its membrane and altering the transport activity; 2, AgNPs and Ag ions penetrate inside the cell and interact with numerous cellular organelles and biomolecules, which can affect corresponding cellular function; 3, AgNPs and Ag ions participate in the generation of reactive oxygen species (ROS) inside the cell leading to a cell damage and; 4, AgNPs and Ag ions induce the genotoxicity.

3.3. Alloy with Other Metals

Alloy NPs exhibit specific properties that are different from their individual NPs. They can be created directly by combining different metallic nanocrystals (NCs) in specific numbers and arrangements [79]. As strong electronic coupling exists between two metals, the bimetallic nanocrystals show more enhanced catalytic, electronic and optical properties compared to monometallic nanocrystals [80]. The properties of alloy NPs are defined by their internal configuration (i.e., arrangement of constituent atoms) and external structures, such as shapes and sizes. The LSPR wavelength of nanocrystals formed from Ag–Au alloys can be tuned by varying the Au:Ag ratio and, therefore, can increase gradually with an increase in the percentage of Au in the alloy nanocrystal. However, metallic atoms are easily bonded in a non-specific manner, lacking the directionality of covalent bonds and equivalence of molecules. The synthesis methods for alloy NPs can be divided into two categories: (i) successive/sequential reduction method and (ii) co-reduction/simultaneous reduction method with metal precursors [81].

Sequential reduction without protective agents is driven thermodynamically and causes the formation of core–shell NPs or other types of hetero-nanostructures. The sequential reduction method involves subsequent seed-mediated growth of NPs with metal precursors and reducing agents over time. The bimetallic colloids with different metals, such as Ag and Au, can be synthesized in several different ways, resulting in Ag-coated Au or Au-coated colloidal particles. The synthesis can be done simply by reducing one metal salt on already-formed counterpart metal NPs—e.g., to synthesize Ag-coated Au colloids, chemically reduce silver salt on the AuNPs [82]. This seed-mediated synthesis method for core–shell and intermetallic structures is widely used in well-defined bimetallic NPs, due to its capability to regulate the size, shape and composition of the final compound [83–85]. Reducing agents also play a vital role in controlling the size distribution. The co-reduction method with different metal precursors to zero-valent atoms has made bimetallic colloids readily accessible [86]. The key advantages lie in the simplicity and versatility of the technique. Using this method, several types of Ag and Au bimetallic core–shell colloids with various shapes have been produced [87]. Bimetallic colloids with gradient metal distribution or with a layered structure are one of the most interesting and promising methods in catalytic applications. In the co-reduction method, however, composition uniformity is a major drawback due to the high prevalence of sequential reduction. For example, the Xia group at Georgia Institute of Technology described nucleation and site-selective growth of Ag on cubic Pd nanocrystal seeds as shown in Figure 5 [80]. Ag atoms were directed to nucleate in a specific-site and then grown on a specific number of faces on a cubic Pd nanocrystal seed by controlling reaction kinetics. This approach allowed the fabrication of bimetallic nanocrystals with well-controlled spatial distributions and tunable LSPR properties. Another example is from Lee and colleagues who demonstrated programmable synthesis of hybrid liposome-metal NPs which allows self-crystallization of metal NPs in liposome [88]. They have synthesized seven types of liposome/monometallic and liposome/bimetallic hybrids with Ag, Au, Pd, Pt, Ag–Au, Au–Pt, and Au–Pd, which were tunable in size and composition. The resulting NPs showed controllable SPR bands in visible and near-infrared spectra as well as better colloidal stability caused by an outer liposome structure. This improved physicochemical property allowed the liposome/NP hybrids to be applied to the intracellular imaging of living cells via SERS. On the other hand, the enhanced catalytic performance of bimetallic Au–Ag (core–shell) colloids on luminol-K_3-Fe(CN)$_6$ chemiluminescence (CL) has been described by Zhang and colleagues [89]. Bimetallic Au–Ag NPs were synthesized by a sequential two-step reduction technique, and subsequently prostate-specific antigen (PSA) specific antibody was conjugated on the Au–Ag NPs. Since the prepared Ag–Au NPs with synergistic catalytic activity significantly enhanced the CL reaction, the PSA antibody-Ag–Au NP immune complex was able to successfully applied to the detection of PSA in human serum sample down to 0.047 pg/mL (S/N = 3).

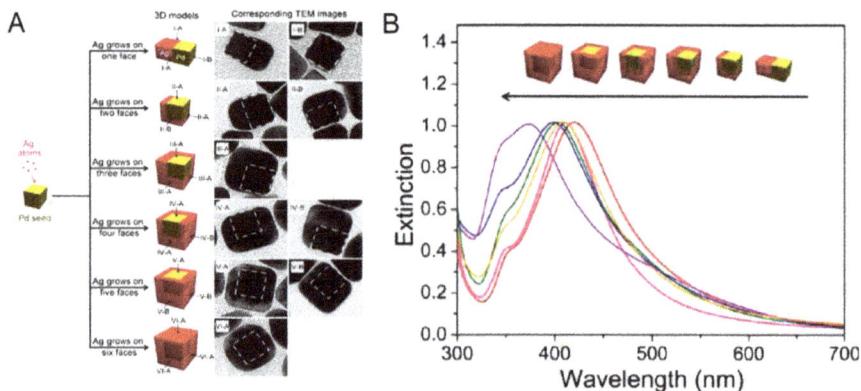

Figure 5. Controlled overgrowth of Ag for bimetallic nanocrystals. (**A**) Schematic illustration showing the site-selective growth of Ag on each cubic seed and corresponding transmission electron microscopy (TEM) images. Well-controlled bimetallic nanocrystals were fabricated along the directed size and number of facets on a cubic Pd seed. The white dashed lines in the TEM indicate the position of the cubic Pd seed. (**B**) Extinction spectra of the Pd–Ag bimetallic nanocrystals with Ag growing on different numbers of faces of the cubic Pd seed. The LSPR peak blue-shifted with the increase in the number of faces involved in the Ag growth. Reprinted with permission from [80]. Copyright 2012 American Chemical Society.

4. Applications of AgNPs

4.1. AgNP-Based Nanomedicine

4.1.1. Plasmonic Nanoantennas

Recently, AgNP has been widely utilized in various subfields of nanomedicine including nanoelectronics, diagnostics, molecular imaging, and biomedicine. These interesting applications are based on utilizing an enhanced electromagnetic field on and near the surface of AgNPs. At the plasmon resonant wavelength, AgNPs act as nanoscale antennas, increasing the intensity of a local electromagnetic field. One spectroscopic technique that benefits from the enhanced electromagnetic field is the Raman spectroscopy, where molecules can be identified by their unique vibrational modes. However, intrinsic Raman scattering of photons from molecules is weak and requires a longer measurement duration to obtain a Raman spectrum. Therefore, surface-enhanced Raman scattering (SERS) from molecules near the surface of plasmonic nanoantenna offers great amplification of Raman signals. Typically, SERS detection involves adsorption of molecules on Ag or Au nanoparticle aggregates or solid substrates with plasmonic nanostructures [90,91]. Strong field enhancement is generated in the nanogaps or interstices known as hot spots within interacting plasmonic nanostructures [92]. The SERS effect can be used to detect critical proteins and biomolecules, such as early cancer biomarkers or drug levels in blood and other body fluids. Up until recently, the SERS effect with hot spots has been the main focus in numerous experimental and theoretical studies, which can enhance the Raman scattering to the factor of 10^8 to 10^{12}, allowing the detection of even a single molecule [93].

Numerous approaches have been made to utilize the plasmonic property of AgNPs. For instance, techniques to control the distance and spacing of hot spots are essential in quantitative SERS covering large areas as shown in Figure 6A,B [94]. Strong enhancement of single hot spots may lead to a false representation of a sample when the signal is mainly determined by a few detection sites. Nanoparticle superlattices have demonstrated a potential to counterbalance the homogeneous distribution in sensing hot-spot bands and to enhance the detection performance of the sensor. Sun and colleagues described a SERS substrate via graphene–AgNPs heterojunction which improved Raman signals [95]. With increasing the density of AgNPs, Raman scattering of graphene-veiled AgNPs heterojunction substrate was significantly enhanced by approximately 67 folds compared to R6G analyte. The cooperative synergy generated by the coupling of graphene and deposited AgNPs can be utilized to create a strong electromagnetic hot spot for an optical sensing platform. Another example is a star-shaped Au/AgNPs SERS substrate on Ge (5 nm)/Ag (25 nm)/GE (75 nm)/glass (germanium–silver multilayers) which was employed via near-infrared (NIR) SERS operation by Lai and colleagues [96]. The hybrid SERS substrate was operated at a 1064 nm excitation and exhibited 30% higher Raman intensity.

AgNPs can be utilized as highly sensitive NP probes for targeting and imaging of small molecules, DNA, proteins, cells tissue, and even tumor in vivo (Figure 6C) [97,98]. AgNPs with stronger and sharper plasmon resonance have been widely used in imaging systems, particularly for cellular imaging with contrast agents functionalized to AgNPs via surface modification. For example, a AgNP-embedded nanoshell structure can be used in cancer imaging and photothermal therapy to explore the location of cancer cells by absorbing light and destroy them via photothermal effect [99]. Kang et al. described NIR-sensitive SERS nanoprobes for an in vivo multiplex molecular imaging to detect aromatic compounds [100]. The NIR SERS probes with plasmonic Au/Ag hollow-shell were assembled onto silica nanospheres which exhibited a red-shift of plasmonic extinction band in the NIR optical window region (700–900 nm). The signals from the NIR-SERS nanoprobe for a single particle detection exhibited a detectable signal from animal tissues that were 8 mm deep [101]. Jun et al. showed Ag-embedded SERS nanoprobes, called M-SERS dot, which have a Raman signature for imaging of target cancer cells as well as strong magnetic properties for identifying desirable cells [102]. AgNP-embedded magnetic nanoparticles (MNPs), which consisted of a magnetic core (18 nm) and a silica shell (16 nm thick) decorated with AgNPs on the surface, were prepared. The M-SERS dots exhibited strong SERS signals originating from diverse encoding materials, such as AgNPs and Raman-labels. The Ag-embedded M-SERS dots with highly sensitive SERS signals enabled targeting, isolation, and imaging of cancer cells. To investigate their specific targeting and sorting abilities, M-SERS conjugated with targeting antibodies were added into multiple cell population, and subsequently, the targeted cancer cells could easily be isolated by an external magnetic field. Hahm et al. described a multilayered core–shell nanoprobe with Ag-embedded silica nanostructure for a SERS-based chemical sensor [64]. The multi-layered nanoprobe consisted of a silica core coated with Raman label, silica shell, and AgNPs. The embedded inner AgNP and Raman label compound in the nanoprobe served as an internal standard for calibrating SERS signals, while the outer AgNPs were utilized as a sensing site for analyte detection. These chemical sensors based on the ratiometric analysis ($I_{Analyte}/I_{Internal\ standard}$) could be applied to various SERS probes for quantitative detection of a wide variety of targets.

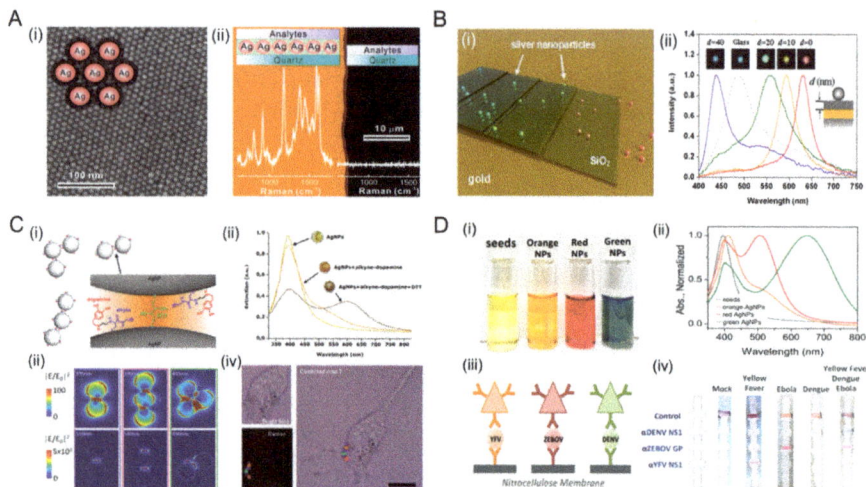

Figure 6. Plasmonic AgNPs for plasmonic nanoantennas and diagnostics. (**A**) Single-layer AgNP surface-enhanced Raman scattering (SERS) film for a large-scale hot spot. (**i**) Scanning electron microscopy (SEM) image of a superlattice of 6 nm. AgNPs were used as a homogeneous single-molecule SERS substrate. Illustration shows an interparticle gap for hot spots, which is regulated by the length of a thiolate chain. (**ii**) Two Raman spectra of single-layered SERS film (left) and quartz surface (right). The enhancement factor was estimated to be larger than 1.2×10^7. Reprinted with permission from [94]. Copyright 2015 American Chemical Society. (**B**) Metal-film induced plasmon resonance tuning of AgNPs. (**i**) Schematic illustration of optical scattering spectra of AgNPs on different substrates. (**ii**) Single AgNP spectra of AgNPs on a silica spacer layer of varying thickness *d* (nm) on a glass substrate with a 50 nm gold film. The inset is a dark-field image of AgNPs with the corresponding color. The dotted lines represent single particle spectra of AgNPs on a plain glass substrate. Reprinted with permission from [103]. Copyright 2010 American Chemical Society. (**C**) SERS-based intracellular imaging using alkyne-AgNPs nanoprobes. (**i**) The structure of colloidal alkyne-AgNP clusters with nano-sized interparticle gaps. (**ii**) Extinction spectra of the alkyne-AgNPs nanoprobe. The resonance peaks at 400 nm shifted around 520 nm after metal functionalization. (**iii**) Computational simulation of the far- and near-field optical responses. Intensity distributions of the single particle mode (upper-panels) and the dimer mode (bottom-panels) (**iv**) Intracellular Raman imaging of a AgNP nanoprobe within the cytoplasmic space of fibroblast. Distinguishable hot spots were highlighted by color-dots related to Raman intensity of the akyne 2045 cm^{-1} band. Reprinted with permission from [104]. Copyright 2018 Nature Publishing Group. (**D**) Multiplexed detection with a tunable wavelength of AgNPs. (**i**) Different colors of AgNPs during a stepwise growth. (**ii**) Corresponding absorption spectra with varying sizes of AgNPs, such as 30, 41, and 47 nm. (**iii**) Individual testing of Yellow Fever virus (YFV) NS1 protein, Zaire Ebola virus (ZEBOV) glycoprotein (GP), and Dengue virus (DENV) NS protein using AgNPs. Orange, red, and green AgNPs were conjugated with monoclonal antibodies specific to YFV NS1, ZEBOV GP, and DENV NS, respectively. (**iv**) Multiplexed detection using different AgNPs-based lateral flow assays. Reprinted with permission from [105]. Copyright 2015 Royal Society of Chemistry.

4.1.2. Diagnostics with Tunable Wavelength

AgNPs can absorb and scatter light with extraordinary efficiency. A large scattering cross-section of the nanospheres allows for an individual AgNP to be imaged under a dark-field microscopy or hyperspectral imaging systems. As mentioned above, AgNPs have been intensively utilized in several applications, including diagnosis and bioimaging of cancer cells [106]. Furthermore, AgNPs have been utilized for the detection of p53 in carcinoma cells [107]. Zhang et al. demonstrated that nanostructures comprising silver cores and a dense layer of Y_2O_3:Er separated by a silica shell is an excellent system model to investigate the interaction between upconversion materials and metals on a nanoscale. Nanoparticles are also potentially promising as fluorescent labels for (single particle) imaging experiments or bioassays, which require low background or tissue penetrating wavelengths [108]. Optical properties of AgNPs can be utilized for multiplexed point-of-care (POC) diagnostics using their size-tunable absorption spectra. As shown in Figure 6D, Yen et al. described a multicolored AgNPs-based multiplexed lateral flow assay (LFA) for multiple pathogen detections [105]. Multiplexed rapid LFA diagnostics has the ability to discriminate among multiple pathogens, thereby facilitating effective investigations for diagnosis. Specifically, triangular plate-shaped AgNPs with varying sizes, such as 30 nm, 41 nm. and 47 nm, have narrow absorbance that are tunable through the visible spectrum, resulting in an easily distinguishable color. The multicolored AgNPs were conjugated with antibodies to recognize dengue virus (DENV) NS protein, Yellow Fever virus (YFV) NS1 protein, and Zaire Ebola virus (ZEBOV) glycoprotein (GP). The limit of detection (LOD) for the biomarkers of each virus was 150 ng/mL in a single channel. Another example is a colorimetric lead detection using AgNPs described by Balakumar and co-workers [109]. The synthesized AgNPs exhibited high sensitivity for the detection of as low as 5.2 nM of Pb^{2+} in the range of 50 to 800 nM and also showed selective recognition even in the presence of interfering metal ions. This approach can be used for a rapid and cost-effective detection of saturnism (lead poisoning) in a water sample.

4.1.3. Surface-Enhanced Fluorescence

Surfaces of metallic nanoparticles can alter the free-space condition of fluorescence with spectral properties which can result in dramatic spectral changes as shown in Figure 7A [110]. This interactions between metal surface and fluorophore have been termed variously as surface-enhanced fluorescence (SEF), metal-enhanced fluorescence (MEF) or radiative decay engineering [111]. The metallic surfaces exhibit the following features as illustrated in Figure 7B: fluorophore quenching within short distances (0–5 nm); spatial disparity of incident light (0–15 nm); and changes in the radiative decay rates (1–20 nm). The enhanced field effect can be leveraged to build an interspace with a shorter distance between a fluorophore and the surface of a metal nanostructure composed of Ag or Au, increasing fluorophore emission rate [112]. The enhanced fluorescence can be attributed to two main factors, which are (i) an enhanced excitation rate by large absorption and scattering cross-section of the plasmonic nanoparticles to incoming light and (ii) a decrease in fluorescence lifetime of the fluorophore that allows an excited state to return to the ground state at a higher frequency. In particular, SEF parameters of the dye/nanoparticle coupled system are related to the distribution of near-field intensity and their distance-dependent decay function. Such effects depend strongly on the overlap of optical properties of the fluorophore and nanosphere and on the physical location of the fluorophore around the particles.

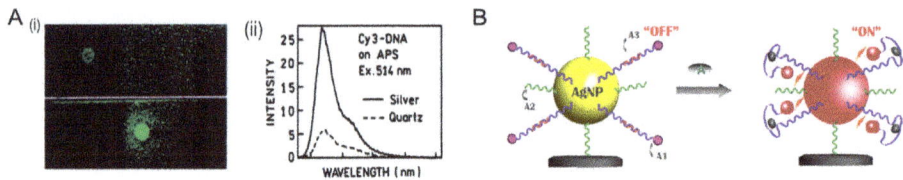

Figure 7. Surface-enhanced fluorescence. (**A**) Metal-enhanced fluorescence on a Ag film. (i) The photograph shows fluorescence spots on quartz (top) and silver (bottom) taken through 530 nm long pass filter for Cy3-DNA. (ii) Emission spectra of Cy3-DNA on APS-treated slides, with (solid line) and without silver island films (dotted line). Reprinted with permission from [113]. Copyright 2003 Future Science Group. (**B**) Schematic illustration of an aptamer-based AgNP nanosensor, showing the 'off' state via fluorophore quenching within short distances (left) and 'on' state via turn-on fluorescence signal (right) based on the spacing distance between the Cyanine 3 and the AgNP surface in the detection of adenosine. Reprinted with permission from [114]. Copyright 2012 Elsevier.

4.2. Biomedical Application of AgNPs

Owing to their intrinsic cytotoxicity, AgNPs have been broadly used as antibacterial and anticancer agents and for biomedical application in the healthcare industry. The degree of toxicity against cells is determined by the surface charges of AgNPs [115]. A positive surface charge of AgNPs renders them more suitable to stay for a longer duration on the tissue surface or luminal side of the blood vessel, which is a major route for administrating anticancer agents [116]. The intrinsic cytotoxic property of AgNPs has been applied against various types of cancer cells, such as hepatocellular carcinoma [117], lung [118] and breast cancer [119,120], and cervical carcinoma [121]. Small sized AgNPs were more efficient in ROS production [122]. Apart from these cellular mechanisms, AgNPs have also shown anti-angiogenic and anti-proliferative properties [123,124]. The anti-proliferative property mediated by AgNPs in cancer cells is due to their ability to damage DNA, break chromosome, produce genomic instability, and disrupt calcium (Ca^{2+}) homeostasis which induces apoptosis and causes cytoskeletal instability. The cytoskeletal injury blocks the cell cycle and division, promoting anti-proliferative activity of cancer cells [89].

For instance, in regard to intracellular transport, Lee et al. characterized the transport of a single AgNP into an in vivo Zebrafish embryo model system and their effects on early embryonic development on a single-nanoparticle resolution in real-time. It was found that a single Ag nanoparticle (5–46 nm) was transported into and out of embryos through chorionic pore channels (CPCs) and exhibited Brownian diffusion (not an active transport). The diffusion coefficient inside the chorionic space (3×10^{-9} cm^2/s) was ~26 times lower than that in egg water (7.7×10^{-8} cm^2/s). Thapa et al. embedded graphene oxide in AgNPs (GO-AgNP) using glucose as a reducing agent. By covalent conjugation of methotrexate (MTX) to Go-AgNP via an amide bond, targeting of folate receptors expressing cancer cells was achieved, and, thus, showing that the combination of anticancer drug and AgNPs could be used synergistically for treatment of cancer [125]. As another example, Azizi and colleagues developed a novel nanocomposite with the aim of developing AgNPs as a new anticancer agent that specifically target tumor cells. Albumin coated AgNPs were synthesized, and their anti-cancerous effects were evaluated against MDA-MB 231, a human breast cancer cell. The cancer cell showed morphological changes, and its DNA agarose gel pattern on gel electrophoresis revealed a cell death process through apoptosis. It was found that AgNPs with a size of 90 nm and with a negative charge of a zeta-potential of about -20 mV could be specifically taken up by tumor cells. The LD$_{50}$ of AgNPs against MDA-MB 231 (5 μM) suggests the AgNPs to be a good candidate as a chemotherapeutic drug [126]. In therapeutics outside oncology, Ayaz and colleagues have described AgNPs conjugated with anti-seizure drugs (as a drug carrier) against brain-eating amoebae (*Naegleria flowleri*) to treat central nervous system (CNS) infection [127]. Anti-seizure drugs which are known to cross the blood–brain barrier (BBB) were attached to the surface AgNPs as capping agents. AgNPs conjugated with drugs, such as diazepam, phenobarbitone, and phyenytoin, exhibited overall anti-amoebic activities against both trophozoite and cyst stages. Moreover, significant enhancement of fungicidal activities was shown against both trophozoite and cyst amoebic stages compared to those of the drugs alone. The researchers suggested that a feasible mechanism of AgNPs-based drugs which can penetrate BBB might lie in their ability to bind to the receptors and ion channels on the cell membrane of amoebae.

The cytotoxic effect of AgNPs has been used extensively in food and healthcare industries, such as food storage, textile, medical device coating, and environmental sensing [128]. Specific toxicity to bacteria has led to the integration of silver in a wide variety of products including wound dressings, packaging materials, and anti-fouling surface coatings. Another interesting approach is AgNP-coated bandages as they can kill harmful microbes and allow better healing at the injured tissue. In addition, silver ions as an antimicrobial agent have been used as composites in dental resin and in coatings of medical instruments [129]. AgNPs have also been utilized in food packaging so that foods can last for longer without contamination.

4.3. Optoelectronics

Diverse silver nanomaterials have been studied as components of nanocomposite due to their high dielectric constants in numerous systems. For example, silver nanowires can be used as conductive coatings in flexible electronics and transparent semiconductors [130]. Similarly, AgNPs have the potential to be applied in silver paste for efficient contact at electronic interfaces because of their high conductivity [10]. In particular, AgNPs can be used as antennas enhancing plasmonic activity for sensing of a specific molecule or in imaging applications. AgNPs can, therefore, be utilized as a sensing material for environmental monitoring [131]. Prosposito et al. reported that negatively-charged AgNPs with -34 mV in zeta-potential exhibited a good response to heavy metals, such as nickel (II) [132]. They synthesized AgNPs with an average diameter of 2.5 nm in water phase using silver nitrate as a precursor, hydrophilic thiol (3-MPS, 3-mercaptopropane sulfonate) as a capping agent, and sodium borohydride as a reducing agent. The SPR spectral changes in the presence of metal ions, such as Ni^{2+}, Cr^{3+}, Nd^{3+}, Cu^{2+}, and Ca^{2+}, were observed. AgNP/3-MPS exhibited the LOD of 0.3 ppm and showed the detection of a low amount of Ni^{2+} ions in water ranging from 0.1 to 1.0 ppm.

In optothermal applications, Hu and colleagues demonstrated a multi-layered bimetallic bactericidal nanoprobe comprising a core–shell–shell (Au–Ag–Au) structure for photothermal heating-mediated controlled release of Ag^+ ions [133]. This bactericidal nanoprobe combined two features of photothermal sterilization based on the outer Au shell as well as the antibacterial effect of the inner Ag shell or Ag^+ ions against surrounding bacteria. The outer shell can be melted even at low-power NIR laser irradiation (785 nm, 50 mW/cm^2). The melting of the shell exposes the inner Ag shell, facilitating the release of antibacterial Ag^+ ions. The bactericidal rate of 100% was observed in *E.coli* O157: H7 at 10 g/mL concentration of nanoprobes under 20-min irradiation. By exploiting the toxicity of Ag, the photothermal approach may alleviate the abuse of broad-spectrum antibiotics. In vivo biomedical application is another avenue that the photothermal method seems promising. A similar approach with Au–Ag core–shell nanospheres and NIR femtosecond laser pulse has been reported, wherein their superior photothermal-induced antibacterial activity was explored [134]. Positively charged Au–Ag nanosphere (19 nm in Au core; 3 nm in Ag shell) were attached to the negatively-charged bacterial surface via electrostatic interaction, forming large clusters on the surface of *S. aureus*. The NIR irradiation of Au–Ag nanospheres generated heat and ROS. As a result, Au–Ag nanospheres exhibited a strong antibacterial activity, as low as 7.5 pM in minimum inhibition concentration (MIC) against *S. aureus*. The result also showed a removal of up to 85% of a notoriously recalcitrant bacterial biofilm within 4 min under NIR irradiation.

Another example is shown by Kamimura and co-workers. The authors described surface-plasmon induced photocatalytic activity based on core@shell (Au–Ag) NPs [135]. Au@Ag NPs and Au–Ag bimetallic NPs were synthesized by multistep citric reduction and photo-reduction methods, respectively. Both types of metallic NPs exhibited strong absorption in the visible wavelength due to localized LSPR of Ag. They both could oxidize 2-propanol to acetone and CO_2 under visible light irradiation (440–800 nm), but Au@Ag exhibited higher turnover rate than Au–Ag NPs. It was found that the improvement in chemical stability of Ag was attributable to the formation of a core@shell structure, which led to the efficient surface plasmon-induced photocatalytic activity.

Metallic nanospheres of Ag or Au has been utilized in optoelectronic light harvesting based on the plasmonic effect [136,137]. In plasmon-assisted solar energy conversion, metal nanostructures are used to scatter solar radiation and better able to couple radiation to semiconductor photovoltaic elements. The efficient extraction of light from LED exploits similar physics where the metal nanostructures play dual roles as a light scatterer and energy-transducing nanoantennas [8]. In this context, the solar cells serve to efficiently couple incident light to the AgNPs, from which optical energy propagates as surface plasmon (SP) polariton [138]. For example, the plasmonic effect triggered by metal NPs was used to enhance the yield of light absorption in solar cells [139,140]. As the photons of incident light were encountered by AgNPs, they caused electron vibration and scattering in the NPs, facilitating more efficient photon absorption. Rho and colleagues have reported dye-sensitized solar cells with AgNPs-decorated TiO_2 nanotube arrays [141,142]. The energy conversion efficiency of the solar cells increased up to 32% by incorporating AgNPs into the TiO_2 film. Another example is polymer optoelectronic devices with carbon dot-supported AgNPs (CD-AgNPs) which were described by Choi and co-workers [143]. The SPR effect of CD-AgNPs allowed additional light absorption and a significant amount of radiative emission in polymer solar cells as well as in polymer light-emitting diodes (LEDs).

5. Conclusions

AgNPs are emerging as a next-generation application in numerous subfields of nanomedicine, and potential benefits of using AgNPs as a prominent nanomaterial in biomedical and industrial sectors have been widely acknowledged. The comprehensive research regarding silver nanomaterials has been explored in this review to understand the synthesis methods and mechanisms, characterization of physicochemical properties, and possible toxicity and to discover more promising applications in oncology, personalized healthcare, and pharmacology. Among the various synthesis methods, biological green synthesis draws our attention as a promising alternative, due to its safety using natural agents and nontoxic chemicals. Diverse applications of AgNPs as plasmonic nanoantenna and biomedical and optoelectronic probes were also highlighted. Lastly, a better understanding of the cytotoxic mechanisms of AgNPs merits future research to broaden their nanomedical applications in diagnostics, therapeutics and pharmaceutics.

Author Contributions: S.H.L. and B.-H.J. conceived the idea and planned for the review article and performed all literature surveys. All authors prepared and reviewed the manuscript.

Funding: This research was supported by basic Science Research Program through the NRF funded by the Ministry of Education (NRF-2018R1D1A1B07045708) and Science, ICT & Future Planning (NRF 2016M3A9B6918892) and funded by the Korean Health Technology R&D Project, Ministry of Health & Welfare (HI17C1264).

Conflicts of Interest: The authors declare no conflict of interest.

References

1. Millstone, J.E.; Hurst, S.J.; Métraux, G.S.; Cutler, J.I.; Mirkin, C.A. Colloidal gold and silver triangular nanoprisms. *Small* **2009**, *5*, 646–664. [CrossRef]
2. Lee, S.H.; Rho, W.-Y.; Park, S.J.; Kim, J.; Kwon, O.S.; Jun, B.-H. Multifunctional self-assembled monolayers via microcontact printing and degas-driven flow guided patterning. *Sci. Rep.* **2018**, *8*, 16763.
3. Lee, S.H.; Sung, J.H.; Park, T.H. Nanomaterial-based biosensor as an emerging tool for biomedical applications. *Ann. Biomed. Eng.* **2012**, *40*, 1384–1397. [CrossRef] [PubMed]

4. Syafiuddin, A.; Salmiati; Salim, M.R.; Kueh, A.B.H.; Hadibarata, T.; Nur, H. A Review of silver nanoparticles: Research trends, global consumption, synthesis, properties, and future Challenges. *J. Clin. Chem. Soc.* **2017**, *64*, 732–756. [CrossRef]

5. Kumar, A.; Vemula, P.K.; Ajayan, P.M.; John, G. Silver-nanoparticle-embedded antimicrobial paints based on vegetable oil. *Nat. Mater.* **2008**, *7*, 236–241. [CrossRef] [PubMed]

6. Desireddy, A.; Conn, B.E.; Guo, J.; Yoon, B.; Barnett, R.N.; Monahan, B.M.; Kirschbaum, K.; Griffith, W.P.; Whetten, R.L.; Landman, U.; et al. Ultrastable silver nanoparticles. *Nature* **2013**, *501*, 399–402. [CrossRef] [PubMed]

7. Sun, Y.; Xia, Y. Shape-controlled synthesis of gold and silver nanoparticles. *Science* **2002**, *298*, 2176–2179. [CrossRef]

8. Atwater, H.A.; Polman, A. Plasmonics for improved photovoltaic devices. *Nat. Mater.* **2010**, *9*, 205–213. [CrossRef]

9. Ouay, B.L.; Stellacci, F. Antibacterial activity of silver nanoparticles: A surface science insight. *Nano Today* **2015**, *10*, 339–354. [CrossRef]

10. Chen, D.; Qiao, X.; Qiu, X.; Chen, J. Synthesis and electrical properties of uniform silver nanoparticles for electronic applications. *J. Mater. Sci.* **2009**, *44*, 1076–1081. [CrossRef]

11. Sun, Y.; Mayers, B.; Herricks, T.; Xia, Y. Polyol synthesis of uniform silver nanowires: a plausible growth mechanism and the supporting evidence. *Nano Lett.* **2003**, *3*, 955–960. [CrossRef]

12. Haes, A.J.; Duyne, R.P.V. A Nanoscale optical biosensor: sensitivity and selectivity of an approach based on the localized surface plasmon resonance spectroscopy of triangular silver nanoparticles. *J. Am. Chem. Soc.* **2002**, *124*, 10596–10604. [CrossRef] [PubMed]

13. Dankovich, T.A.; Gray, D.G. Bactericidal paper impregnated with silver nanoparticles for point-of-use water treatment. *Environ. Sci. Technol.* **2011**, *45*, 1992–1998. [CrossRef] [PubMed]

14. Zhang, X.-F.; Liu, Z.-G.; Shen, W.; Gurunathan, S. Silver nanoparticles: Synthesis, characterization, properties, applications, and therapeutic approaches. *Int. J. Mol. Sci.* **2016**, *17*, 1534. [CrossRef] [PubMed]

15. Wang, L.; Zhang, T.; Li, P.; Huang, W.; Tang, J.; Wang, P.; Liu, J.; Yuan, Q.; Bai, R.; Li, B.; et al. Use of synchrotron radiation-analytical techniques to reveal chemical origin of silver-nanoparticle cytotoxicity. *ACS Nano.* **2015**, *9*, 6532–6547. [CrossRef] [PubMed]

16. El-Nour, K.M.M.A.; Eftaiha, A.; Al-Warthan, A.; Ammar, R.A.A. Synthesis and applications of silver nanoparticles. *Arab. J. Chem.* **2010**, *3*, 135–140. [CrossRef]

17. Heiligtag, F.J.; Niederberger, M. The fascinating world of nanoparticle research. *Mater. Today.* **2013**, *16*, 262–271. [CrossRef]

18. Wei, L.; Lu, J.; Xu, H.; Patel, A.; Chen, Z.-S.; Chen, G. Silver nanoparticles: Synthesis, properties, and therapeutic applications. *Drug. Discov. Today.* **2015**, *20*, 595–601. [CrossRef]

19. Burduşel, A.-C.; Gherasim, O.; Grumezescu, A.M.; Mogoantă, L.; Ficai, A.; Andronescu, E. Biomedical applications of silver nanoparticles: An up-to-date overview. *Nanomaterials* **2018**, *8*, 681. [CrossRef]

20. Chugh, H.; Sood, D.; Chandra, I.; Tomar, V.; Dhawan, G.; Chandra, R. Role of gold and silver nanoparticles in cancer nano-medicine. *Artif. Cell. Nanomed. Biotechnol.* **2018**, *46*, 1210–1220. [CrossRef]

21. Amendola, V.; Meneghetti, M. Laser ablation synthesis in solution and size manipulation of noble metal nanoparticles. *Phys. Chem. Chem. Phys.* **2009**, *11*, 3805–3821. [CrossRef] [PubMed]

22. Iravani, S.; Korbekandi, H.; Mirmohammadi, S.V.; Zolfaghari, B. Synthesis of silver nanoparticles: Chemical, physical and biological methods. *Res. Pharm. Sci.* **2014**, *9*, 385–406. [PubMed]

23. Kruis, F.E.; Fissan, H.; Rellinghaus, B. Sintering and evaporation characteristics of gas-phase synthesis of size-selected PbS nanoparticles. *Mater. Sci. Eng. B* **2000**, *69*, 329–334. [CrossRef]

24. Jung, J.H.; Oh, H.C.; Noh, H.S.; Ji, J.H.; Kim, S.S. Metal nanoparticle generation using a small ceramic heater with a local heating area. *J. Aerosol. Sci.* **2006**, *37*, 1662–1670. [CrossRef]

25. Chen, Y.-H.; Yeh, C.-S. Laser ablation method: Use of surfactants to form the dispersed Ag nanoparticles. *Colloids Surf. A* **2002**, *197*, 133–139. [CrossRef]

26. Kinnear, C.; Moore, T.L.; Rodriguez-Lorenzo, L.; Rothen-Rutishauser, B.; Petri-Fink, A. Form follows function: Nanoparticle shape and its implications for nanomedicine. *Chem. Rev.* **2017**, *117*, 11476–11521. [CrossRef] [PubMed]

27. Pillai, Z.S.; Kamat, P.V. What factors control the size and shape of silver nanoparticles in the citrate ion reduction method? *J. Phys. Chem. B* **2004**, *108*, 945–951. [CrossRef]

28. Turkevich, J.; Kim, G. Palladium: Preparation and catalytic properties of particles of uniform size. *Science* **1970**, *169*, 873–879. [CrossRef]

29. Turkevich, J. Colloidal gold. Part I. *Gold. Bull.* **1985**, *18*, 86–91. [CrossRef]

30. Brust, M.; Walker, M.; Bethell, D.; Schiffrin, D.J.; Whyman, R. Synthesis of thiol-derivatised gold nanoparticles in a two-phase liquid-liquid system. *J. Chem. Soc. Chem. Commun.* **1994**, 801–802. [CrossRef]

31. Dakal, T.C.; Kumar, A.; Majumdar, R.S.; Yadav, V. Mechanistic basis of antimicrobial actions of silver nanoparticles. *Front. Microbiol.* **2016**, *7*, 1831. [CrossRef] [PubMed]

32. Evanoff, D.D., Jr.; Chumanov, G. Synthesis and optical properties of silver nanoparticles and arrays. *Chem. Phys. Chem.* **2005**, *6*, 1221–1231. [CrossRef] [PubMed]

33. Goulet, P.J.G.; Lennox, R.B. New insights into Brust—Schiffrin metal nanoparticle synthesis. *J. Am. Chem. Soc.* **2010**, *132*, 9582–9584. [CrossRef] [PubMed]

34. Dong, X.; Ji, X.; Wu, H.; Zhao, L.; Li, J.; Yang, W. Shape control of silver nanoparticles by stepwise citrate reduction. *J. Phys. Chem. B* **2009**, *113*, 6573–6576. [CrossRef]

35. Bai, J.; Li, Y.; Du, J.; Wang, S.; Zheng, J.; Yang, Q.; Chen, X. One-pot synthesis of polyacrylamide-gold nanocomposite. *Mater. Chem. Phys.* **2007**, *106*, 412–415. [CrossRef]

36. Oliveira, M.M.; Ugarte, D.; Zanchet, D.; Zarbina, A.J.G. Influence of synthetic parameters on the size, structure, and stability of dodecanethiol-stabilized silver nanoparticles. *J. Colloid Interface Sci.* **2005**, *292*, 429–435. [CrossRef]

37. Xue, C.; Métraux, G.S.; Millstone, J.E.; Mirkin, C.A. Mechanistic study of photomediated triangular silver nanoprism growth. *J. Am. Chem. Soc.* **2008**, *130*, 8337–8344. [CrossRef]

38. Fayaz, A.M.; Balaji, K.; Girilal, M.; Yadav, R.; Kalaichelvan, P.T.; Venketesan, R. Biogenic synthesis of silver nanoparticles and their synergistic effect with antibiotics: A study against gram-positive and gram-negative bacteria. *Nanomedicine* **2010**, *6*, 103–109. [CrossRef]

39. Ahmed, S.; Ikram, S. Silver nanoparticles: One pot green synthesis using *Terminalia arjuna* extract for biological application. *J. Nanosci. Nanotechnol.* **2015**, *6*, 1000309.

40. Klaus-Joerger, T.; Joerger, R.; Olsson, E.; Granqvist, C.-G. Bacteria as workers in the living factory: Metal-accumulating bacteria and their potential for materials science. *Trends. Biotechnol.* **2001**, *19*, 15–20. [CrossRef]

41. Arokiyaraj, S.; Vincent, S.; Saravanan, M.; Lee, Y.; Oh, Y.K.; Kim, K.H. Green synthesis of silver nanoparticles using *Rheum palmatum* root extract and their antibacterial activity against *Staphylococcus aureus* and *Pseudomonas aeruginosa*. *Artif. Cells Nanomed. Biotechnol.* **2016**, *45*, 372–379. [CrossRef] [PubMed]

42. Patra, S.; Mukherjee, S.; KumarBarui, A.; Ganguly, A.; Sreedhar, B.; Patra, C.R. Green synthesis, characterization of gold and silver nanoparticles and their potential application for cancer therapeutics. *Mater. Sci. Eng. C* **2015**, *53*, 298–309. [CrossRef] [PubMed]

43. Gajbhiye, M.; Kesharwani, J.; Ingle, A.; Gade, A.; Rai, M. Fungus-mediated synthesis of silver nanoparticles and their activity against pathogenic fungi in combination with fluconazole. *Nanomedicine* **2009**, *5*, 382–386. [CrossRef] [PubMed]

44. Mukherjee, P.; Ahmad, A.; Mandal, D.; Senapati, S.; Sainkar, S.R.; Khan, M.I.; Parishcha, R.; Ajaykumar, P.V.; Alam, M.; Kumar, R.; et al. Fungus-mediated synthesis of silver nanoparticles and their immobilization in the Mycelial matrix: A novel biological approach to nanoparticle synthesis. *Nano Lett.* **2001**, *1*, 515–519. [CrossRef]

45. Rauwel, P.; Küünal, S.; Ferdov, S.; Rauwel, E. A review on the green synthesis of silver nanoparticles and their morphologies studied via TEM. *Adv. Mater. Sci. Eng.* **2015**, *2015*, 682749. [CrossRef]

46. Lengke, M.F.; Fleet, M.E.; Southam, G. Biosynthesis of silver nanoparticles by filamentous cyanobacteria from a silver (I) nitrate complex. *Langmuir* **2007**, *27*, 2694–2699. [CrossRef] [PubMed]

47. Kalimuthu, K.; Babu, R.S.; Venkataraman, D.; Bilal, M.; Gurunathan, S. Biosynthesis of silver nanocrystals by Bacillus licheniformis. *Colloids Surf. B Biointerfaces* **2008**, *65*, 150–153. [CrossRef]

48. Kumar, S.A.; Abyaneh, M.K.; Gosavi, S.W.; Kulkarni, S.K.; Pasricha, R.; Ahmad, A.; Khan, M.I. Nitrate reductase-mediated synthesis of silver nanoparticles from $AgNO_3$. *Biotechnol. Lett.* **2007**, *29*, 439–445. [CrossRef]

49. Vaidyanathan, R.; Gopalram, S.; Kalishwaralal, K.; Deepak, V.; Ram, S.; Pandian, K.; Gurunathan, S. Enhanced silver nanoparticle synthesis by optimization of nitrate reductase activity. *Colloids Surf. B Biointerfaces* **2010**, *75*, 335–341. [CrossRef]

50. Li, G.; He, D.; Qian, Y.; Guan, B.; Gao, S.; Cui, Y.; Yokoyama, K.; Wang, L. Fungus-mediated green synthesis of silver nanoparticles using *Aspergillus terreus*. *Int. J. Mol. Sci.* **2012**, *13*, 466–476. [CrossRef]

51. Klaus, T.; Joerger, R.; Olsson, E.; Granqvist, C.-G. Silver-based crystalline nanoparticles, microbially fabricated. *Proc. Natl. Acad. Sci. USA* **1999**, *96*, 13611–13614. [CrossRef] [PubMed]

52. Xin, H.; Namgung, B.; Lee, L.P. Nanoplasmonic optical antennas for life sciences and medicine. *Nat. Rev. Mater.* **2018**, *3*, 228–243. [CrossRef]

53. Lee, S.H.; Ko, H.J.; Park, T.H. Real-time monitoring of odorant-induced cellular reactions using surface plasmon resonance. *Biosens. Bioelectron.* **2009**, *25*, 55–60. [CrossRef] [PubMed]

54. Agnihotri, S.; Mukherji, S.; Mukher, S. Size-controlled silver nanoparticles synthesized over the range 5-100 nm using the same protocol and their antibacterial efficacy. *RSC Adv.* **2014**, *4*, 3974–3983. [CrossRef]

55. Aherne, D.; Ledwith, D.M.; Gara, M.; Kelly, J.M. Optical properties and growth aspects of silver nanoprisms produced by a highly reproducible and rapid synthesis at room temperature. *Adv. Funct. Mater.* **2008**, *18*, 2005–2016. [CrossRef]

56. Jin, R.; Cao, Y.; Mirkin, C.A.; Kelly, K.L.; Schatz, G.C.; Zheng, J.G. Photoinduced conversion of silver nanospheres to nanoprisms. *Science* **2001**, *294*, 1901–1903. [CrossRef] [PubMed]

57. Sönnichsen, C.; Reinhard, B.M.; Liphardt, J.; Alivisatos, A.P. A molecular ruler based on plasmon coupling of single gold and silver nanoparticles. *Nat. Biotechnol.* **2005**, *23*, 741. [CrossRef] [PubMed]

58. Wiley, B.; Herricks, T.; Sun, Y.; Xia, Y. Polyol synthesis of silver nanoparticles: Use of chloride and oxygen to promote the formation of single-crystal, truncated cubes and tetrahedrons. *Nano Lett.* **2004**, *4*, 1733–1739. [CrossRef]

59. Hanisch, M.; Mackovic', M.; Taccardi, N.; Spiecker, E.; Taylor, R.N.K. Synthesis of silver nanoparticle necklaces without explicit addition of reducing or templating agent. *Chem. Commun.* **2012**, *48*, 4287–4289. [CrossRef]

60. Wiley, B.J.; Chen, Y.; McLellan, J.M.; Xiong, Y.; Li, Z.-Y.; Ginger, D.; Xia, Y. Synthesis and optical properties of silver nanobars and nanorice. *Nano Lett.* **2007**, *7*, 1032–1036. [CrossRef]

61. Xue, B.; Wang, D.; Zuo, J.; Kong, X.; Zhang, Y.; Liu, X.; Tu, L.; Chang, Y.; Li, C.; Wu, F.; et al. Towards high quality triangular silver nanoprisms: Improved synthesis, six-tip based hot spots and ultra-high local surface plasmon resonance sensitivity. *Nanoscale* **2015**, *7*, 8048. [CrossRef] [PubMed]

62. Wiley, B.J.; Xiong, Y.; Li, Z.-Y.; Yin, Y.; Xi, Y. Right bipyramids of silver: A new shape derived from single twinned seeds. *Nano Lett.* **2006**, *6*, 765–768. [CrossRef] [PubMed]

63. Tran, H.N.; Nghiem, T.H.L.; Vu, T.T.D.; Chu, V.H.; Le, Q.H.; Hoang, T.M.N.; Nguyen, L.T.; Pham, D.M.; Tong, K.T.; Do, Q.H.; et al. Optical nanoparticles: Synthesis and biomedical application. *Adv. Nat. Sci. Nanosci. Nanotechnol.* **2015**, *6*, 023002.

64. Hahm, E.; Cha, M.G.; Kang, E.J.; Pham, X.-H.; Lee, S.H.; Kim, H.-M.; Kim, D.-E.; Lee, Y.-S.; Jeong, D.-H.; Jun, B.-H. Multilayer Ag-embedded silica nanostructure as a surface-enhanced raman scattering-based chemical sensor with dual-function internal standards. *ACS Appl. Mater. Interfaces* **2018**, *10*, 40748–40755. [CrossRef] [PubMed]

65. Park, J.; Cha, S.-H.; Cho, S.; Park, Y. Green synthesis of gold and silver nanoparticles using gallic acid: Catalytic activity and conversion yield toward the 4-nitrophenol reduction reaction. *J. Nanopart. Res.* **2016**, *18*, 166. [CrossRef]

66. Lee, K.-S.; El-Sayed, M.A. Dependence of the enhanced optical scattering efficiency relative to that of absorption for gold metal nanorods on aspect ratio, size, end-cap shape, and medium refractive index. *J. Phys. Chem. B* **2005**, *109*, 20331–20338. [CrossRef]

67. Sriram, M.I.; Kalishwaralal, K.; Barathmanikanth, S.; Gurunathani, S. Size-based cytotoxicity of silver nanoparticles in bovine retinal endothelial cells. *Nanosci. Methods* **2012**, *1*, 56–77. [CrossRef]

68. Stoehr, L.C.; Gonzalez, E.; Stampfl, A.; Casals, E.; Duschl, A.; Puntes, V.; Oostingh, G.J. Shape matters: Effects of silver nanospheres and wires on human alveolar epithelial cells. *Part. Fiber Toxicol.* **2011**, *8*, 36. [CrossRef]

69. Li, W.-R.; Xie, X.-B.; Shi, Q.-S.; Zeng, H.-Y.; OU-Yang, Y.-S.; Chen, Y.-B. Antibacterial activity and mechanism of silver nanoparticles on *Escherichia coli*. *Appl. Microbiol. Biotechnol.* **2010**, *85*, 1115–1122. [CrossRef]

70. Greulich, C.; Braun, D.; Peetsch, A.; Diendorf, J.; Siebers, B.; Epple, M.; Köller, M. The toxic effect of silver ions and silver nanoparticles towards bacteria and human cells occurs in the same concentration range. *RSC Adv.* **2012**, *2*, 6981–6987. [CrossRef]

71.   Abuayyash, A.; Ziegler, N.; Gessmann, J.; Sengstock, C.; Schildhauer, T.A.; Ludwig, A.; Köller, M. Antibacterial efficacy of sacrifical anode thin films combining silver with platinum group elements within a bacteria-containing human plasma clot. *Adv. Eng. Mater.* **2018**, *20*, 1700493. [CrossRef]

72.   Maurer, L.L.; Meyer, J.N. A systematic review of evidence for silver nanoparticle-induced mitochondrial toxicity. *Environ. Sci. Nano.* **2016**, *3*, 311–322. [CrossRef]

73.   Kim, S.-H.; Lee, H.-S.; Ryu, D.-S.; Choi, S.-J.; Lee, D.-S. Antibacterial activity of silver-nanoparticles against *Staphylococcus aureus* and *Escherichia coli. Korean J. Microbiol. Biotechnol.* **2011**, *39*, 77–85.

74.   Tamayo, L.A.; Zapata, P.A.; Vejar, N.D.; Azócar, M.I.; Gulppi, M.A.; Zhou, X.; Thompson, G.E.; Rabagliati, F.M.; Páez, M.A. Release of silver and copper nanoparticles from polyethylene nanocomposites and their penetration into Listeria monocytogenes. *Mat. Sci. Eng.* **2014**, *40*, 24–31. [CrossRef]

75.   Abbaszadegan, A.; Ghahramani, Y.; Gholami, A.; Hemmateenejad, B.; Dorostkar, S.; Nabavizadeh, M.; Sharghi, H. The effect of charge at the surface of silver nanoparticles on antimicrobial activity against gram-positive and gram-negative bacteria: A preliminary study. *J. Nanomater.* **2015**, *720654.* [CrossRef]

76.   Loo, C.-Y.; Rohanizadeh, R.; Young, P.M.; Traini, D.; Cavaliere, R.; Whitchurch, C.B.; Lee, W.-H. Combination of silver nanoparticles and curcumin nanoparticles for enhanced anti-biofilm activities. *J. Agric. Food Chem.* **2016**, *64*, 2513–2522. [CrossRef]

77.   Pařil, P.; Baar, J.; Čermák, P.; Rademacher, P.; Prucek, R.; Sivera, M.; Panáček, A. Antifungal effects of copper and silver nanoparticles against white and brown-rot fungi. *J. Mater. Sci.* **2017**, *52*, 2720–2729. [CrossRef]

78.   Tang, S.; Zheng, J. Antibacterial activity of silver nanoparticles: Structural effects. *Adv. Healthc. Mater.* **2018**, *7*, 1701503. [CrossRef] [PubMed]

79.   Gu, J.; Zhang, Y.-W.; Tao, F. Shape control of bimetallic nanocatalysts through well-designed colloidal chemistry approaches. *Chem. Soc. Rev.* **2012**, *41*, 8050–8065. [CrossRef] [PubMed]

80.   Zhu, C.; Zeng, J.; Tao, J.; Johnson, M.C.; Schmidt-Krey, I.; Blubaugh, L.; Zhu, Y.; Gu, Z.; Xia, Y. Kinetically controlled overgrowth of Ag or Au on Pd nanocrystal seeds: From hybrid dimers to nonconcentric and concentric bimetallic nanocrystals. *J. Am. Chem. Soc.* **2012**, *134*, 15822–15831. [CrossRef] [PubMed]

81.   Srinoi, P.; Chen, Y.-T.; Vittur, V.; Marquez, M.D.; Lee, T.R. Bimetallic nanoparticles: Enhanced magnetic and optical properties for emerging biological applications. *Appl. Sci.* **2018**, *8*, 1106. [CrossRef]

82.   Kim, T.H.; Pham, X.H.; Rho, W.Y.; Kim, H.M.; Hahm, E.; Ha, Y.; Son, B.S.; Lee, S.H.; Jun, B.H. Ag and Ag- Au introduced silica-coated magnetic beads. *B Korean Chem. Soc.* **2018**, *39*, 250–256. [CrossRef]

83.   Pham, X.-H.; Hahm, E.; Kang, E.; Ha, Y.N.; Lee, S.H.; Rho, W.-Y.; Lee, Y.-S.; Jeong, D.H.; Jun, B.-H. Gold-silver bimetallic nanoparticles with a raman labeling chemical assembled on silica nanoparticles as an internal-standard-containing nanoprobe. *J. Alloy. Compd.* **2018**, *779*, 360–366. [CrossRef]

84.   Pham, X.-H.; Lee, M.; Shim, S.; Jeong, S.; Kim, H.-M.; Hahm, E.; Lee, S.H.; Lee, Y.-S.; Jeong, D.H.; Jun, B.-H. Highly sensitive and reliable SERS probes based on nanogap control of Au-Ag alloy on silica nanoparticle. *RSC Adv.* **2017**, *7*, 7015–7021. [CrossRef]

85.   Xia, Y.; Gilroy, K.D.; Peng, H.C.; Xia, X. Seed-mediated growth of colloidal metal nanocrystals. *Angew. Chem. Int. Ed.* **2017**, *56*, 60–95. [CrossRef] [PubMed]

86.   Bönnemann, H.; Richards, R.M. Nanoscopic metal particles-synthetic methods and potential applications. *Eur. J. Inorg. Chem.* **2001**, 2455–2480. [CrossRef]

87.   Skrabalak, S.E.; Chen, J.; Sun, Y.; Lu, X.; Au, L.; Cobley, C.M.; Xia, Y. Gold nanocages: Synthesis, properties, and applications. *Acc. Chem. Res.* **2008**, *41*, 1587–1595. [CrossRef]

88.   Lee, J.-H.; Shin, Y.; Lee, W.; Whang, K.; Kim, D.; Lee, L.P.; Choi, J.-W.; Kang, T. General and programmable synthesis of hybrid liposome/metal nanoparticles. *Sci. Adv.* **2016**, *2*, e1601838. [CrossRef]

89.   Zhang, W.-S.; Cao, J.-T.; Dong, Y.-X.; Wang, H.; Ma, S.-H.; Liu, Y.-M. Enhanced chemiluminescence by Au-Ag core-shell nanoparticles: A general and practical biosensing platform for tumor marker detection. *J. Lumin.* **2018**, *201*, 163–169. [CrossRef]

90.   Lee, S.J.; Morrill, A.R.; Moskovits, M. Hot spots in silver nanowire bundles for Surface-enhanced raman spectroscopy. *J. Am. Chem. Soc.* **2006**, *128*, 2200–2201. [CrossRef]

91.   Kneipp, J.; Kneipp, H.; McLaughlin, M.; Brown, D.; Kneipp, K. In vivo molecular probing of cellular compartments with gold nanoparticles and nanoaggregates. *Nano Lett.* **2006**, *6*, 2225–2231. [CrossRef] [PubMed]

92. Camden, J.P.; Dieringer, J.A.; Wang, Y.; Masiello, D.J.; Marks, L.D.; Schatz, G.C.; Duyne, R.P.V. Probing the structure of single-molecule surface-enhanced raman scattering hot spots. *J. Am. Chem. Soc.* **2008**, *130*, 12616–12617. [CrossRef] [PubMed]

93. Kleinman, S.L.; Frontiera, R.R.; Henry, A.-I.; Dieringer, J.A.; Duyne, R.P.V. Creating, characterizing, and controlling chemistry with SERS hot spots. *Phys. Chem. Chem. Phys.* **2013**, *15*, 21–36. [CrossRef] [PubMed]

94. Chen, H.-Y.; Lin, M.-H.; Wang, C.-Y.; Chang, Y.-M.; Gwo, S. Large-scale hot spot engineering for quantitative SERS at the single-molecule scale. *J. Am. Chem. Soc.* **2015**, *137*, 13698–13705. [CrossRef] [PubMed]

95. Sun, H.-B.; Fu, C.; Xia, Y.-J.; Zhang, C.-W.; Du, J.-H.; Yang, W.-C.; Guo, P.-F.; Xu, J.-Q.; Wang, C.-L.; Jia, Y.-L. Enhanced Raman scattering of graphene by silver nanoparticles with different densities and locations. *Mater. Res. Express* **2017**, *4*, 025012. [CrossRef]

96. Lai, C.-H.; Wang, G.-A.; Ling, T.-K.; Wang, T.-J.; Chiu, P.-K.; Chau, C.Y.-F.; Huang, C.-C.; Chiang, H.-P. Near infrared surface-enhanced Raman scattering based on starshaped gold/silver nanoparticles and hyperbolic metamaterial. *Sci. Rep.* **2017**, *7*, 5546. [CrossRef] [PubMed]

97. Mulvaney, S.P.; Musick, M.D.; Keating, C.D.; Natan, M.J. Glass-coated, analyte-tagged nanoparticles: A new tagging system based on detection with surface-enhanced raman scattering. *Langmuir* **2003**, *19*, 4784–4790. [CrossRef]

98. Doering, W.E.; Piotti, M.E.; Natan, M.J.; Freeman, R.G. SERS as a foundation for nanoscale, optically detected biological labels. *Adv. Mater.* **2007**, *19*, 3100–3108. [CrossRef]

99. Loo, C.; Lowery, A.; Halas, N.; West, J.; Drezek, R. Immunotargeted nanoshells for integrated cancer imaging and therapy. *Nano Lett.* **2005**, *5*, 709–711. [CrossRef]

100. Kang, H.; Jeong, S.; Park, Y.; Yim, J.; Jun, B.-H.; Kyeong, S.; Yang, J.-K.; Kim, G.; Hong, S.; Lee, L.P.; et al. Near-infrared SERS nanoprobes with plasmonic Au/Ag hollow-shell assemblies for in vivo multiplex detection. *Adv. Funct. Mater.* **2013**, *23*, 3719–3727. [CrossRef]

101. Kelkar, S.S.; Reineke, T.M. Theranostics: Combining imaging and therapy. *Bioconjug. Chem.* **2011**, *22*, 1879–1903. [CrossRef] [PubMed]

102. Jun, B.H.; Noh, M.S.; Kim, J.; Kim, G.; Kang, H.; Kim, M.S.; Seo, Y.T.; Baek, J.; Kim, J.H.; Park, J.; et al. Multifunctional silver-embedded magnetic nanoparticles as SERS nanoprobes and their applications. *Small* **2010**, *6*, 119–125. [CrossRef] [PubMed]

103. Hu, M.; Ghoshal, A.; Marquez, M.; Kik, P.G. Single particle spectroscopy study of metal-film-induced tuning of silver nanoparticle plasmon resonances. *J. Phys. Chem. C* **2010**, *114*, 7509–7514. [CrossRef]

104. Ardini, M.; Huang, J.-A.; Sánchez, C.S.; Mousavi, M.Z.; Caprettini, V.; Maccaferri, N.; Melle, G.; Bruno, G.; Pasquale, L.; Garoli, D.; et al. Live intracellular biorthogonal imaging by surface enhanced raman spectroscopy. *Sci. Rep.* **2018**, *8*, 12652.

105. Yen, C.-W.; de Puig, H.; Tam, J.; Gómez-Márquez, J.; Bosch, I.; Hamad-Schifferli, K.; Gehrkea, L. Multicolored silver nanoparticles for multiplexed disease diagnostics: Distinguishing dengue, yellow fever, and ebola viruses. *Lab Chip.* **2015**, *15*, 1638–1641. [CrossRef] [PubMed]

106. Liu, J.; Wang, Z.; Liu, F.D.; Kane, A.B.; Hurt, R.H. Chemical transformations of nanosilver in biological environment. *ACS Nano* **2012**, *6*, 9887–9899. [CrossRef] [PubMed]

107. Zhou, W.; Ma, Y.; Yang, H.; Ding, Y.; Luo, X. A label-free biosensor based on silver nanoparticles array for clinical detection of serum p53 in head and neck squamous cell carcinoma. *Int. J. Nanomed.* **2011**, *6*, 381–386. [CrossRef] [PubMed]

108. Zhang, F.; Braun, G.B.; Shi, Y.; Zhang, Y.; Sun, X.; Reich, N.O.; Zhao, D.; Stucky, G. Fabrication of Ag@SiO2@Y$_2$O$_3$:Er nanostructures for bioimaging: Tuning of the upconversion fluorescence with silver nanoparticles. *J. Am. Chem. Soc.* **2010**, *132*, 2850–2851. [CrossRef]

109. Bala, V.; Peria, k.; Prakash, k.; Muthupandi, K.; Rajan, A. Nanosilver for selective and sensitive sensing of saturnism. *Sens. Actuators B* **2017**, *241*, 814–820.

110. Dragan, A.I.; Bishop, E.S.; Casas-Finet, J.R.; Strouse, R.J.; McGivney, J.; Schenerman, M.A.; Geddes, C.D. Distance dependence of metal-enhanced fluorescence. *Plasmonics* **2012**, *7*, 739–744. [CrossRef]

111. Aslan, K.; Gryczynski, I.; Malicka, J.; Matveeva, E.; Lakowicz, J.R.; Geddes, C.D. Metal-enhanced fluorescence: An emerging tool in biotechnology. *Curr. Opin. Biotechnol.* **2005**, *16*, 55–62. [CrossRef] [PubMed]

112. Stockman, M. Nanoplasmonics: The physics behind the applications. *Phys. Today* **2011**, *64*, 39. [CrossRef]

113. Lakowicz, J.R.; Malicka, J.; Gryczynski, I. Silver particles enhance emission of fluorescent DNA oligomers. *Biotechniques* **2003**, *34*, 62–68. [CrossRef] [PubMed]

114. Wang, Y.; Li, Z.; Li, H.; Vuki, M.; Xu, D.; Chen, H.-Y. A novel aptasensor based on silver nanoparticle enhanced fluorescence. *Biosens. Bioelectron.* **2012**, *32*, 76–81. [CrossRef] [PubMed]

115. Suresh, A.K.; Pelletier, D.A.; Wang, W.; Morrell-Falvey, J.L.; Gu, B.; Doktycz, M.J. Cytotoxicity induced by engineered silver nanocrystallites is dependent on surface coatings and cell types. *Langmuir* **2012**, *28*, 2727–2735. [CrossRef] [PubMed]

116. Schlinkert, P.; Casals, E.; Boyles, M.; Tischler, U.; Hornig, E.; Tran, N.; Zhao, J.; Himly, M.; Riediker, M.; Oostingh, G.J.; et al. The oxidative potential of differently charged silver and gold nanoparticles on three human lung epithelial cell types. *J. Nanobiotechnol.* **2015**, *13*, 1. [CrossRef] [PubMed]

117. Kawata, K.; Osawa, M.; Okabe, S. In vitro toxicity of silver nanoparticles at noncytotoxic doses to HepG2 human hepatoma cells. *Environ. Sci. Technol.* **2009**, *43*, 6046–6051. [CrossRef]

118. Foldbjerg, R.; Dang, D.A.; Autrup, H. Cytotoxicity and genotoxicity of silver nanoparticles in the human lung cancer cell line, A549. *Arch. Toxicol.* **2011**, *85*, 743–750. [CrossRef]

119. Jeyaraj, M.; Sathishkumar, G.; Sivanandhan, G.; MubarakAli, D.; Rajesh, M.; Arun, R.; Kapildev, G.; Manickavasagam, M.; Thajuddin, N.; Premkumar, K.; et al. Biogenic silver nanoparticles for cancer treatment: An experimental report. *Colloids Surf. B Biointerfaces* **2013**, *106*, 86–92. [CrossRef]

120. Gurunathan, S.; Han, J.W.; Eppakayala, V.; Jeyaraj, M.; Kim, J.-H. Cytotoxicity of biologically synthesized silver nanoparticles in MDA-MB-231 human breast cancer cells. *Biomed. Res. Int.* **2013**, *2013*, 535796. [CrossRef]

121. Vasanth, K.; Ilango, K.; MohanKumar, R.; Agrawal, A.; Dubey, G.P. Anticancer activity of *Moringa oleifera* mediated silver nanoparticles on human cervical carcinoma cells by apoptosis induction. *Colloids Surf. B Biointerfaces* **2014**, *117*, 354–359. [CrossRef] [PubMed]

122. Carlson, C.; Hussain, M.; Schrand, A.M.; Braydich-Stolle, L.K.; Hess, K.L.; Jones, R.L.; Schlager, J.J. Unique cellular interaction of silver nanoparticles: Size-dependent generation of reactive oxygen species. *J. Phys. Chem. B* **2008**, *112*, 13608–13619. [CrossRef] [PubMed]

123. AshaRani, P.; Hande, M.P.; Valiyaveettil, S. Anti-proliferative activity of silver nanoparticles. *BMC Cell Biol.* **2009**, *10*, 65. [CrossRef] [PubMed]

124. Gurunathan, S.; Lee, K.-J.; Kalishwaralal, K.; Sheikpranbabu, S.; Vaidyanathan, R.; Eom, S.H. Antiangiogenic properties of silver nanoparticles. *Biomaterials* **2009**, *30*, 6341–6350. [CrossRef] [PubMed]

125. Thapa, R.K.; Kim, J.H.; Jeong, J.-H.; Shin, B.S.; Choi, H.-G.; Yong, C.S.; Kim, J.O. Silver nanoparticle-embedded graphene oxide-methotrexate for targeted cancer treatment. *Colloids Surf. B* **2017**, *153*, 95–103. [CrossRef]

126. Azizi, M.; Ghourchian, H.; Yazdian, F.; Bagherifam, S.; Bekhradnia, S.; Nyström, B. Anti-cancerous effect of albumin coated silver nanoparticles on MDA-MB 231 human breast cancer cell line. *Sci. Rep.* **2017**, *7*, 5178. [CrossRef] [PubMed]

127. Anwar, A.; Rajendran, K.; Siddiqui, R.; Shah, M.R.; Khan, N.A. Clinically approved drugs against CNS diseases as potential therapeutic agents to target brain-eating amoebae. *ACS Chem. Neurosci.* **2018**. [CrossRef]

128. Sharma, V.K.; Yngard, R.A.; Lin, Y. Silver nanoparticles: Green synthesis and their antimicrobial activities. *Adv. Colloid. Interface. Sci.* **2009**, *145*, 83–96. [CrossRef]

129. Sondi, I.; Salopek-Sondi, B. Silver nanoparticles as antimicrobial agent: A case study on *E. coli* as a model for Gram-negative bacteria. *J. Colloid Interface Sci.* **2004**, *275*, 177–182. [CrossRef]

130. Triyana, J.K.; Suharyadi, H.E. High-performance silver nanowire film on flexible substrate prepared by meyer-rod coating. *Mater. Sci. Eng.* **2017**, *202*, 012055.

131. Pham, X.-H.; Hahm, E.; Kim, T.H.; Kim, H.-M.; Lee, S.H.; Lee, Y.-S.; Jeong, D.H.; Jun, B.-H. Enzyme-catalyzed Ag growth on Au nanoparticle-assembled structure for highly sensitive colorimetric immunoassay. *Sci. Rep.* **2018**, *8*, 6290. [CrossRef] [PubMed]

132. Prosposito, P.; Mochi, F.; Ciotta, E.; Casalboni, M.; Matteis, F.D.; Venditti, I.; Fontana, L.; Testa, G.; Fratoddi, I. Hydrophilic silver nanoparticles with tunable optical properties: Application for the detection of heavy metals in water. *Beilstein. J. Nanotechnol.* **2016**, *7*, 1654–1661. [CrossRef] [PubMed]

133. Hu, B.; Wang, N.; Hana, L.; Chen, M.-L.; Wang, J.-H. Core-shell-shell nanorods for controlled release of silver that can serve as a nanoheater for photothermal treatment on bacteria. *Acta Biomater.* **2015**, *11*, 511–519. [CrossRef] [PubMed]

134. Ding, X.; Yuan, P.; Gao, N.; Zhu, H.; Yang, Y.Y.; Xu, Q.-H. Au-Ag core-shell nanoparticles for simultaneous bacterial imaging and synergistic antibacterial activity. *Nanomed. Nanotechnol. Biol. Med.* **2017**, *13*, 297–305. [CrossRef] [PubMed]

135. Kamimura, S.; Yamashita, S.; Abe, S.; Tsubota, T.; Ohno, T. Effect of core@shell (Au@Ag) nanostructure on surface plasmon-induced photocatalytic activity under visible light irradiation. *Appl. Catal. B Environ.* **2017**, *211*, 11–17. [CrossRef]

136. Lu, L.; Luo, Z.; Xu, T.; Yu, L. Cooperative plasmonic effect of Ag and Au nanoparticles on enhancing performance of polymer solar cells. *Nano Lett.* **2013**, *13*, 59–64. [CrossRef]

137. Song, D.H.; Kim, H.-Y.; Kim, H.-S.; Suh, J.S.; Jun, B.-H.; Rho, W.-Y. Preparation of plasmonic monolayer with Ag and Au nanoparticles for dye-sensitized solar cells. *Chem. Phys. Lett.* **2017**, *687*, 152–157. [CrossRef]

138. Angelis, F.D.; Das, G.; Candeloro, P.; Patrini, M.; Galli, M.; Bek, A.; Lazzarino, M.; Maksymov, I.; Liberale, C.; Andreani, L.C.; et al. Nanoscale chemical mapping using three-dimensional adiabatic compression of surface plasmon polaritons. *Nat. Nanotechnol.* **2009**, *5*, 67–72. [CrossRef]

139. Chen, X.; Jia, B.; Saha, J.K.; Cai, B.; Stokes, N.; Qiao, Q.; Wang, Y.; Shi, Z.; Gu, M. Broadband enhancement in thin-film amorphous silicon solar cells enabled by nucleated silver nanoparticles. *Nano Lett.* **2012**, *12*, 2187–2192. [CrossRef]

140. Rho, W.-Y.; Chun, M.-H.; Kim, H.-S.; Kim, H.-M.; Suh, J.S.; Jun, B.-H. Ag nanoparticle-functionalized open-ended freestanding TiO$_2$ nanotube arrays with a scattering layer for improved energy conversion efficiency in dye-sensitized solar cells. *Nanomaterials* **2016**, *6*, 117. [CrossRef]

141. Kim, H.-S.; Chun, M.-H.; Suh, J.S.; Jun, B.-H.; Rho, W.-Y. Dual functionalized freestanding TiO$_2$ nanotube arrays coated with Ag nanoparticles and carbon materials for dye-sensitized solar cells. *Appl. Sci.* **2017**, *7*, 576. [CrossRef]

142. Rho, W.-Y.; Kim, H.-S.; Lee, S.H.; Jung, S.; Suh, J.S.; Hahn, Y.-B.; Jun, B.-H. Front-illuminated dye-sensitized solar cells with Ag nanoparticle-functionalized freestanding TiO$_2$ nanotube arrays. *Chem. Phys. Lett.* **2014**, *614*, 78–81. [CrossRef]

143. Choi, H.; Ko, S.-J.; Choi, Y.; Joo, P.; Kim, T.; Lee, B.R.; Jung, J.-W.; Choi, H.J.; Cha, M.; Jeong, J.-R.; et al. Versatile surface plasmon resonance of carbon-dot-supported silver nanoparticles in polymer optoelectronic devices. *Nat. Photonics* **2013**, *7*, 732–738. [CrossRef]

International Journal of
Molecular Sciences

MDPI

Article

Functionalized β-Cyclodextrin Immobilized on Ag-Embedded Silica Nanoparticles as a Drug Carrier

Eun Ji Kang [1], Yu Mi Baek [1], Eunil Hahm [1], Sang Hun Lee [1], Xuan-Hung Pham [1], Mi Suk Noh [2], Dong-Eun Kim [1] and Bong-Hyun Jun [1,*]

[1] Department of Bioscience and Biotechnology, Konkuk University, Seoul 05029, Korea;
    ejkang@konkuk.ac.kr (E.J.K.); undine1213@naver.com (Y.M.B.); greenice@konkuk.ac.kr (E.H.);
    shlee.ucb@gmail.com (S.H.L.); phamricky@gmail.com (X.-H.P.); kimde@konkuk.ac.kr (D.-E.K.)
[2] Bio-Health Convergence Institute, Korea Testing Certification, Gunpo 15809, Korea; pourlady@ktc.re.kr
* Correspondence: bjun@konkuk.ac.kr; Tel.: +82-450-0521

Received: 18 December 2018; Accepted: 12 January 2019; Published: 14 January 2019

Abstract: Cyclodextrins (CDs) have beneficial characteristics for drug delivery, including hydrophobic interior surfaces. Nanocarriers with β-CD ligands have been prepared with simple surface modifications as drug delivery vehicles. In this study, we synthesized β-CD derivatives on an Ag-embedded silica nanoparticle (NP) (SiO$_2$@Ag NP) structure to load and release doxorubicin (DOX). Cysteinyl-β-CD and ethylenediamine-β-CD (EDA-β-CD) were immobilized on the surface of SiO$_2$@Ag NPs, as confirmed by transmission electron microscopy (TEM), ultraviolet-visible (UV-Vis) spectrophotometry, and Fourier transform infrared (FTIR) spectroscopy. DOX was introduced into the β-CD on the SiO$_2$@Ag NPs and then successfully released. Neither cysteinyl-β-CD and EDA-β-CD showed cytotoxicity, while DOX-loaded cysteinyl-β-CD and EDA-β-CD showed a significant decrease in cell viability in cancer cells. The SiO$_2$@Ag NPs with β-CD provide a strategy for designing a nanocarrier that can deliver a drug with controlled release from modified chemical types.

Keywords: cyclodextrin; doxorubicin (DOX); drug delivery

---

1. Introduction

Nanomaterial-based carriers have been widely studied as transport vehicles for various substances, such as drugs, due to their ability to increase local accessibility to the target and enhance bioavailability [1–4]. It is important to select a suitable nanoparticle (NP) for the fabrication of a nanocarrier as well as a ligand, which will be immobilized on the NP, to capture and release the target drug.

In β-cyclodextrin (β-CD), the inner cavity is hydrophobic, and the outside is hydrophilic, which is beneficial for incorporating hydrophobic materials into the cavity. The formation of an inclusion complex between β-CD and hydrophobic materials can result in the dissolution of insoluble materials in water [5–7]. A variety of studies in the food, pharmaceutical, medical, and cosmetic industries have evaluated the complexation ability of β-CD [8–10]. In the medical field, the inclusion complex between β-CD and doxorubicin (DOX) was first reported in the 1990s [11], and subsequent studies have examined its complex-forming ability [12–14]. Surface modification is essential for the introduction of β-CD as a stable ligand onto NPs (unpublished). In addition, the affinity and materials that can be captured differ depending on the kind of β-CD functional group. Moreover, little is known about the amounts of loaded and released DOX using β-CD.

Metal-embedded silica NPs (SiO$_2$ NPs) have been prepared for various applications and provide several key advantages over other NPs, such as the potential for plasmon tuning for deep tissue imaging [15] and photothermal therapy [16], the strong plasmonic property for sensitive detection [17–22], and easy handling and surface modification [17]. Moreover, metal NPs have affinity to ligands with unshared electron pairs, such as thiol and amine groups [18,23]. Using the SiO$_2$@Ag NP structure, a target material can be detected by introducing β-CD, which is used as a capture ligand. Functionalized β-CD included amine (ethylenediamine) and thiol (cysteinyl) group was introduced on the metal surface [24,25]. Detection using the assembled structure and β-CD has been reported, but subsequent analyses of drug delivery are lacking. In addition, drug release using functionalized β-CD has not been studied.

In this study, we used SiO$_2$@Ag NPs coated with cysteinyl-β-CD and ethylenediamine (EDA)-β-CD as ligands to capture DOX. The rates of loaded and released DOX depended on the kind of β-CD. The drug release kinetics differed depending on the kind of immobilized β-CD on the SiO$_2$@Ag NPs. In addition, we treated breast cancer cells with DOX-loaded SiO$_2$@Ag@β-CD NPs to assess the cytotoxicity of DOX-loaded cysteinyl-β-CD and EDA-β-CD. Our results suggest that β-CD derivatives could be used for drug capture and release, and cysteinyl-β-CD might be useful as a nanocarrier in drug delivery systems.

2. Results and Discussion

As illustrated in Figure 1a, SiO$_2$@Ag NPs, which have the advantage of facile handling and surface modifications, were used to immobilize β-CD while maintaining the SiO$_2$@Ag NP structure. Then, the functionalized β-CD was immobilized on SiO$_2$@Ag NPs, and DOX was loaded onto the NPs to investigate its loading and release.

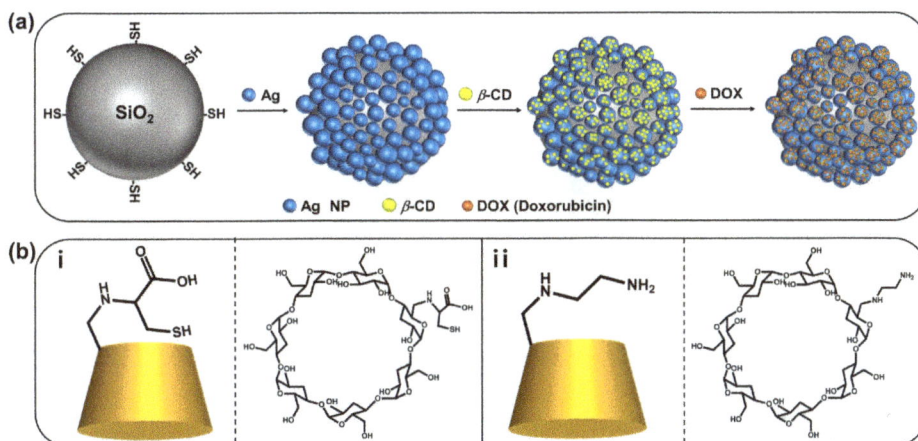

Figure 1. (a) Procedure for synthesizing SiO$_2$@Ag NPs and introducing β-CD derivatives and DOX. (b) Chemical structures and illustration of (i) cysteinyl-β-CD and (ii) EDA-β-CD.

First, SiO$_2$ NPs with diameters of 178 ± 6.1 nm were synthesized by the well-known Stöber method [26,27]. The SiO$_2$ NPs were then functionalized with (3-mercaptopropyl)trimethoxysilane (MPTS) to introduce thiol groups, which have high affinity to Ag NPs. To allow the stable and dense immobilization of β-CD derivatives, Ag NPs were embedded on the SiO$_2$ NP surface by the reduction of silver nitrate with octylamine. Three kinds of ligands (cysteinyl-β-CD, EDA-β-CD, and methoxypoly(ethylene glycol)sulfhydryl [m-PEG-SH]), prepared following previously reported methods [25,28], were added after the synthesis of SiO$_2$@Ag NPs. Cysteinyl-β-CD and EDA-β-CD (Figure 1b) were selected as functionalized β-CDs because they are known to effectively capture

DOX and the thiol group of cysteinyl-β-CD and diamine of EDA-β-CD have strong affinity to metal surfaces [23].

To confirm the shape of the produced NPs, transmission electron microscopy (TEM) images were recorded. As shown in Figure 2a, uniform SiO$_2$@Ag NPs were synthesized, and Ag NPs with diameters of 16 ± 7 nm were densely immobilized on the SiO$_2$ surface following the reduction reaction. Following coating with cysteinyl-β-CD or EDA-β-CD, the structure of the SiO$_2$@Ag NPs was maintained, as shown in Figure 2b,c. However, when the m-PEG-SH solution was used as a ligand, the Ag NPs detached from the SiO$_2$ NPs, as shown in Figure 2d. The ultraviolet-visible (UV-Vis) spectra for SiO$_2$@Ag@cysteinyl-β-CD NPs and SiO$_2$@Ag@EDA-β-CD NPs were similar to that for SiO$_2$@Ag NPs, which absorbed a broad wavelength range from 395 nm to 1000 nm, as shown in Figure 3. The UV-Vis spectrum for SiO$_2$@Ag-PEG NPs, which had a peak at 401 nm, was narrower than that for SiO$_2$@Ag NPs and similar to that for Ag NPs [29]. In addition, the solution color of SiO$_2$@Ag NPs incorporated with β-CDs was not significantly different from that of the SiO$_2$@Ag NPs, while the solution of SiO$_2$@Ag-PEG NPs turned yellow. The introduction of m-PEG-SH might cause Ag NPs to detach from SiO$_2$ NPs, and the synthesis of the assembled structure is not easy to control. Therefore, among the three ligands immobilized on nanostructures, the SiO$_2$@Ag@cysteinyl-β-CD NPs and SiO$_2$@Ag@EDA-β-CD NPs were used to further analyze DOX loading and release.

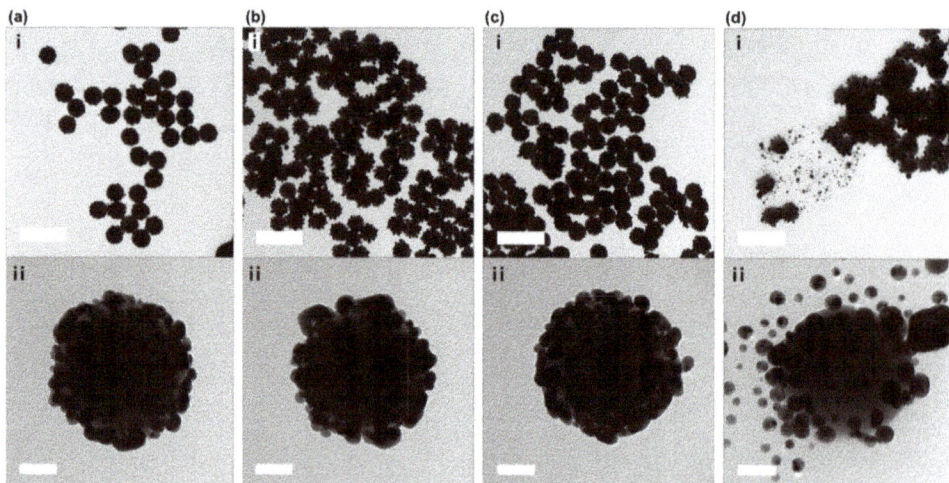

Figure 2. TEM images of SiO$_2$@Ag NPs immobilized with three kinds of ligands. (**a**) Bare SiO$_2$@Ag NPs, (**b**) SiO$_2$@Ag@cysteinyl-β-CD NPs, (**c**) SiO$_2$@Ag@EDA-β-CD NPs, and (**d**) SiO$_2$@Ag-PEG NPs. (i) Low-magnification TEM images of the overall morphology of NPs after the addition of ligands. (ii) High-magnification TEM images of single NPs. Scale bars, (i) 500 nm and (ii) 50 nm.

To confirm that the SiO$_2$@Ag NPs were successfully coated with β-CDs and loaded with DOX, attenuated total reflection-Fourier transform infrared (ATR-FTIR) spectra of the synthesized NPs were recorded after each step, as shown in Figure 4 (see also Figure S1). To compare the FTIR spectra of our synthesized materials, we normalized the signal of our material at ~3800 cm^{-1} (background signal). The IR spectra of the SiO$_2$@Ag NPs coated with two types of β-CDs were similar to that of the SiO$_2$@Ag NPs; however, two observations corroborated that β-CDs adhered to the SiO$_2$@Ag NPs. The IR spectra of the two kinds of β-CDs are shown in Figure S2a. The intensity of the band at 1627 cm^{-1} in the IR spectrum for SiO$_2$@Ag@cysteinyl-β-CD NPs increased due to the N–H bending vibration of β-CD derivatives, as shown in Figure 4a. In addition, the intensity of the band at 1635 cm^{-1}, which represents the N–H bending vibration of β-CD derivatives, was different from that of SiO$_2$@Ag@EDA-β-CD NPs, as shown in Figure 4b. In particular, the peak around 1000 cm^{-1} is larger than that in the other spectra.

This peak is attributed to bonds in the SiO$_2$ NPs such as Si–O–Si and Si–OH. However, the peak intensity decreases after modification with DOX. In addition, when the cysteinyl-β-CD was introduced onto the SiO$_2$@Ag surface, some part of the Ag NPs could detach or move to another Ag or thiol group on SiO$_2$, as shown in Figure 2b. The thiol group included in cysteinyl-β-CD has a higher affinity to Ag NPs than the amine group. As a result, the surface of SiO$_2$@Ag@cysteinyl-β-CD NPs exhibits more SiO$_2$, and the FTIR spectra could be larger than the other spectra. This observation can be taken as evidence that DOX was loaded onto the surface of the NPs.

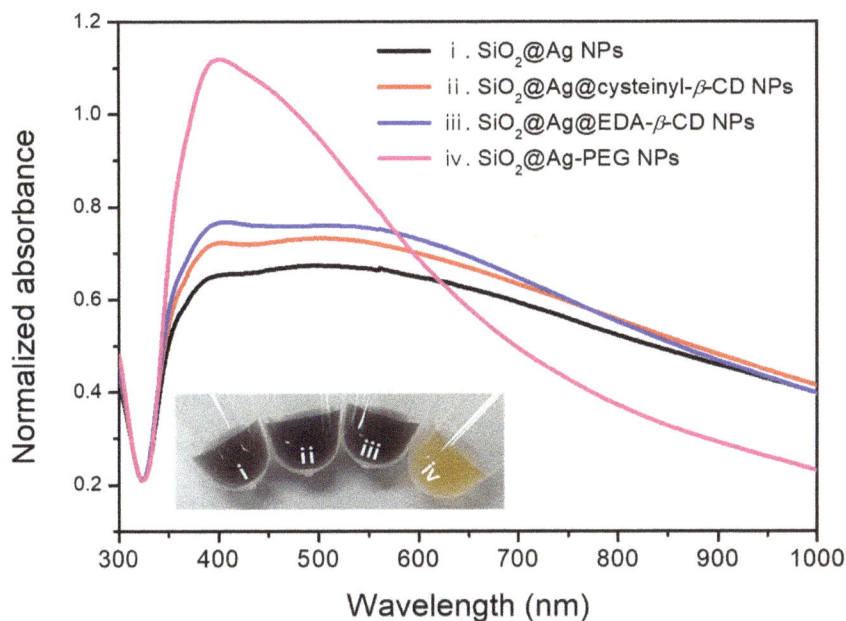

Figure 3. UV-Vis spectra for SiO$_2$@Ag NPs, SiO$_2$@Ag@cysteinyl-β-CD NPs, SiO$_2$@Ag@EDA-β-CD NPs, and SiO$_2$@Ag-PEG NPs. (i) SiO$_2$@Ag NPs, (ii) SiO$_2$@Ag@cysteinyl-β-CD NPs, (iii) SiO$_2$@Ag@EDA-β-CD NPs, and (iv) SiO$_2$@Ag-PEG NPs. (Inset: Photograph of synthesized NP solutions using four kinds of ligands, showing the color change).

A band at 1424 cm^{-1} was assigned to the CH$_2$ bending vibration from β-CD derivatives (Figure 4ii). Although a CH$_2$ group was included in the thiol-functionalized SiO$_2$@Ag NPs, CH$_2$ bending does not appear in Figure 4i. This is presumably because MPTS is covered with Ag NPs [30]. After loading DOX onto SiO$_2$@Ag@cysteinyl-β-CD, new bands at 1608 cm^{-1} and 1574 cm^{-1}, which were assigned to the aromatic C=C stretching vibration in Figure 4a, clearly indicate the presence of DOX. Moreover, bands at 1718 cm^{-1} and 1403 cm^{-1} appeared, which were assigned to ketone group stretching and methyl group bending vibrations, respectively. The IR spectra for SiO$_2$@Ag@EDA-β-CD NPs with DOX showed aromatic C=C stretching vibration bands at 1611 cm^{-1} and 1575 cm^{-1}. Additionally, ketone group stretching and methyl group bending vibration bands of DOX at 1724 cm^{-1} and 1409 cm^{-1}, respectively, were detected. The IR spectra of DOX is shown in Figure S2b. Thus, we confirmed that the β-CDs were immobilized on the SiO$_2$@Ag NPs, and DOX was loaded onto the NPs.

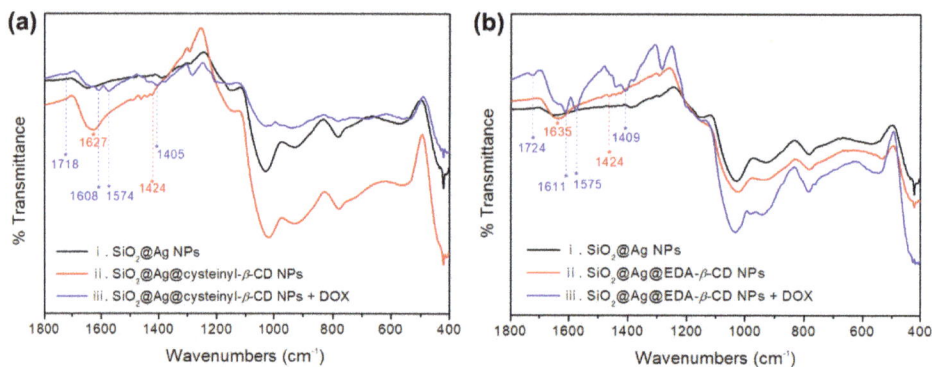

Figure 4. FTIR spectra confirming the introduction of cysteinyl-β-CD, EDA-β-CD, and DOX on SiO$_2$@Ag NPs. (**a**) Immobilized cysteinyl-β-CD (N–H bending vibration at 1627 cm^{-1}, CH$_2$ bending vibration at 1424 cm^{-1}, red) on the SiO$_2$@Ag NPs and DOX (ketone group stretching at 1718 cm^{-1}, C=C stretching vibration at 1608 cm^{-1} and 1574 cm^{-1}, and methyl group bending at 1403 cm^{-1}, blue) on SiO$_2$@Ag@cysteinyl-β-CD NPs. (**b**) Immobilized EDA-β-CD (N–H bending vibration at 1635 cm^{-1}, CH$_2$ bending vibration at 1424 cm^{-1}, red) on SiO$_2$@Ag NPs and DOX (ketone group stretching at 1724 cm^{-1}, C=C stretching vibration at 1611 cm^{-1} and 1575 cm^{-1}, methyl group bending at 1409 cm^{-1}, blue) on SiO$_2$@Ag@EDA-β-CD NPs. (i) SiO$_2$@Ag NPs, (ii) SiO$_2$@Ag@cysteinyl-β-CD NPs and SiO$_2$@Ag@EDA-β-CD NPs, (iii) NPs after loading DOX onto SiO$_2$@Ag@cysteinyl-β-CD NPs and SiO$_2$@Ag@EDA-β-CD NPs.

To evaluate the amount of DOX loaded onto each NP, a DOX solution of 50 μmol/mL (44 μg/mL) was separately added to 1 mg/mL solutions of SiO$_2$@Ag@cysteinyl-β-CD NP and SiO$_2$@Ag@EDA-β-CD NP. After vortexing for 12 h to mix the solutions, the supernatant was separated by centrifugation at 13,000 rpm for 15 min.

The absorbance of the supernatant was measured at 483 nm, and a DOX calibration curve (Figure S2) was used to evaluate the quantity of DOX loaded onto the synthesized NPs. Figure 5a shows the percentage of DOX loaded onto two kinds of synthesized NPs. The quantity of DOX loaded onto SiO$_2$@Ag@cysteinyl-β-CD NPs and SiO$_2$@Ag@EDA-β-CD NPs were 34.4 μg/mg and 32.3 μg/mg, corresponding to 78.2 and 73.6% of the initial DOX, respectively. Thus, slightly more DOX was loaded onto SiO$_2$@Ag@cysteinyl-β-CD NPs than onto SiO$_2$@Ag@EDA-β-CD NPs. According to Hassan et al., β-CD is a basket-shaped oligosaccharide with a thinner and broader ring's edge [14]. The exterior part of β-CD is hydrophilic, and its internal cavity is relatively non-polar. Due to this construction, non-polar guest can be encapsulated by β-CD to form the inclusion complex between CDs and guest by generating hydrogen bonding, hydrophobic interaction, van der Waals interaction. In the same way, DOX can interacts with β-CD to generate the supramolecular complex by host-guest inclusion. However, it is not easy to explain the mechanism in which DOX were loaded and released with different rates in SiO$_2$@Ag@cysteinyl-β-CD NPs and SiO$_2$@Ag@EDA-β-CD NPs. However, the loading of DOX depends on types of the functional group of β-CD previously reported by our group. Ethylenediamine β-CD derivative can captured various flavonoids [25]. We believed that not only cavity but also the exterior part of β-CD can interact selectively with DOX.

To confirm the release behavior, the DOX release from each NP was monitored over time by measuring the UV-Vis absorbance of the supernatant after 1, 6, 12, 24, and 48 h at room temperature. The supernatant was collected at each time point, and then, a new solution was added for the next release step. The results are shown as the accumulated values of released DOX, as calculated using the DOX calibration curve.

Figure 5b shows the release profiles of the two kinds of NPs as a function of time. The SiO$_2$@Ag@EDA-β-CD NPs released more DOX than the SiO$_2$@Ag@cysteinyl-β-CD NPs during the early stage of release as well as more DOX overall than the SiO$_2$@Ag@cysteinyl-β-CD NPs, which released the lowest levels DOX throughout the release process. Over 48 h, the SiO$_2$@Ag@EDA-β-CD NPs and SiO$_2$@Ag@cysteinyl-β-CD NPs released 2.15 µg (7.40%) and 1.57 µg (5.48%) of DOX, respectively. The percentage of released DOX is slightly lower than the reported value of ~10% [31], but the treatment concentration of DOX was lower than those of previous studies [32,33] to avoid side effects. The release behavior of DOX from β-CD was reported by Viale et al. [34]. These results reveal the release kinetics of β-CD with an amine group, i.e., it releases DOX more slowly when it is loaded onto cationic oligomeric β-CD. This released quantity of DOX from SiO$_2$@Ag@EDA-β-CD NPs could be explained by the positive charge owing to NH$_2$ at neutral pH.

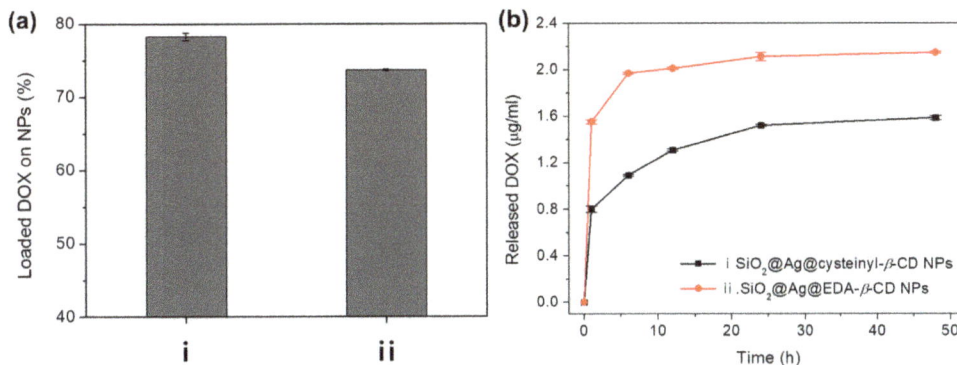

Figure 5. (**a**) The percentage of DOX loaded onto each NP; (i) SiO$_2$@Ag@cysteinyl-β-CD NPs and (ii) SiO$_2$@Ag@EDA-β-CD NPs after adding a DOX solution (44 µg/mL) to the NPs. (**b**) The quantity of DOX released from the NPs at room temperature over 48 h. DOX was released more slowly from SiO$_2$@Ag@cysteinyl-β-CD NPs.

Figure 5 shows that the highest amount of DOX was introduced onto the SiO$_2$@Ag@cysteinyl-β-CD NPs, and less DOX was released by the SiO$_2$@Ag@cysteinyl-β-CD NPs. Thus, we believe that DOX is captured by cysteinyl-β-CD in the SiO$_2$@Ag@cysteinyl-β-CD NPs. These results indicate that the loading and release of DOX on NPs depends on the kind of β-CD ligand. Furthermore, the observed release behavior indicates that among the two types of NPs examined, the SiO$_2$@Ag@cysteinyl-β-CD NPs could be a good candidate for capturing and sequestering DOX in drug delivery systems.

To assess cysteinyl-β-CD and EDA-β-CD NPs as an anticancer drug carrier, the cell viability was measured in cancer cells treated with cysteinyl-β-CD and EDA-β-CD with or without DOX (Figure 6). The cytotoxic effects of SiO$_2$@Ag, SiO$_2$@Ag@cysteinyl-β-CD, and SiO$_2$@Ag@EDA-β-CD NPs on breast cancer cells (MCF-7 cells) were negligible up to a concentration of 5 µg/mL (Figure 6a). This result indicates that our Ag-embedded SiO$_2$ NPs were biocompatible and not significantly cytotoxic. In contrast, when DOX-loaded cysteinyl-β-CD and EDA-β-CD nanocarriers were incubated with MCF-7 cells at 37 °C with increasing incubation time, the viability of MCF-7 cells significantly decreased from 1 h to 48 h (Figure 6b). The rate of cytotoxicity with DOX-loaded SiO$_2$@Ag@EDA-β-CD was faster than that of DOX-loaded SiO$_2$@Ag@cysteinyl-β-CD. In addition, the cell viability of cancer cells dropped sharply to ~60% after 12 h if incubation with both DOX-loaded cysteinyl-β-CD and EDA-β-CD nanocarriers, which was maintained until 48 h. This result was consistent with the DOX release time (Figure 5b) and demonstrated that DOX was completely released from the nanocarriers after 12 h. The cytotoxicity of DOX loaded onto the nanocarriers was also investigated, as shown in Figure 6c. The cell viability decreased with increasing DOX concentrations. At low DOX concentrations (<200 nM), the cell viability was ~80%, which decreased to 60% at 1000 nM. The concentration of DOX needed to attain 50% cell viability (GI50) was calculated by fitting with a hyperbolic equation, providing GI50 values of 1.323 µM and 1.154 µM for SiO$_2$@Ag@cysteinyl-β-CD and SiO$_2$@Ag@EDA-β-CD, respectively.

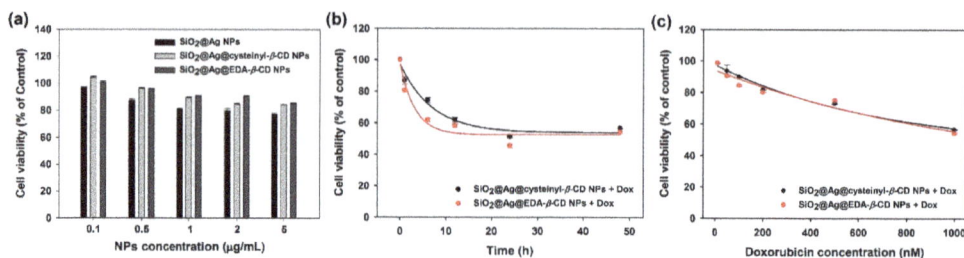

Figure 6. Assessment of cytotoxicity in cancer cells treated with Ag-embedded SiO$_2$ nanocarriers. (**a**) MCF-7 cancer cells were treated with each type of NP at increasing concentrations (0.1–5 µg/mL) for 48 h. The cell viability was measured using the WST-1 assay. (**b**) Cell viability after treatment with cysteinyl-β-CD and EDA-β-CD loaded with 1 µM DOX with increasing incubation time (1, 6, 12, 24 and 48 h). The cytotoxicity rate was obtained by fitting with an exponential equation (lines). (**c**) Cell viability after treatment with various concentrations of DOX-loaded cysteinyl-β-CD and EDA-β-CD NPs for 48 h. The decrease in cell viability with increasing DOX was fit to the hyperbolic equation (lines).

3. Materials and Methods

3.1. Materials

To synthesize the NPs, tetraethyl orthosilicate (TEOS), ethylene glycol, polyvinylpyrrolidone (PVP, Mw ≈ 40,000), silver nitrate (AgNO$_3$, 99.99%), octylamine, and MPTS were purchased from Sigma-Aldrich (St. Louis, MO, USA) and used without any purification. Ethyl alcohol (EtOH) and aqueous ammonium hydroxide (NH$_4$OH, 27%) were purchased from Daejung (Siheung, Korea). Designed cysteinyl-β-CD and EDA-β-CD were obtained from the Microbial Carbohydrate Resource Bank (MCRB) at Konkuk University, Korea [25,28], and m-PEG-SH (M.W. 5000) was purchased from Sunbio (Anyang, Korea). Deionized (DI) water was used in all experiments.

3.2. Synthesis of SiO$_2$@Ag NPs

The SiO$_2$@Ag NPs were synthesized according to a previously reported method [24]. SiO$_2$ NPs were prepared by a modified Stöber method [26] using 40 mL of 99.9% EtOH, 3 mL of NH$_4$OH, and 1.6 mL of TEOS. The solution was stirred vigorously for 20 h at room temperature and then washed with 95% EtOH three times. To synthesize the embedded Ag NPs, PVP in 25 mL of ethylene glycol was mixed with the thiol-functionalized silica solution (30 mg/mL). AgNO$_3$ in ethylene glycol and octylamine were added in sequence. After a reaction time of 1 h, the sample was washed with 95% EtOH several times. To obtain a 10 mg/mL SiO$_2$@Ag NP solution, it was dispersed in 3 mL of absolute EtOH. The NPs were examined with an energy-filtering transmission electron microscope (LIBRA 120; Carl Zeiss, Oberkochen, Germany) operated at an accelerating voltage of 120 kV.

3.3. Preparation of SiO$_2$@Ag@cysteinyl-β-CD NPs, SiO$_2$@Ag@EDA-β-CD NPs, and SiO$_2$@Ag-PEG NPs

To prepare SiO$_2$@Ag@cysteinyl-β-CD NPs, the cysteinyl-β-CD solution (1 mmol in DI water) was supplemented with 1 mg of SiO$_2$@Ag NPs. The mixture was vortexed vigorously for 12 h at 25 °C. The suspension was washed one time with 95% EtOH by centrifugation and dispersed in 1 mL of absolute EtOH. The SiO$_2$@Ag@EDA-β-CD NPs were synthesized using the same method as that for SiO$_2$@Ag@cysteinyl-β-CD NPs. To synthesize SiO$_2$@Ag-PEG NPs, m-PEG-SH solution (1 mmol in DI water) was added to 1 mg of SiO$_2$@Ag NPs. The mixture was vortexed vigorously for 12 h at room temperature and then washed once with 95% EtOH by centrifugation. Finally, it was dispersed in 99% EtOH.

Int. J. Mol. Sci. **2019**, *20*, 315

3.4. Loading of DOX on SiO₂@Ag NPs with Ligands (β-CD Derivatives and PEG)

3.4. Loading of DOX on SiO$_2$@Ag NPs with Ligands (β-CD Derivatives and PEG)

A DOX solution was added to SiO$_2$@Ag@β-CD derivative NP and SiO$_2$@Ag-PEG NP (1 mg/mL) solutions. The concentration of DOX dispersed in DI water was 50 μmol/mL. The mixture was vigorously shaken at room temperature for 12 h in the dark. Free DOX was removed by centrifugation at 13,000 rpm for 15 min. To determine the amount of DOX introduced, a calibration curve was obtained based on absorbance at 483 nm using various concentrations of DOX in 50% EtOH, as shown in Figure S3.

3.5. DOX Release

DOX release was observed at room temperature without additional or external factors. To monitor the release of DOX, the supernatant of DOX-loaded SiO$_2$@Ag@β-CD derivative NPs was measured at 1, 6, 12, 24, and 48 h. The supernatant of each NP was harvested after centrifugation at 13,000 rpm for 15 min. To measure the released amount of DOX, the absorbance of the supernatant was measured at 483 nm using a UV spectrophotometer (OPTIZEN POP; Mecasys, Daejeon, Korea).

3.6. Cell Culture and Cell Viability Assay

The cells used in this study were MCF-7 human breast cancer cells purchased from ATCC (American Type Culture Collection; code no ATCC® HTB-22™). MCF-7 cells were cultured in DMEM (Dulbecco's Modified Eagle Medium) culture medium (HyClone Laboratories, Logan, UT, USA) supplemented with 10% fetal bovine serum (HyClone Laboratories) and 1% of penicillin/streptomycin (Welgene, Daegu, Korea). MCF-7 cells were seeded onto 96-well plates at a density of 5.0×10^3 cells/well and incubated at 37 °C for 24 h. The WST-1 assay was then performed according to the manufacturer's instructions 48 h after treatment with NPs of DOX-loaded NPs at the indicated concentrations. The absorbance was measured by VICTOR X3 multi-label plate reader (PerkinElmer, Waltham, MA, USA) at 450 nm.

4. Conclusions

Functionalized β-CD derivatives, namely, cysteinyl-β-CD and EDA-β-CD, were successfully immobilized on SiO$_2$@Ag NPs to load DOX. The percentages of DOX introduced onto each NP were similar; however, the release behavior differed. In comparison to SiO$_2$@Ag@EDA-β-CD NPs, the SiO$_2$@Ag@cysteinyl-β-CD NPs captured relatively more and released less DOX. Moreover, the cell viability was decreased by increasing the concentration of NPs with DOX. These features indicate that β-CDs, in particular SiO$_2$@Ag@cysteinyl-β-CD NPs, are useful candidate materials for drug capture and show promise for the development of bioapplications and nanomedicine, with particular potential for drug delivery systems.

Supplementary Materials: Supplementary materials can be found at http://www.mdpi.com/1422-0067/20/2/315/s1.

Author Contributions: Conceptualization, B.-H.J.; data curation, E.J.K. and Y.M.B.; formal analysis, E.J.K.; investigation, E.H.; methodology, M.S.N.; project administration, B.-H.J.; resources, E.H.; visualization, E.J.K., Y.M.B.; writing—original draft, E.J.K.; writing—review & editing, B.-H.J., S.H.L., X.-H.P., M.S.N., and D.-E.K.

Funding: This work was supported by Konkuk University in 2016.

Acknowledgments: Microbial Carbohydrate Resource Bank (MCRB, Seoul, Korea) is kindly acknowledged for providing the carbohydrate materials.

Conflicts of Interest: This author declares no conflict of interest.

References

1. Yu, X.; Trase, I.; Ren, M.; Duval, K.; Guo, X.; Chen, Z. Design of nanoparticle-based carriers for targeted drug delivery. *J. Nanomater.* **2016**, *2016*. [CrossRef] [PubMed]

2. Hillaireau, H.; Couvreur, P. Nanocarriers' entry into the cell: Relevance to drug delivery. *Cell Mol. Life Sci.* **2009**, *66*, 2873–2896. [CrossRef] [PubMed]

3. Kumari, P.; Ghosh, B.; Biswas, S. Nanocarriers for cancer-targeted drug delivery. *J. Drug. Target.* **2016**, *24*, 179–191. [CrossRef] [PubMed]

4. Ruiz-Gatón, L.; Espuelas, S.; Larrañeta, E.; Reviakine, I.; Yate, L.A.; Irache, J.M. Pegylated poly (anhydride) nanoparticles for oral delivery of docetaxel. *Eur. J. Pharm. Sci.* **2018**, *118*, 165–175. [CrossRef] [PubMed]

5. Van De Manakker, F.; Vermonden, T.; Van Nostrum, C.F.; Hennink, W.E. Cyclodextrin-based polymeric materials: Synthesis, properties, and pharmaceutical/biomedical applications. *Biomacromolecules* **2009**, *10*, 3157–3175. [CrossRef] [PubMed]

6. Lavoine, N.; Givord, C.; Tabary, N.; Desloges, I.; Martel, B.; Bras, J. Elaboration of a new antibacterial bio-nano-material for food-packaging by synergistic action of cyclodextrin and microfibrillated cellulose. *Innov. Food Sci. Emerg.* **2014**, *26*, 330–340. [CrossRef]

7. Yallapu, M.M.; Jaggi, M.; Chauhan, S.C. B-cyclodextrin-curcumin self-assembly enhances curcumin delivery in prostate cancer cells. *Colloid Surf. B-Biointerfaces* **2010**, *79*, 113–125. [CrossRef] [PubMed]

8. Del Valle, E.M. Cyclodextrins and their uses: A review. *Process Biochem.* **2004**, *39*, 1033–1046. [CrossRef]

9. Huarte, J.; Espuelas, S.; Lai, Y.; He, B.; Tang, J.; Irache, J.M. Oral delivery of camptothecin using cyclodextrin/poly (anhydride) nanoparticles. *Int. J. Pharm.* **2016**, *506*, 116–128. [CrossRef]

10. Calleja, P.; Espuelas, S.; Corrales, L.; Pio, R.; Irache, J.M. Pharmacokinetics and antitumor efficacy of paclitaxel–cyclodextrin complexes loaded in mucus-penetrating nanoparticles for oral administration. *Nanomedicine* **2014**, *9*, 2109–2121. [CrossRef]

11. Bekers, O.; Kettenes, J.J.V.D.B.; Van Helden, S.P.; Seijkens, D.; Beijnen, J.H.; Bulti, A.; Underberg, W.J. Inclusion complex formation of anthracycline antibiotics with cyclodextrins; a proton nuclear magnetic resonance and molecular modelling study. *J. Inclus. Phenom. Mol.* **1991**, *11*, 185–193. [CrossRef]

12. Swiech, O.; Mieczkowska, A.; Chmurski, K.; Bilewicz, R. Intermolecular interactions between doxorubicin and β-cyclodextrin 4-methoxyphenol conjugates. *J. Phys. Chem. B* **2012**, *116*, 1765–1771. [CrossRef]

13. Anand, R.; Manoli, F.; Manet, I.; Daoud-Mahammed, S.; Agostoni, V.; Gref, R.; Monti, S. β-cyclodextrin polymer nanoparticles as carriers for doxorubicin and artemisinin: A spectroscopic and photophysical study. *Photochem. Photobiol. Sci.* **2012**, *11*, 1285–1292. [CrossRef]

14. Yousef, T.; Hassan, N. Supramolecular encapsulation of doxorubicin with β-cyclodextrin dendrimer: In vitro evaluation of controlled release and cytotoxicity. *J. Incl. Phenom. Macrocycl. Chem.* **2017**, *87*, 105–115. [CrossRef]

15. Kang, H.; Jeong, S.; Park, Y.; Yim, J.; Jun, B.H.; Kyeong, S.; Yang, J.K.; Kim, G.; Hong, S.; Lee, L.P. Near-infrared sers nanoprobes with plasmonic au/ag hollow-shell assemblies for in vivo multiplex detection. *Adv. Funct. Mater.* **2013**, *23*, 3719–3727. [CrossRef]

16. Noh, M.S.; Lee, S.; Kang, H.; Yang, J.-K.; Lee, H.; Hwang, D.; Lee, J.W.; Jeong, S.; Jang, Y.; Jun, B.-H. Target-specific near-ir induced drug release and photothermal therapy with accumulated au/ag hollow nanoshells on pulmonary cancer cell membranes. *Biomaterials* **2015**, *45*, 81–92. [CrossRef]

17. Jun, B.H.; Kim, G.; Jeong, S.; Noh, M.S.; Pham, X.H.; Kang, H.; Cho, M.H.; Kim, J.H.; Lee, Y.S.; Jeong, D.H. Silica core-based surface-enhanced raman scattering (sers) tag: Advances in multifunctional sers nanoprobes for bioimaging and targeting of biomarkers#. *Bull. Korean Chem. Soc.* **2015**, *36*, 963–978.

18. Chang, H.; Kang, H.; Ko, E.; Jun, B.-H.; Lee, H.-Y.; Lee, Y.-S.; Jeong, D.H. Psa detection with femtomolar sensitivity and a broad dynamic range using sers nanoprobes and an area-scanning method. *ACS Sens.* **2016**, *1*, 645–649. [CrossRef]

19. Noh, M.S.; Jun, B.-H.; Kim, S.; Kang, H.; Woo, M.-A.; Minai-Tehrani, A.; Kim, J.-E.; Kim, J.; Park, J.; Lim, H.-T. Magnetic surface-enhanced raman spectroscopic (m-sers) dots for the identification of bronchioalveolar stem cells in normal and lung cancer mice. *Biomaterials* **2009**, *30*, 3915–3925. [CrossRef] [PubMed]

20. Kim, J.-H.; Kim, J.-S.; Choi, H.; Lee, S.-M.; Jun, B.-H.; Yu, K.-N.; Kuk, E.; Kim, Y.-K.; Jeong, D.H.; Cho, M.-H. Nanoparticle probes with surface enhanced raman spectroscopic tags for cellular cancer targeting. *Anal. Chem.* **2006**, *78*, 6967–6973. [CrossRef] [PubMed]

21. Jun, B.H.; Noh, M.S.; Kim, J.; Kim, G.; Kang, H.; Kim, M.S.; Seo, Y.T.; Baek, J.; Kim, J.H.; Park, J. Multifunctional silver-embedded magnetic nanoparticles as sers nanoprobes and their applications. *Small* **2010**, *6*, 119–125. [CrossRef] [PubMed]
22. Cha, M.G.; Kim, H.-M.; Kang, Y.-L.; Lee, M.; Kang, H.; Kim, J.; Pham, X.-H.; Kim, T.H.; Hahm, E.; Lee, Y.-S. Thin silica shell coated ag assembled nanostructures for expanding generality of sers analytes. *PLoS ONE* **2017**, *12*, e0178651. [CrossRef] [PubMed]
23. Ueno, R.; Kim, B. Reliable transfer technique of gold micro heater through different affinities of thiol (sh) and amine (nh2) groups. *Microelectron. Eng.* **2017**, *171*, 6–10. [CrossRef]
24. Hahm, E.; Jeong, D.; Cha, M.G.; Choi, J.M.; Pham, X.-H.; Kim, H.-M.; Kim, H.; Lee, Y.-S.; Jeong, D.H.; Jung, S. β-cd dimer-immobilized ag assembly embedded silica nanoparticles for sensitive detection of polycyclic aromatic hydrocarbons. *Sci. Rep.* **2016**, *6*, 26082. [CrossRef] [PubMed]
25. Choi, J.M.; Hahm, E.; Park, K.; Jeong, D.; Rho, W.-Y.; Kim, J.; Jeong, D.H.; Lee, Y.-S.; Jhang, S.H.; Chung, H.J. Sers-based flavonoid detection using ethylenediamine-β-cyclodextrin as a capturing ligand. *Nanomaterials* **2017**, *7*, 8. [CrossRef] [PubMed]
26. Stöber, W.; Fink, A.; Bohn, E. Controlled growth of monodisperse silica spheres in the micron size range. *J. Colloid Interface Sci.* **1968**, *26*, 62–69. [CrossRef]
27. Liberman, A.; Mendez, N.; Trogler, W.C.; Kummel, A.C. Synthesis and surface functionalization of silica nanoparticles for nanomedicine. *Surf. Sci. Rep.* **2014**, *69*, 132–158. [CrossRef]
28. Kim, H.; Yiluo, H.; Park, S.; Lee, J.Y.; Cho, E.; Jung, S. Characterization and enhanced antioxidant activity of the cysteinyl β-cyclodextrin-baicalein inclusion complex. *Molecules* **2016**, *21*, 703. [CrossRef]
29. Bastús, N.G.; Merkoçi, F.; Piella, J.; Puntes, V. Synthesis of highly monodisperse citrate-stabilized silver nanoparticles of up to 200 nm: Kinetic control and catalytic properties. *Chem. Mater.* **2014**, *26*, 2836–2846. [CrossRef]
30. Zhang, S.; Zhang, Y.; Liu, J.; Xu, Q.; Xiao, H.; Wang, X.; Xu, H.; Zhou, J. Thiol modified fe3o4@ sio2 as a robust, high effective, and recycling magnetic sorbent for mercury removal. *Chem. Eng. J.* **2013**, *226*, 30–38. [CrossRef]
31. Liu, T.; Li, X.; Qian, Y.; Hu, X.; Liu, S. Multifunctional ph-disintegrable micellar nanoparticles of asymmetrically functionalized β-cyclodextrin-based star copolymer covalently conjugated with doxorubicin and dota-gd moieties. *Biomaterials* **2012**, *33*, 2521–2531. [CrossRef] [PubMed]
32. Al-Ahmady, Z.S.; Al-Jamal, W.T.; Bossche, J.V.; Bui, T.T.; Drake, A.F.; Mason, A.J.; Kostarelos, K. Lipid–peptide vesicle nanoscale hybrids for triggered drug release by mild hyperthermia in vitro and in vivo. *ACS Nano* **2012**, *6*, 9335–9346. [CrossRef] [PubMed]
33. Dabbagh, A.; Mahmoodian, R.; Abdullah, B.J.J.; Abdullah, H.; Hamdi, M.; Abu Kasim, N.H. Low-melting-point polymeric nanoshells for thermal-triggered drug release under hyperthermia condition. *Int. J. Hyperthermia* **2015**, *31*, 920–929. [CrossRef] [PubMed]
34. Viale, M.; Giglio, V.; Monticone, M.; Maric, I.; Lentini, G.; Rocco, M.; Vecchio, G. New doxorubicin nanocarriers based on cyclodextrins. *Investig. New Drugs* **2017**, *35*, 539–544. [CrossRef] [PubMed]

International Journal of
Molecular Sciences

MDPI

Article

Polydopamine-Assisted Silver Nanoparticle Self-Assembly on Sericin/Agar Film for Potential Wound Dressing Application

Liying Liu [1], Rui Cai [2], Yejing Wang [1,2,*], Gang Tao [1], Lisha Ai [1], Peng Wang [2], Meirong Yang [1], Hua Zuo [3], Ping Zhao [1,4] and Huawei He [1,4,*]

[1] State Key Laboratory of Silkworm Genome Biology, Southwest University, Chongqing 400715, China; l3341345@email.swu.edu.cn (L.L.); taogang@email.swu.edu.cn (G.T.); als123@email.swu.edu.cn (L.A.); yangmeirong@email.swu.edu.cn (M.Y.); zhaop@swu.edu.cn (P.Z.)
[2] College of Biotechnology, Southwest University, Chongqing 400715, China; cairui0330@email.swu.edu.cn (R.C.); modelsums@email.swu.edu.cn (P.W.)
[3] College of Pharmaceutical Sciences, Southwest University, Chongqing 400715, China; zuohua@swu.edu.cn
[4] Chongqing Key Laboratory of Sericultural Science, Chongqing Engineering and Technology Research Center for Novel Silk Materials, Southwest University, Chongqing 400715, China
* Correspondence: yjwang@swu.edu.cn (Y.W.); hehuawei@swu.edu.cn (H.H.); Tel.: +86-23-6825-1575 (Y.W. & H.H.)

Received: 23 August 2018; Accepted: 19 September 2018; Published: 21 September 2018

Abstract: Silver nanoparticles (AgNPs) are extensively applied for their broad-spectrum and excellent antibacterial ability in recent years. Polydopamine (PDA) has great advantages for synthesizing large amounts of AgNPs, as it has multiple sites for silver ion binding and phenolic hydroxyl structure to reduce silver ions to AgNPs. Here, we mixed sericin and agar solution and dried at 65 °C to prepare a sericin (SS)/Agar composite film, and then coated polydopamine (PDA) on the surface of SS/Agar film by soaking SS/Agar film into polydopamine solution, subsequently synthesizing high-density AgNPs with the assistance of PDA to yield antibacterial AgNPs-PDA- SS/Agar film. Scanning electron microscope (SEM), Fourier transform infrared spectroscopy (FT-IR) and X-ray diffraction (XRD) spectra indicated the successful synthesis of high-density AgNPs on the surface of PDA-SS/Agar film. PDA coating and AgNPs modification did not affect the structure of sericin and agar. Furthermore, water contact angle, water absorption and mechanical property analysis showed that AgNPs-PDA-SS/Agar film had excellent hydrophilicity and proper mechanical properties. Inhibition zone and growth curve assays suggested the prepared film had excellent and long-lasting antibacterial ability. In addition, it had excellent cytocompatibility on the fibroblast NIH/3T3 cells. The film shows great potential as a novel kind of wound dressing.

Keywords: polydopamine; silver nanoparticle; sericin; antimicrobial activity; cytocompatibility

1. Introduction

The antibacterial ability of a material surface is crucial to inhibiting the growth of bacteria on and around the material, which has great potential in food packaging and biomedical application [1]. Recently, nanomaterials have received increasing interest for their specific properties and applications in different fields. Silver nanoparticle (AgNP) is a brilliant nanomaterial, as it has a broad inhibitive effect against a variety of bacteria and fungus [2]. Surface immobilization of AgNPs is one of the most effective ways to increase the antibacterial property of materials [3].

Silk sericin (SS) is a natural hydrophilic macromolecular protein derived from silkworm cocoon. Sericin makes up 25–30% of silkworm cocoon, and it wraps silk fibroin fiber with a continuous, viscous layer that helps the formation of cocoon [4]. Sericin is a polymer protein with a molecular weight ranging from 10 to over 300 kDa. Sericin has high contents of serine and aspartate, accounting for about 33.4% and 16.7%, respectively [4]. Serine and aspartic acid have strong polar side chains. Thus, sericin can easily copolymerize and blend with other macromolecules to produce biocompatible materials with enhanced properties [5,6]. Sericin is considered to be one of the skin's important natural moisturizing factors due to its excellent hydrophilicity and hygroscopicity [7]. In addition, sericin has a great deal of excellent properties, such as biocompatibility, oxidation resistance, and anticoagulation [8]. The moisturizing property, the ability of promoting epithelial cell growth and oxidation resistance mean that sericin possesses great potential in biomedical applications. Silk-based materials have been attracting increasing interest for biomedical materials and tissue engineering applications in recent years. Chlapanidas et al. studied the biological properties of silk fibroin from 20 strains, and then picked the most promising strains in which sericin was developed for cosmetic and dermatological applications [9]. They also showed that sericin microspheres loaded with tumor necrosis factor-α (TNF-α) blockers contribute to the down-regulation of cytokines [10]. In addition, a great deal of research on sericin and silk fibroin-based biomaterials, including the use of silk fibroin microspheres as a promoting wound healing material or a local coagulant [11–13], and sericin as a natural carrier for drug delivery, has been documented [14–16]. However, sericin contains a large amount of random coil structures, resulting in the formation of amorphous and brittle sericin materials which are not suitable for biomaterial application [17]. Therefore, sericin is usually cross-linked or blended with other polymers to enhance its mechanical performance. Agar is a macromolecular polysaccharide with hydrophilic, biocompatible and biodegradable ability [18]. In our previous study, we developed sericin/agar composite film modified with AgNPs to expand the application of sericin-based biomaterials [19,20]. AgNPs are synthesized with the assistance of ultraviolet (UV) light irradiation. However, long-term exposure under UV light may cause damage to sericin. Also, the aggregation of AgNPs on sericin is a major drawback of the UV-assisted method, as well as other methods currently available. In addition, the high-density synthesis of AgNPs is another key issue to be considered in the preparation of AgNPs functionalized materials. Recently, kinds of polymers such as poly (vinyl pyrrolidone) and polyamide network have been used as three-dimensional substrates for high-density growth of AgNPs [21,22]. However, most polymers are hydrophobic and not suitable for biomedical application. Therefore, it is important to find a substance which could not only improve the density of AgNPs, but also increase the hydrophilicity of the material surface for biomaterial application. Dopamine (DA) is a small molecule, and is the main component of viscous proteins secreted by mussels and can self-polymerize to polydopamine (PDA) under alkaline conditions and adhere to almost any substrate [23,24]. PDA contains several hydrophilic groups, such as -OH, -COOH and -NH$_2$ [25]. PDA coating is an effective method used in recent years to improve the hydrophilicity and biocompatibility of materials [26,27]. PDA can not only provide sites for metal ions binding, but also reduce silver ions to AgNPs with its phenolic hydroxyl groups. PDA has been proved to be non-toxic to cells [25,28]. Thus, PDA is a very promising candidate for AgNPs synthesis. The issue of whether PDA can produce biologically active dopamine in vivo has not been well addressed in the literature on PDA-modified materials.

In this work, we utilized PDA to assist the synthesis of high-density AgNPs on the surface of PDA-SS/Agar film to yield AgNPs-PDA-SS/Agar film. Scanning electron microscopy (SEM), Fourier transform infrared spectroscopy (FT-IR), X-ray diffraction (XRD) confirmed the high-density synthesis of AgNPs on the surface of the blend film. In addition, the novel film exhibited excellent hydrophilicity and proper mechanical property. Antibacterial tests indicated that the fabricated film had excellent antibacterial activity against *Escherichia coli* (*E. coli*) and *Staphylococcus aureus* (*S. aureus*). Cell viability assay indicated the composite film had excellent cytocompatibility on the fibroblast NIH/3T3 cells. AgNPs-PDA-SS/Agar film shows great prospects in novel wound dressing, artificial skin, tissue engineering and antibacterial packaging. In addition, SS/Agar composite can be prepared into three-dimensional scaffold and gel materials to expand its application in bone repair and injectable hydrogel materials.

2. Results and Discussion

Here, we used PDA to direct the synthesis of the antibacterial AgNPs on the surface of PDA-coated SS/Agar film. The principle and procedure are briefly shown in Figure 1. The procedure of the preparation of the films is shown in Figure 1a. Agar solution (2%, *w/v*) was added into sericin solution (2%, *w/v*) to become sericin/Agar mixture, and then dried at 65 °C to form SS/Agar composite film. Next, dopamine hydrochloride powder was dissolved in Tris-HCl buffer (pH 8.5) to become 2% (*w/v*) polydopamine solution. Then, SS/Agar film was immersed into dopamine solution to produce PDA-coated SS/Agar film. Furthermore, PDA-SS/Agar film was soaked in AgNO$_3$ solution at room temperature for 4 h. The PDA layer acts as a secondary reaction platform, which can not only provide sites for metal ions binding, but also reduce silver ions to AgNPs with its phenolic hydroxyl groups to promote the synthesis of high-density AgNPs on the PDA-SS/Agar film. The prepared AgNPs-PDA-SS/Agar film was expected to have excellent antibacterial ability and cytocompatibility for wound dressing applications.

Figure 1b shows the mechanism of dopamine polymerization and AgNPs synthesis with the assistance of PDA. First, dopamine was oxidized to form dopamine quinone, and then dopamine quinone was converted to leukodopaminechrome by cyclization. Leukodopaminechrome was oxidized to form dopaminechrome. Finally, dopaminechrome was transformed to PDA by means of rearrangement and polymerization. Almost all dopamine can be converted to PDA under alkaline conditions. PDA contains active phenolic hydroxyl groups, which can react with silver ions to reduce them to AgNPs.

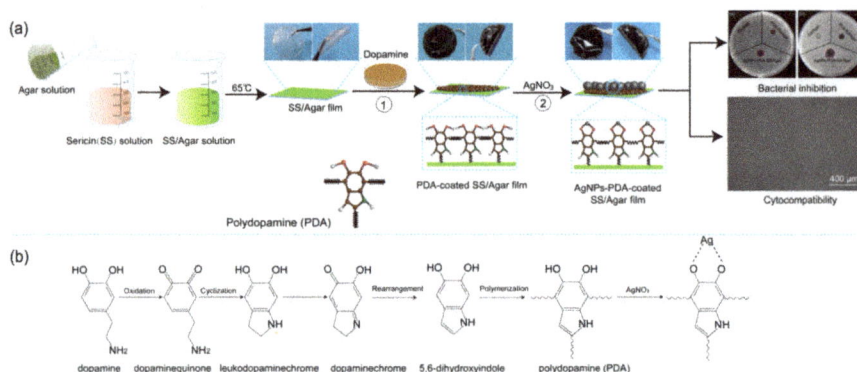

Figure 1. A diagram of the preparation of antibacterial AgNPs-PDA-SS/Agar film (**a**); schematic diagram of dopamine polymerization and AgNPs synthesis with the assistance of PDA (**b**).

2.1. Scanning Electron Microscope (SEM), Energy Dispersive Spectroscopy (EDS)

The SEM images of SS/Agar, PDA-SS/Agar, AgNPs-SS/Agar and AgNPs-PDA-SS/Agar films are shown in Figure 2. SS/Agar film exhibited a uniform and smooth surface (Figure 2a), indicating that sericin and agar were well blended. Figure 2b shows the surface of PDA-SS/Agar film, which was covered with a layer of ridge-like substance. SS/Agar film was soaked into PDA solution at room temperature for 12 h to ensure PDA coating on its surface. Thus, we deduced that the ridge-like substance covering the surface of SS/Agar film was the coated PDA. Figure 2c shows the SEM image of AgNPs-SS/Agar film without PDA. High-density AgNPs (marked by red arrows) were observed in the SEM image of AgNPs-PDA-SS/Agar film (Figure 2d). The number of AgNPs in Figure 2d were much greater than in Figure 2c, which indirectly indicates the existence of PDA on the SS/Agar film, and suggests that PDA could effectively promote the synthesis of large amounts of AgNPs. Particle size analysis revealed that the size of the synthesized AgNPs was concentrated at 300–500 nm (Figure 2e). Most of the particles were spherical in shape. In some cases, AgNPs seemed to be merged, as the density of the particles was too high (Figure 2d). Furthermore, EDS confirmed the presence of silver in the AgNPs-PDA-SS/Agar composite film (Figure 2f).

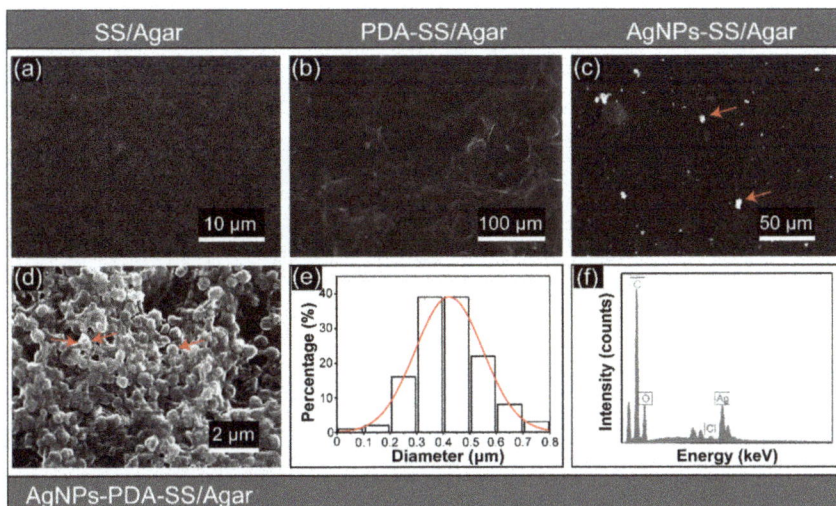

Figure 2. Surface morphologies of SS/Agar (**a**); PDA-SS/Agar (**b**); AgNPs-SS/Agar (**c**); AgNPs-PDA-SS/Agar films (**d**); (**e**) is the particle size analysis of (**d**); EDS spectrum of AgNPs-PDA-SS/Agar film (**f**). Red arrows indicate high-density AgNPs.

2.2. Fourier Transform Infrared Spectroscopy (FT-IR)

FT-IR spectra were collected to characterize the structure of different films. The results are shown in Figure 3. The peaks that appeared at 1037 cm^{-1} and 926 cm^{-1} in agar film were characteristic peaks of agar, corresponding to the C=O stretching vibration of 3,6-anhydrogalactose [29]. The two characteristic peaks at 1614 and 1516 cm^{-1} in the sericin film corresponded to amid I and II [30], respectively. Four characteristic peaks at 926, 1037, 1516 and 1614 cm^{-1} occurred in the spectra of SS/Agar, PDA-SS/Agar and AgNPs-PDA-SS/Agar films, indicating the presence of sericin and agar in the blend films, and that the structure of sericin and agar was not affected after blending. After PDA coating, two additional characteristic peaks at 1510 and 1601 cm^{-1} were observed in the spectra of PDA-SS/Agar and AgNPs-PDA-SS/Agar films, which corresponded to the C=C stretching and N-H bending vibrations of the indoline or indole structures in PDA, respectively [31,32]. The appearance of these two peaks demonstrated the successful PDA coating on the SS/Agar film, which was consistent with the observation of SEM (Figure 2b). The spectra of AgNPs-PDA-SS/Agar film showed similar characteristic peaks of sericin and agar with that of SS/Agar film, but differed from that of PDA-SS/Agar film. The possible reason was that the characteristic peak of PDA was close to that of sericin; thus, PDA coating affected the characteristic peaks of sericin and agar, and resulted in the change of SS/Agar spectrum. However, when PDA was used to reduce Ag$^+$ to AgNPs, the structure of PDA was changed; thus, the characteristic peak of PDA disappeared and could no longer affect the characteristic peaks of sericin and agar. AgNPs alone did not affect the structure of sericin and agar, and could not change the spectrum of SS/Agar film.

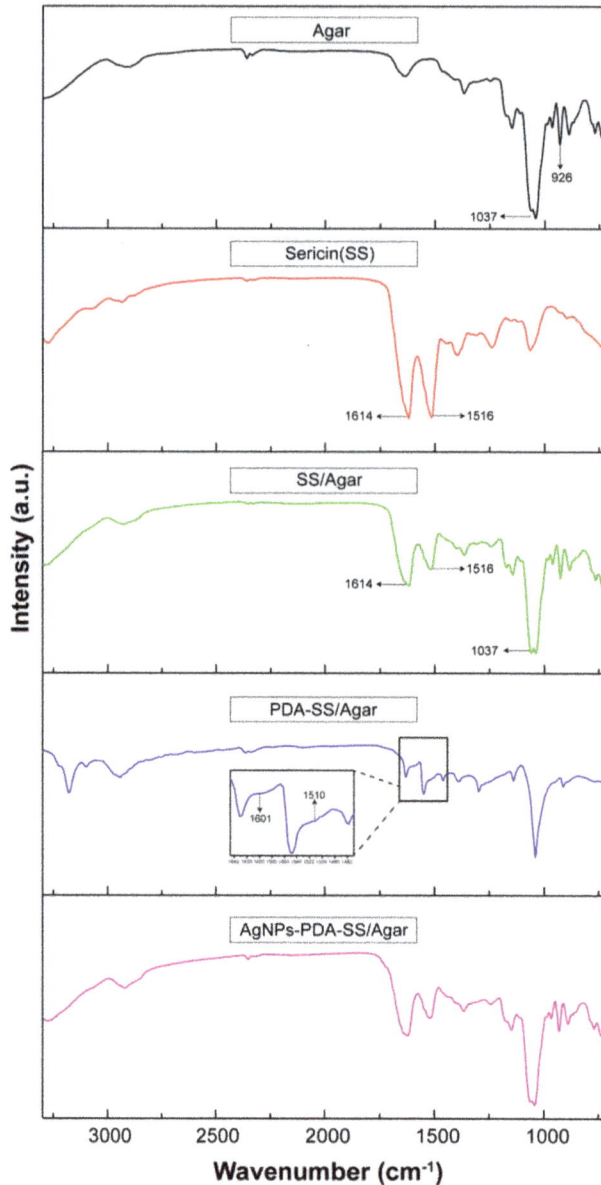

Figure 3. FT-IR spectra of Agar, Sericin (SS), SS/Agar, PDA-SS/Agar and AgNPs-PDA-SS/Agar films.

2.3. X-ray Diffraction (XRD)

The XRD patterns of AgNPs-PDA-SS/Agar, PDA-SS/Agar, SS/Agar, sericin and agar films are shown in Figure 4. The peak located at $2\theta = 19.2°$ corresponded to the silk II structure of sericin protein. The peak at $2\theta = 14.9°$ was the characteristic peak of agar [33]. The peak at $2\theta = 13.3°$ appeared in all XRD patterns except sericin, and was ascribed to the characteristic peak of agar with a slight deviation. The peak at $2\theta = 20.3°$ appeared in all composite films, which may be due to the deviation of the characteristic peak of silk II at $2\theta = 19.3°$ after blending with agar [34]. Our results showed that the blending of sericin and agar slightly changed the characteristic peak of agar and sericin. After PDA coating, the XRD patterns of SS/Agar film did not change, indicating that PDA did not affect the crystal structure of sericin and agar. The diffraction peak at 38.1° and 32.4° could be assigned to the (111) and (122) planes of the face-centered cubic structure of Ag [35,36], demonstrating the presence of AgNPs in the AgNPs-PDA-SS/Agar film.

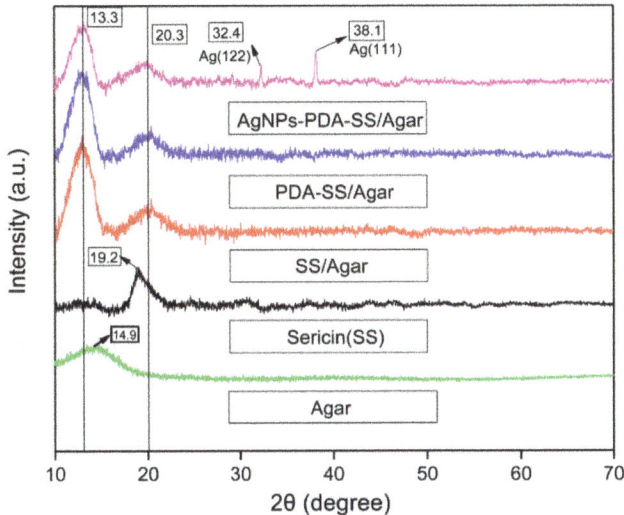

Figure 4. XRD patterns of Agar, Sericin (SS), SS/Agar, PDA-SS/Agar and AgNPs-PDA-SS/Agar films.

2.4. Wettability and Water Uptake Ability

The water contact angle of SS/Agar, PDA-SS/Agar and AgNPs-PDA-SS/Agar films are shown in Figure 5. The water contact angle of SS/Agar was 78.4°, indicating the surface of SS/Agar was hydrophilic. After PDA coating, the water contact angle decreased to 62.3°, indicating that the wettability of the composite film increased. Figure 5c shows that the water contact angle of AgNPs-PDA-SS/Agar film was 81.3°, suggesting AgNPs modification slightly reduced the wettability of the material surface. The reason may be AgNPs are hydrophobic substances, and the uniform distribution of AgNPs on the PDA-SS/Agar composite film would reduce the hydrophilicity of the film. The water contact angle of AgNPs-PDA-SS/Agar film was still less than 90°, indicating that the prepared material was hydrophilic and potentially useful for biomaterial application.

To further explain the hydrophilicity, the swelling property of the material was tested, as illustrated in Figure 5d,e. The result showed the swelling of AgNPs-PDA-SS/Agar, PDA-SS/Agar and SS/Agar films in 60 seconds. It is obvious that in two seconds, the composite film absorbed a lot of water, indicating that the surface of the composite film had excellent hydrophilicity. The water absorption capacity of PDA-SS/Agar film in a short period of time was better than that of AgNPs-PDA-SS/Agar and SS/Agar films. After 12–48 h, the swelling ratios of different films were in the range of 150% to 250% (Figure 5e), indicating that the prepared film had excellent water uptake ability. According to the moist wound healing theory, moisture promotes wound healing, reduces the pain of dressing removal and scarring and does not destroy freshly formed tissue. Therefore, the excellent hydrophilicity and wetting properties of the prepared composite film are advantageous for wound dressing or other potential biomedical applications.

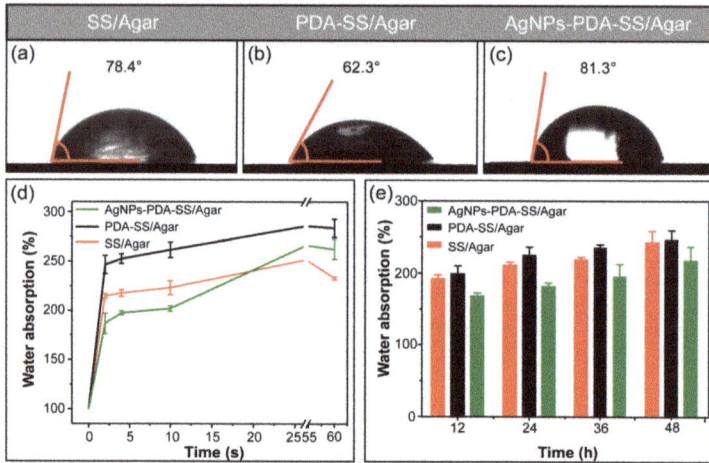

Figure 5. Water contact angle of SS/Agar (**a**), PDA-SS/Agar (**b**), AgNPs-PDA-SS/Agar films (**c**) and water absorption of different films (**d**,**e**).

2.5. Mechanical Properties

Figure 6a,b shows that SS/Agar and PDA-SS/Agar films had tensile strength exceeding 40 MPa. It is known that sericin is fragile and lacks mechanical properties, while agar has high tensile strength [37]. Therefore, the blending of sericin and agar improved the tensile strength of sericin film. The incorporation of AgNPs into PDA-SS/Agar film resulted in the reduction of tensile strength to about 25 MPa, when compared with SS/Agar and PDA-SS/Agar films. This may be because the synthesis of AgNPs on the PDA-SS/Agar film increased the film's thickness and roughness, which resulted in a reduction in tensile strength. However, this strength was still competent for some applications such as wound caring and food packaging. Elongation at break reflects the flexibility of a material [38]. The elongation at break of SS/Agar was less than 5%. PDA coating increased the elongation at break of SS/Agar film to about 7%, probably due to an increase in the thickness of the composite film. Compared with PDA-SS/Agar film, the elongation at break of AgNPs-PDA-SS/Agar film slightly increased to 8%, indicating the enhanced flexibility of AgNPs-PDA-SS/Agar film. Similarly, thickness and roughness were likely responsible for the increase of the elongation at break of AgNPs- PDA-SS/Agar film. The flexible nature of AgNPs-PDA-SS/Agar film would be beneficial for wound dressing and other potential applications.

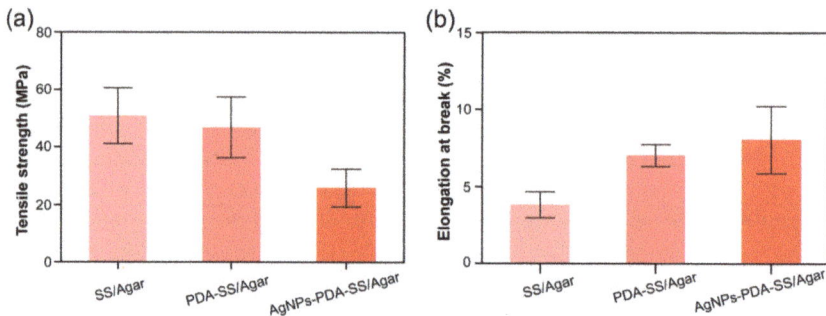

Figure 6. Mechanical properties of different films: (**a**) tensile strength, and (**b**) elongation at break.

2.6. Inhibition Zone Assay

Bacterial infection impedes wound healing. So antibacterial ability is necessary for wound dressing. Figure 7 shows the inhibition zones of AgNPs and PDA-modified or unmodified SS/Agar film against two common bacteria found in wound infections (*S. aureus*, *E. coli*). No inhibition zone appeared for SS/Agar and PDA-SS/Agar films toward the two types of bacteria. An obvious inhibition zone occurred in the presence of AgNPs-PDA-SS/Agar film toward *E. coli* and *S. aureus*, demonstrating that the fabricated AgNPs-PDA-SS/Agar film had excellent antibacterial ability. The diameters of the inhibition zones are shown in Table 1.

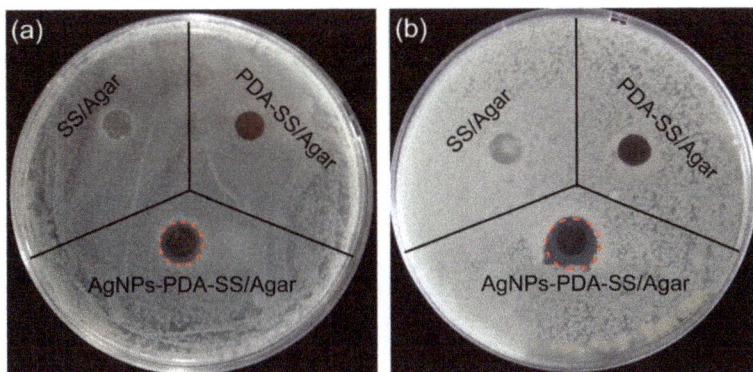

Figure 7. The inhibition zones of SS/Agar, PDA-SS/Agar, AgNPs-PDA-SS/Agar films against *E. coli* (**a**) and *S. aureus* (**b**). Red dotted circle represents the edge of the inhibition zone.

Table 1. Diameters of the inhibition zones of SS/Agar, PDA-SS/Agar and AgNPs-PDA-SS/Agar films against *E. coli* (**a**) and *S. aureus* (**b**).

Bacteria	SS/Agar (cm)	PDA-SS/Agar (cm)	AgNPs-PDA-SS/Agar (cm)
E. coli	1.10 ± 0.00	1.10 ± 0.00	1.63 ± 0.03
S. aureus	1.10 ± 0.00	1.10 ± 0.00	1.91 ± 0.11

2.7. Bacterial Growth Curve

A bacterial growth curve experiment was carried out to assess the inhibition effect of AgNPs and PDA treated and untreated SS/Agar film on bacterial growth. Figure 8 shows the growth curves of *E. coli* (Figure 8a) and *S. aureus* (Figure 8b) in the presence of different films, respectively. The growth of *E. coli* and *S. aureus* in the presence of SS/Agar and PDA-SS/Agar films was similar to the control, indicating that SS/Agar and PDA-SS/Agar films did not have bacteriostatic activity. Compared with the control, AgNPs-PDA-SS/Agar significantly inhibited bacterial growth up to 20 h, suggesting that AgNPs-PDA-SS/Agar film had a long-term and efficient inhibition effect on bacterial growth.

2.8. Antimicrobial Stability

AgNPs-PDA-SS/Agar film was treated at different pH (4.0, 7.4, 10.0) for 24 h, and then the inhibitory effect of the treated film against *E. coli* and *S. aureus* was determined. As shown in Figure 8c,d, in the absence of AgNPs, there was no significant difference in bacterial growth between SS/Agar and the control at different time points, indicating SS/Agar film had no bacteriostasis ability. Compared with the control, the bacterial growth was obviously inhibited in the presence of AgNPs-PDA-SS/Agar film after treatment with different pH, suggesting AgNPs-PDA-SS/Agar film had stable and long-term antibacterial ability, which was advantageous for wound dressing and other potential applications.

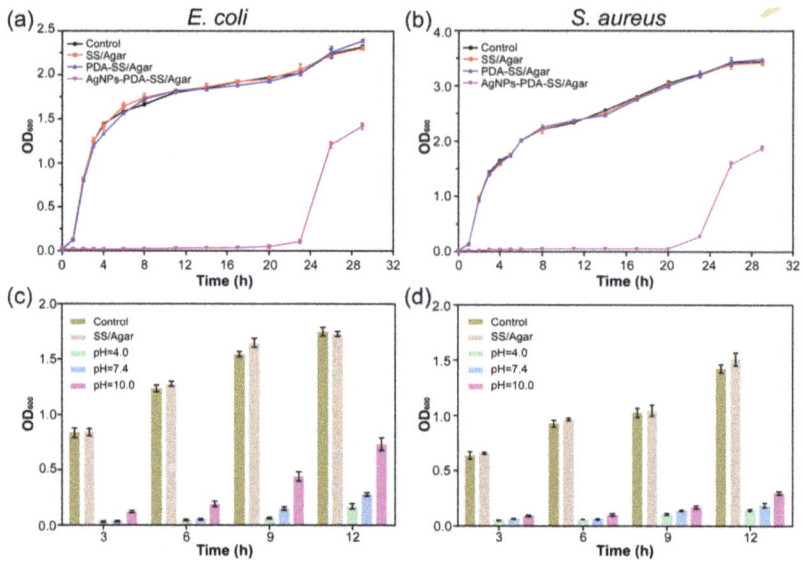

Figure 8. Bacterial growth curve of *E. coli* (**a**) and *S. aureus* (**b**) in the presence of different films, and antimicrobial stability analysis of AgNPs-PDA-SS/Agar film under different pH conditions (**c,d**).

2.9. Cytocompatibility

To evaluate the cytotoxicity of SS/Agar, PDA-SS/Agar and AgNPs-PDA-SS/Agar films, cell counting kit-8 (CCK-8) assay was performed to examine the cells treated with different films. In the test, the metabolically active cells react with the tetrazolium salt in the CCK-8 solution to produce a soluble formaldehyde nitrogen dye with maximum absorbance at 450 nm [39]. Optical density (OD) reflects cell survival and living cells [40]. The results showed there was no significant difference in cell viability between the control and the experimental group treated with AgNPs-PDA-SS/Agar film (Figure 9). Notably, the cell viability when treated with PDA-SS/Agar film was higher than that of the control, indicating PDA was not only non-toxic on cells, but also could promote cell proliferation to improve cell viability. In addition, the cell morphology under different treatments almost did not change after 24 h (Figure 10), suggesting that the prepared films had excellent cytocompatibility on the fibroblast NIH/3T3 cells, which is beneficial for its application in biomaterials.

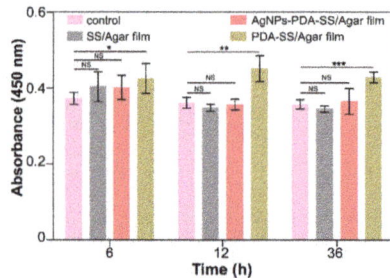

Figure 9. CCK-8 assay of the cytocompatibility of different films on NIH/3T3 cells. The statically significant values are expressed by "NS" (not significant), "★" (*p* < 0.05), "★★" (*p* < 0.01) and "★★★" (*p* < 0.001).

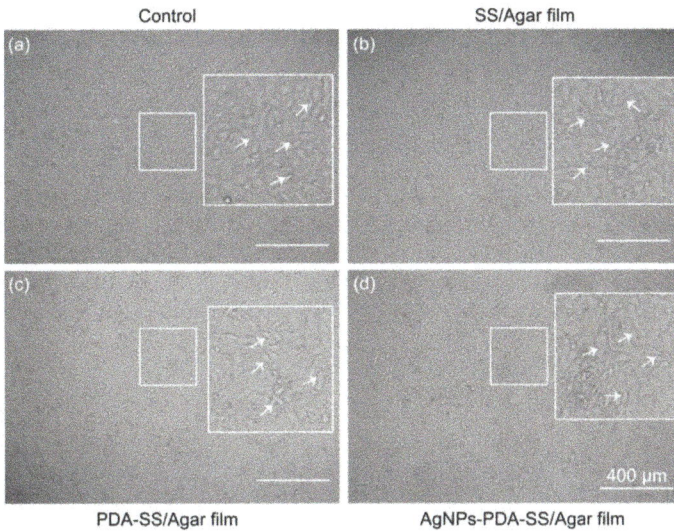

Figure 10. Microscopic observation of NIH/3T3 cells morphology with control (**a**), in the presence of SS/Agar film (**b**), PDA-SS/Agar film (**c**) and AgNPs-PDA-SS/Agar film (**d**). Small box represents a selected area, big box represents the enlarged image in the small box. White arrows indicate the observed fibroblast NIH/3T3 cells. The scale bar is 400 μm.

To better visualize the effects of the prepared films on NIH/3T3 cells viability, a living/dead cell staining assay was performed. In this assay, living cells are stained green, while dead cells are red. After being treated with different films for 24 h, the fluorescence images clearly showed almost all cells were stained green, a very few cells (<1‰) were stained red (marked with white arrows, Figure 11), indicating the excellent cytocompatibility of the films on NIH/3T3 cells. This result was in good agreement with that of CCK-8 assay and the microscopic observation on cell morphology.

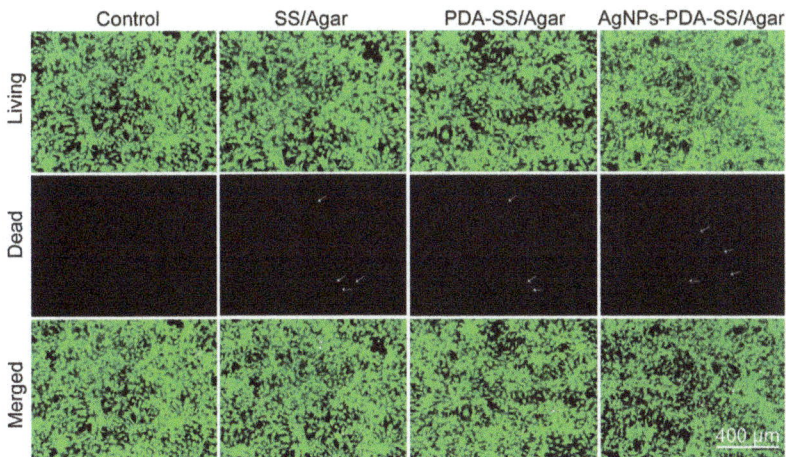

Figure 11. Living/dead cell staining assay of NIH/3T3 cells after being treated with different films. White arrows indicate a very few cells (<1‰) were stained red.

3. Materials and Methods

3.1. Materials and Chemicals

The strain of silkworm we used in this work was a commercial silkworm strain 872. Silkworms were reared in our laboratory with fresh mulberry leaves at 25 °C and 75% relative humidity under 12 h photoperiod. Fresh silkworm cocoons were collected for sericin preparation. Dopamine hydrochloride and silver nitrate ($AgNO_3$) were purchased from Aladdin (Shanghai, China). Hydrochloric acid and Tris (hydroxymethyl) aminomethane (Tris) were from Sangon Biotech (Shanghai, China). Ultrapure water made by a MilliQ water purification system (Millipore, Billerica, MA, USA) was used in the experiment. The Cell counting kit-8 (CCK-8) used in the experiment was bought from Beyotime (Beijing, China). LIVE/DEAD cell viability kit was bought from Thermo Fisher Scientific (Waltham, MA, USA). NIH3T3 (mouse embryonic fibroblast) cell lines were received from the China Infrastructure of Cell Line Resources. The Dulbecco's modified Eagle's medium (DMEM), Penicillin/Streptomycin, Fetal Bovine Serum (FBS), and Trypsin-EDTA were bought from Gibco BRL (Gaithersburg, MD, USA).

3.2. Preparation of AgNPs-PDA-SS/Agar Film

Sericin was obtained from silkworm cocoons through high temperature. Silkworm cocoons were cut into pieces and treated at 121 °C and 0.1 MPa for 15 min. Then, sericin was extracted into solution and separated from silk fibroin by filtration, which is descripted in our previous report [41–43]. Agar was dissolved in water with agitation at 90 °C to a final concentration of 2% (*w/v*). Sericin solution (2%, *w/v*) and agar solution (2%, *w/v*) were well mixed as a ratio of 1:1 at 60 °C. Subsequently, the mixture was dried at 65 °C overnight to form SS/Agar film. Next, dopamine hydrochloride was dissolved in Tris-HCl buffer (pH 8.5) to form polydopamine solution (2%, *w/v*), the chemical process was shown in Figure 1b. SS/Agar film was directly placed in freshly prepared polydopamine solution with stirring at 37 °C for 12 h. Then, the PDA-treated blend film was removed, washed by water for three times, and dried at 25 °C for 12 h to obtain PDA-coated SS/Agar composite film. Next, PDA-SS/Agar film was immersed into $AgNO_3$ solution (10 mM) at 25 °C for 4 h. After multiple washes, the film was dried at 25 °C to yield AgNPs modified PDA-SS/Agar film. The average thickness of the SS/Agar, PDA-SS/Agar and AgNPs-PDA-SS/Agar films determined by microscope were 145.47 μm, 153.29 μm and 158.90 μm, respectively.

3.3. Material Characteristics

The surface morphologies of SS/Agar, PDA-SS/Agar, AgNPs-SS/Agar and AgNPs-PDA-SS/Agar films were characterized by JEOL scanning electron microscopy JSM-6510LV (Tokyo, Japan). The films were cut into strips with a dimension of 1 cm × 1 cm (length × width), dried, and sputtered with gold prior to SEM test. The accelerating voltage for the test was 20 kV. The working distance was 10 mm. Energy dispersive spectra (EDS) were collected on Oxford INCA X-Max 250 (Abingdon, UK) during SEM test to analyze the chemical elements. XRD spectra were recorded on PANalytical x'pert (Almelo, Netherland) within 10–80°. Nicolet iz10 FT-IR spectrometer (Thermofisher Scientific, Waltham, MA, USA) was used to obtain FT-IR spectra in the wavenumber of 4000–800 cm^{-1}.

3.4. Hydrophilicity

The hydrophilicity of the prepared films was measured by surveying the sessile drip contact angle using a KRÜSS DSA100 contact angle analyzer (Hamburg, Germany) at 25 °C. Water droplets (4 μL) were dispensed on the surface of the film and the water contact angle was measured. Five points are measured for each sample and averaged.

3.5. Water Absorption Ratio and Moisture Retention Capacity

Water absorption ratio was used to characterize the swelling property of the film. The dry films (3 cm × 3 cm, length × width) were weighed, and then immersed into water. Afterwards, the swollen samples were removed from water at different intervals and weighed after the removal of water on the surface by filter paper. Swelling property was defined as follows:

$$\text{Water absorption ratio (\%)} = (m_2 - m_1)/m_1 \times 100\% \tag{1}$$

where m_1 and m_2 were the weights of dry and swollen films, respectively. Three replications per sample were made to ensure the accuracy of the experiment.

3.6. Mechanical Properties

For film materials, the tensile strength and elongation at break are two common indicators for mechanical properties. The tensile strength and elongation at break of the films were studied by SHIMADZU AG-X plus (Tokyo, Japan), which is a common instrument for mechanical properties analysis. For the test, the films with a dimension of 4 cm × 1 cm (length × width) were measured with a crosshead speed of 3 mm/min. The length and width of the samples were determined according to the mold we used, and the stretching rate was suggested by the manufacturer's protocol and validated by previous reports [44–46]. At least 8 replicates of an individual film were examined.

3.7. Inhibition Zone Assay

The method presented by Schillinger and Lücke was applied to test anti-bacterial inhibition zone of the prepared films [47]. Briefly, *E. coli* and *S. aureus* were inoculated into Luria-bertani (LB) medium (pH 7.4) with shaking at 37 °C until optical density (OD) value at 600 nm (OD_{600}) reached 1.0. Then, bacterial suspension (200 µL) was uniformly spread on agar plate in the presence of SS/Agar, PDA-SS-SS/Agar and AgNPs-PDA-SS-SS/Agar films and incubated at 37 °C for 24 h. The diameters of inhibition zone around the samples were measured.

3.8. Growth Curve Assay

The growth curve analysis was carried out based on Pal's protocol [48]. Bacteria were inoculated into LB medium and cultured at 37 °C with constant shaking (220 rpm) in the presence of different films. Then, bacterial suspension (0.5 mL) was collected at different time intervals to measure OD_{600}. All samples were tested in triplicate.

3.9. Antimicrobial Stability

To test the antibacterial stability of AgNPs-PDA-SS/Agar film, the films (1 cm × 1 cm, length × width) were soaked in PBS buffers (pH 4.0, 7.4, 10.0). After 24 h, the films were dried at 25 °C and then added into 10 mL bacterial suspensions with the same initial OD_{600}. Bacterial suspensions (0.5 mL) were collected at different intervals and OD_{600} was measured to determine the antibacterial stability of the prepared films.

3.10. Cytotoxicity

The fibroblast NIH/3T3 cells were cultured in Dulbecco's modified eagle medium at 37 °C with 5% CO_2 and 95% relative humidity. Prior to the cytotoxicity test, the circular SS/Agar, PDA-SS/Agar and AgNPs-PDA-SS/Agar films (diameter = 7 mm) were irradiated by UV light overnight to ensure the sterility of the materials. NIH/3T3 cells (100 µL) were loaded in a 96-well plate at a density of 1×10^4 cells/well and incubated at 37 °C for 12 h in the presence of the sterile films. Untreated cells were set as controls. CCK-8 kit was used to detect cells viability after treated with different films. After various time, the films were removed from the medium, and CCK-8 solution (10 µL) was added into each well and incubated at 37 °C. After 1 h, the optical density (OD) were measured at 450 nm on a Tecan Infinite M200 Pro microplate reader (Männedorf, Switzerland). Cell viability was the percentage of OD values in the treated and control groups. Each experimental group was tested for at least three replications. After culture for 24 h, NIH/3T3 cells morphology were observed on a fluorescence microscope.

In addition, a living/dead cells staining assay was performed to further assess the effect of the films on NIH/3T3 cells. NIH/3T3 cells were cultured at 37 °C as mentioned above. After being treated with SS/Agar, PDA-SS/Agar and AgNPs-PDA-SS/Agar films for 24 h, the staining solution (30 μL) was added into each well and incubated with the cells at 37 °C for 15 min. The fluorescence images were recorded on an Invitrogen EVOS FL Auto Cell Imaging System (Waltham, MA, USA). Each sample was tested in triplicate.

3.11. Statistics

All experiments were performed at least in triplicate, and the results were presented as average ± standard deviation (SD). Student's *t*-test, together with variance analysis, was carried out to determine the statistical significance between two groups. The statically significant values were expressed by "★" ($p < 0.05$), "★★" ($p < 0.01$) and "★★★" ($p < 0.001$).

4. Conclusions

In this study, we successfully synthesize high-density AgNPs on PDA-SS/Agar composite film with the assistance of PDA. AgNPs-PDA-SS/Agar composite film exhibits good hydrophilicity and proper mechanical properties. In addition, AgNPs-PDA-SS/Agar film has high efficiency and durable antibacterial ability, and excellent compatibility on the fibroblast NIH/3T3 cells. These excellent properties facilitate the potential applications of AgNPs-PDA-SS/Agar film in wound dressing, tissue engineering and antibacterial packaging.

Author Contributions: L.L., Y.W. and H.H. conceived and designed the experiments; L.L., R.C., G.T., L.A., M.Y. and P.W. performed the experiments; L.L., R.C. and G.T. analyzed the data; H.Z. and P.Z. contributed reagents/materials/analysis tools; L.L. wrote the draft; Y.W. and H.H. supervised the research and revised the manuscript.

Funding: This research was funded by National Natural Science Foundation of China (grant number 31572465), Chongqing Research Program of Basic Research, Frontier Technology (grant number cstc2015jcyjBX0035), State Key Program of the National Natural Science of China (grant number 31530071), Fundamental Research Funds for the Central Universities (grant number XDJK2018B010, XDJK2018C063), Graduate Research and Innovation Project of Chongqing (grant number CYB17069, CYS18123) and Open Project Program of Chongqing Engineering and Technology Research Center for Novel Silk Materials (grant number silkgczx2016003).

Conflicts of Interest: The authors declare no conflict of interest. The founding sponsors had no role in the design of the study; in the collection, analyses, or interpretation of data; in the writing of the manuscript, and in the decision to publish the results.

References

1. Ouattar, B.; Simard, R.E.; Piett, G.; Bégin, A.; Holley, R.A. Inhibition of surface spoilage bacteria in processed meats by application of antimicrobial films prepared with chitosan. *Int. J. Food. Microbiol.* **2000**, *62*, 139–148. [CrossRef]

2. Sondi, I.; Salopeksondi, B. Silver nanoparticles as antimicrobial agent: A case study on *E. Coli* as a model for gram-negative bacteria. *J. Colloid Interf. Sci.* **2004**, *275*, 177–182. [CrossRef] [PubMed]

3. Huang, L.; Zhao, S.; Wang, Z.; Wu, J.; Wang, J.; Wang, S. In situ immobilization of silver nanoparticles for improving permeability, antifouling and anti-bacterial properties of ultrafiltration membrane. *J. Membr. Sci.* **2016**, *499*, 269–281. [CrossRef]

4. Zhang, Y.Q. Applications of natural silk protein sericin in biomaterials. *Biotechnol. Adv.* **2003**, *20*, 91–100. [CrossRef]

5. Yang, M.; Wang, Y.; Cai, R.; Tao, G.; Chang, H.; Ding, C.; Zuo, H.; Shen, H.; Zhao, P.; He, H. Preparation and characterization of silk sericin/glycerol films coated with silver nanoparticles for antibacterial application. *Sci. Adv. Mater.* **2018**, *10*, 1–8. [CrossRef]

6. Aramwit, P.; Siritientong, T.; Kanokpanont, S.; Srichana, T. Formulation and characterization of silk sericin-pva scaffold crosslinked with genipin. *Int. J. Biol. Macromol.* **2010**, *47*, 668–675. [CrossRef] [PubMed]

7. Zhaorigetu, S.; Yanaka, N.; Sasaki, M.; Watanabe, H.; Kato, N. Inhibitory effects of silk protein, sericin on uvb-induced acute damage and tumor promotion by reducing oxidative stress in the skin of hairless mouse. *J. Photochem. Photobiol. B Biol.* **2003**, *71*, 11–17. [CrossRef]

8. Kundu, B.; Kundu, S.C. Silk sericin/polyacrylamide in situ forming hydrogels for dermal reconstruction. *Biomaterials* **2012**, *33*, 7456–7467. [CrossRef] [PubMed]

9. Chlapanidas, T.; Faragò, S.; Lucconi, G.; Perteghella, S.; Galuzzi, M.; Mantelli, M.; Avanzini, M.A.; Tosca, M.C.; Marazzi, M.; Vigo, D. Sericins exhibit ROS-scavenging, anti-tyrosinase, anti-elastase, and in vitro immunomodulatory activities. *Int. J. Biol. Macromol.* **2013**, *58*, 47–56. [CrossRef] [PubMed]

10. Chlapanidas, T.; Perteghella, S.; Leoni, F.; Faragò, S.; Marazzi, M.; Rossi, D.; Martino, E.; Gaggeri, R.; Collina, S. TNF-α blocker effect of naringenin-loaded sericin microparticles that are potentially useful in the treatment of psoriasis. *Int. J. Mol. Sci.* **2014**, *15*, 13624–13636. [CrossRef] [PubMed]

11. Faragò, S.; Lucconi, G.; Perteghella, S.; Vigani, B.; Tripodo, G.; Sorrenti, M.; Catenacci, L.; Boschi, A.; Faustini, M.; Vigo, D. A dry powder formulation from silk fibroin microspheres as a topical auto-gelling device. *Pharm. Dev. Technol.* **2016**, *21*, 453–462. [CrossRef] [PubMed]

12. Bari, E.; Arciola, C.R.; Vigani, B.; Crivelli, B.; Moro, P.; Marrubini, G.; Sorrenti, M.; Catenacci, L.; Bruni, G.; Chlapanidas, T. In vitro effectiveness of microspheres based on silk sericin and chlorella vulgaris or arthrospira platensis for wound healing applications. *Materials* **2017**, *10*, 983. [CrossRef] [PubMed]

13. Perteghella, S.; Martella, E.; De Girolamo, L.; Perucca Orfei, C.; Pierini, M.; Fumagalli, V.; Pintacuda, D.V.; Chlapanidas, T.; Viganò, M.; Faragò, S. Fabrication of innovative silk/alginate microcarriers for mesenchymal stem cell delivery and tissue regeneration. *Int. J. Mol. Sci.* **2017**, *18*, 1829. [CrossRef] [PubMed]

14. Crivelli, B.; Perteghella, S.; Bari, E.; Sorrenti, M.; Tripodo, G.; Chlapanidas, T.; Torre, M.L. Silk nanoparticles: From inert supports to bioactive natural carriers for drug delivery. *Soft Matter* **2018**, *14*, 546–557. [CrossRef] [PubMed]

15. Bari, E.; Perteghella, S.; Faragò, S.; Torre, M.L. Association of silk sericin and platelet lysate: Premises for the formulation of wound healing active medications. *Int. J. Biol. Macromol.* **2018**, *119*, 37–47. [CrossRef] [PubMed]

16. Bari, E.; Perteghella, S.; Marrubini, G.; Sorrenti, M.; Catenacci, L.; Tripodo, G.; Mastrogiacomo, M.; Mandracchia, D.; Trapani, A.; Faragò, S. In vitro efficacy of silk sericin microparticles and platelet lysate for intervertebral disk regeneration. *Int. J. Biol. Macromol.* **2018**, *118*, 792–799. [CrossRef] [PubMed]

17. Cai, R.; Tao, G.; He, H.; Song, K.; Zuo, H.; Jiang, W.; Wang, Y. One-step synthesis of silver nanoparticles on polydopamine-coated sericin/polyvinyl alcohol composite films for potential antimicrobial applications. *Molecules* **2017**, *22*, 721. [CrossRef] [PubMed]

18. Reddy, J.P.; Rhim, J.W. Characterization of bionanocomposite films prepared with agar and paper-mulberry pulp nanocellulose. *Carbohydr. Polym.* **2014**, *110*, 480–488. [CrossRef] [PubMed]

19. Wang, Y.; Cai, R.; Tao, G.; Wang, P.; Zuo, H.; Zhao, P.; Umar, A.; He, H. A novel AgNPs/sericin/agar film with enhanced mechanical property and antibacterial capability. *Molecules* **2018**, *23*, 1821. [CrossRef] [PubMed]

20. Liu, L.; Cai, R.; Wang, Y.; Tao, G.; Ai, L.; Wang, P.; Yang, M.; Zuo, H.; Zhao, P.; Shen, H. Preparation and characterization of AgNPs in situ synthesis on polyelectrolyte membrane coated sericin/agar film for antimicrobial applications. *Materials* **2018**, *11*, 1205. [CrossRef] [PubMed]

21. Aramwit, P.; Kanokpanont, S.; Nakpheng, T.; Srichana, T. The effect of sericin from various extraction methods on cell viability and collagen production. *Int. J. Mol. Sci.* **2010**, *11*, 2200–2211. [CrossRef] [PubMed]

22. He, H.; Rui, C.; Wang, Y.; Gang, T.; Guo, P.; Hua, Z.; Chen, L.; Liu, X.; Ping, Z.; Xia, Q. Preparation and characterization of silk sericin/pva blend film with silver nanoparticles for potential antimicrobial application. *Int. J. Biol. Macromol.* **2017**, *104*, 457. [CrossRef] [PubMed]

23. Seo, D.; Kim, J. Effect of the molecular size of analytes on polydiacetylene chromism. *Adv. Funct. Mater.* **2010**, *20*, 1397–1403. [CrossRef]

24. Cui, J.; Wang, Y.; Postma, A.; Hao, J.; Hosta-Rigau, L.; Caruso, F. Monodisperse polymer capsules: Tailoring size, shell thickness, and hydrophobic cargo loading via emulsion templating. *Adv. Funct. Mater.* **2010**, *20*, 1625–1631. [CrossRef]

25. Zhu, L.P.; Yu, J.Z.; Xu, Y.Y.; Xi, Z.Y.; Zhu, B.K. Surface modification of pvdf porous membranes via poly(dopa) coating and heparin immobilization. *Colloid Surface B* **2009**, *69*, 152–155. [CrossRef] [PubMed]

26. Peng, F.; Wang, Q.; Shi, R.; Wang, Z.; You, X.; Liu, Y.; Wang, F.; Gao, J.; Mao, C. Fabrication of sesame sticks-like silver nanoparticles/polystyrene hybridnanotubes and their catalytic effects. *Sci. Rep.* **2016**, *6*, 39502. [CrossRef] [PubMed]

27. Xu, H.; Zhang, G.; Xu, K.; Wang, L.; Yu, L.; Xing, M.; Qiu, X. Mussel-inspired dual-functional peg hydrogel inducing mineralization and inhibiting infection in maxillary bone reconstruction. *Mat. Sci. Eng. A-Struct.* **2018**, *90*, 379–386. [CrossRef] [PubMed]

28. Ke, W.; Yun, Y.; Zhang, Y.; Deng, J.; Lin, C. Antimicrobial activity and cytocompatibility of silver nanoparticles coated catheters via a biomimetic surface functionalization strategy. *Int. J. Nanomed.* **2015**, *10*, 7241–7252.

29. Tacora Cauna, R.L.; Luna Mercado, G.I.; Bravo Portocarrero, R.; Mayta Hancco, J.; Choque Yucra, M.; Ibañez Quispe, V. Effect ofthe expansion pressure process by explosion and toast temperature on some functional and physicochemical characteristics on two varieties of cañihua (chenopodium pallidicaule aellen). *Journal De Ciencia Y Tecnologia Agraria* **2010**, *10*, 188–198.

30. Lu, Q.; Hu, X.; Wang, X.; Kluge, J.A.; Lu, S.; Cebe, P.; Kaplan, D.L. Water-insoluble silk films with silk i structure. *Acta Biomater.* **2010**, *6*, 1380–1387. [CrossRef] [PubMed]

31. Dreyer, D.R.; Miller, D.J.; Freeman, B.D.; Paul, D.R.; Bielawski, C.W. Elucidating the structure of poly(dopamine). *Langmuir* **2012**, *28*, 6428. [CrossRef] [PubMed]

32. Jin, Y.; Deng, D.; Cheng, Y.; Kong, L.; Xiao, F. Annealing-free and strongly adhesive silver nanowire networks with long-term reliability by introduction of a nonconductive and biocompatible polymer binder. *Nanoscale* **2014**, *6*, 4812–4818. [CrossRef] [PubMed]

33. Gómezordóñez, E.; Rupérez, P. Ftir-atr spectroscopy as a tool for polysaccharide identification in edible brown and red seaweeds. *Food Hydrocolloid* **2011**, *25*, 1514–1520. [CrossRef]

34. Tao, W.; Li, M.; Xie, R. Preparation and structure of porous silk sericin materials. *Macromol. Mater. Eng.* **2010**, *290*, 188–194. [CrossRef]

35. Jiang, Y.; Lu, Y.; Zhang, L.; Liu, L.; Dai, Y.; Wang, W. Preparation and characterization of silver nanoparticles immobilized on multi-walled carbon nanotubes by poly(dopamine) functionalization. *J. Nanopart. Res.* **2012**, *14*, 1–10. [CrossRef]

36. Roy, K.; Sarkar, C.K.; Ghosh, C.K. Rapid colorimetric detection of hg2+ ion by green silver nanoparticles synthesized using dahlia pinnata leaf extract. *Green. Process. Synth.* **2015**, *4*, 455–461. [CrossRef]

37. Kanmani, P.; Rhim, J.W. Antimicrobial and physical-mechanical properties of agar-based films incorporated with grapefruit seed extract. *Carbohydr. Polym.* **2014**, *102*, 708–716. [CrossRef] [PubMed]

38. Kumar, P.T.; Lakshmanan, V.K.; Anilkumar, T.V.; Ramya, C.; Reshmi, P.; Unnikrishnan, A.G.; Nair, S.V.; Jayakumar, R. Flexible and microporous chitosan hydrogel/nano zno composite bandages for wound dressing: In vitro and *in vivo* evaluation. *ACS Appl Mater Interfaces* **2012**, *4*, 2618–2629. [CrossRef] [PubMed]

39. Yang, M.; Wang, Y.; Gang, T.; Rui, C.; Peng, W.; Liu, L.; Ai, L.; Hua, Z.; Ping, Z.; Umar, A. Fabrication of sericin/agrose gel loaded lysozyme and its potential in wound dressing application. *Nanomaterials* **2018**, *8*, 235. [CrossRef] [PubMed]

40. Yao, Y.; Liu, H.; Ding, X.; Jing, X.; Gong, X.; Zhou, G.; Fan, Y. Preparation and characterization of silk fibroin/poly(l-lactide-co-caprolactone) nanofibrous membranes for tissue engineering applications. *J. Bioact. Compat. Pol.* **2015**, *30*, 14825–14829. [CrossRef]

41. Cai, R.; Tao, G.; He, H.; Guo, P.; Yang, M.; Ding, C.; Zuo, H.; Wang, L.; Zhao, P.; Wang, Y. In situ synthesis of silver nanoparticles on the polyelectrolyte-coated sericin/pva film for enhanced antibacterial application. *Materials* **2017**, *10*, 967. [CrossRef] [PubMed]

42. Wu, J.H.; Wang, Z.; Xu, S.Y. Preparation and characterization of sericin powder extracted from silk industry wastewater. *Food Chem.* **2007**, *103*, 1255–1262. [CrossRef]

43. He, H.; Tao, G.; Wang, Y.; Cai, R.; Guo, P.; Chen, L.; Zuo, H.; Zhao, P.; Xia, Q. In situ green synthesis and characterization of sericin-silver nanoparticle composite with effective antibacterial activity and good biocompatibility. *Mat. Sci. Eng. C-Mater.* **2017**, *80*, 509. [CrossRef] [PubMed]

44. Karim, Z.; Mathew, A.P.; Grahn, M.; Mouzon, J.; Oksman, K. Nanoporous membranes with cellulose nanocrystals as functional entity in chitosan: Removal of dyes from water. *Carbohydr. Polym.* **2014**, *112*, 668–676. [CrossRef] [PubMed]
45. Tao, G.; Cai, R.; Wang, Y.; Song, K.; Guo, P.; Zhao, P.; Zuo, H.; He, H. Biosynthesis and characterization of AgNPs –silk/pva film for potential packaging application. *Materials* **2017**, *10*, 667. [CrossRef] [PubMed]
46. Janisha, J.; Rosamma, A. Deproteinised natural rubber latex grafted poly(dimethylaminoethyl methacrylate)–poly(vinyl alcohol) blend membranes: Synthesis, properties and application. *Int. J. Biol. Macromol.* **2017**, *107*, 1821.
47. Schillinger, U.; Lücke, F.K. Antibacterial activity of lactobacillus sake isolated from meat. *Appl. Environ. Microb.* **1989**, *55*, 1901.
48. Hwang, J.J.; Ma, T.W. Preparation, morphology, and antibacterial properties of polyacrylonitrile/montmorillonite/silver nanocomposites. *Mater. Chem. Phys.* **2012**, *136*, 613–623. [CrossRef]

International Journal of
Molecular Sciences

MDPI

Article

Studies on Silver Ions Releasing Processes and Mechanical Properties of Surface-Modified Titanium Alloy Implants

Aleksandra Radtke [1,2]**, Marlena Grodzicka** [1,2]**, Michalina Ehlert** [1,2]**, Tadeusz M. Muzioł** [1]**, Marek Szkodo** [3]**, Michał Bartmański** [3] **and Piotr Piszczek** [1,2,]*

[1] Faculty of Chemistry, Nicolaus Copernicus University in Toruń, Gagarina 7, 87-100 Toruń, Poland; aradtke@umk.pl (A.R.); marlena.grodzicka@doktorant.umk.pl (M.G.); m.ehlert@doktorant.umk.pl (M.E.); tmuziol@chem.umk.pl (T.M.M.)
[2] Nano-Implant Ltd. Jurija Gagarina 5/102, 87-100 Toruń, Poland
[3] Faculty of Mechanical Engineering, Gdańsk University of Technology, ul. Gabriela Narutowicza 11/12, 80-233 Gdańsk, Poland; marek.szkodo@pg.edu.pl (M.S.); michal.bartmanski@pg.edu.pl (M.B.)
* Correspondence: piszczek@umk.pl; Tel.: +48-607-883-357

Received: 17 November 2018; Accepted: 6 December 2018; Published: 9 December 2018

Abstract: Dispersed silver nanoparticles (AgNPs) on the surface of titanium alloy (Ti6Al4V) and titanium alloy modified by titania nanotube layer (Ti6Al4V/TNT) substrates were produced by the chemical vapor deposition method (CVD) using a novel precursor of the formula $[Ag_5(O_2CC_2F_5)_5(H_2O)_3]$. The structure and volatile properties of this compound were determined using single crystal X-ray diffractometry, variable temperature IR spectrophotometry (VT IR), and electron inducted mass spectrometry (EI MS). The morphology and the structure of the produced Ti6Al4V/AgNPs and Ti6Al4V/TNT/AgNPs composites were characterized by scanning electron microscopy (SEM) and atomic force microscopy (AFM). Moreover, measurements of hardness, Young's modulus, adhesion, wettability, and surface free energy have been carried out. The ability to release silver ions from the surface of produced nanocomposite materials immersed in phosphate-buffered saline (PBS) solution has been estimated using inductively coupled plasma mass spectrometry (ICP-MS). The results of our studies proved the usefulness of the CVD method to enrich of the Ti6Al4V/TNT system with silver nanoparticles. Among the studied surface-modified titanium alloy implants, the better nano-mechanical properties were noticed for the Ti6Al4V/TNT/AgNPs composite in comparison to systems non-enriched by AgNPs. The location of silver nanoparticles inside of titania nanotubes caused their lowest release rate, which may indicate suitable properties on the above-mentioned type of the composite for the construction of implants with a long term antimicrobial activity.

Keywords: titanium alloy; silver nanoparticles; surface morphology; mechanical properties; surface free energy; silver ions release

1. Introduction

The design and the manufacture of customized implants using innovative technologies is one of the main directions in modern implantology development [1,2]. New generation implants fabrication besides their anatomic fit [3,4] requires the development of new alloys and composite coatings, which provide them suitable biointegration properties, but also improve their mechanical properties, durability, hydrophilicity, etc. The implant, in order to be effective, must not only restore the function of the organ, but also ensure the patient's comfort, and protect him from negative effects, e.g., formation of inflammation or allergic reactions. The customization of implants in the patient anatomy is associated with the development of the numeric image techniques and such three-dimensional (3D) printing technology as selective laser sintering (SLS) and selective laser melting (SLM). Both above-mentioned techniques allow for the formation of three-dimensional metal structures by selective melting of metal powder in a layer-by-layer manner, which enable the formation of products with new chemical properties, differing from their macroscopic equivalents [5–7]. The response of the tissues to the implant is largely controlled by the implant surface morphology and properties. When compared to smooth surfaces, textured implants surfaces exhibit larger surface area for integrating with bone, via osseointegration process. Improved bone bonding and accelerated bone formation appears to be possible with roughened surfaces that are modified with certain treatments, which can be classified into mechanical, physical, chemical, and electrochemical methods [8]. Our earlier works on the modification of titanium and Ti6Al4V implants have shown that the fabrication of TiO$_2$ nanotube (TNT) layers of strictly defined tubes diameter on their surface had an impact on the cell adhesion and proliferation improvement [9,10].

Another problem, which should be taken into account during the designing new generation of implants, is the appearance of complications after implant introduction—a possible bacterial infections. Infections that are related to foreign body are difficult to treat because the bacteria, which cause these infections, live in well-developed biofilms. In this way, there are protected against the action of antimicrobials [11,12]. The providing the antimicrobial activity of implant surface is a complicated issue due to the necessity of the antimicrobial coatings use, which should be universal versus different strains of bacteria that are present in the organism and/or introduced with the implant [13]. Surface-modified implants with a layer of titanium dioxide can be enriched with biocides: antibiotics or other antibacterial agents. Antibiotics can be used for this purpose, however, due to bacterial resistance and concerns about their long-term effectiveness, they may not produce the desired effects [14]. Silver is the most well known antimicrobial agent of low toxicity to mammalian cells and it is effective against more than 650 pathogens. According results of previous studies, it should be noted that AgNPs are one of the most viable alternatives to antibiotics and they may be used in a wide range of applications [15–20]. AgNPs can be synthesized using the sol-gel method, electrophoretic deposition from aqueous suspensions, physical vapor deposition (PVD), chemical vapor deposition (CVD), and atomic layer deposition (ALD) [21–24]. In our works, we have focused on the use of CVD methods, which allow the formation of dispersed AgNPs on the substrate surface. The shape, size, and dispersion level of silver nanoparticles can be controlled, by optimizing the deposition parameters and also by the type of chemical compound that is used as a precursor [25]. Silver(I) complexes with fluorinated or non-fluorinated β-diketonates/carboxylates and tertiary phosphines are the most commonly used as precursors in these techniques [26,27]. Also, the selected silver(I) carboxylates can be applied as a solid source of metallic particles in CVD techniques, within silver(I) pentafluoropropionate (Ag(O$_2$CC$_2$F$_5$), as an example. The advantage of this compound is suitable volatility, low decomposition temperature at low vacuum, and a short deposition time. Moreover, it is characterized by simple and cheap synthesis [28]. In this paper, we present the results of the use of trihydrate of the above-mentioned compound as a new silver CVD precursor. The carried out studies concern the optimization of a CVD method for the production of dispersed AgNPs on the surface of Ti6Al4V implants that were manufactured by the SLS method, as obtained and modified by titanium dioxide nanotubes of different diameters. The important part of our works was the estimation of wettability, surface roughness, and mechanical properties of the produced implants. The results concerning the above-mentioned issues are not widely discussed in previous reports. Moreover, the silver ions releasing from the surface is discussed in the presented paper. It is especially important for the potential application of Ti6Al4V/AgNPs and T6Al4V/TNT/AgNPs composite materials in the construction of customized implants.

2. Results

2.1. The Chemical Vapor Deposition of Silver Nanoparticles

2.1.1. Precursor—The Structure and Thermal Properties of $[Ag_5(O_2CC_2F_5)_5(H_2O)_3]$

Simple and inexpensive synthesis of silver(I) pentafluoropropionate, and the especially suitable properties of this compound as a silver CVD precursor decided its choice for all of our experiments related to the enrichment of Ti6Al4V implants by AgNPs [28,29]. The purification of $Ag(O_2CC_2F_5)$ by the slow recrystallization of this compound from anhydrous ethanol enabled obtaining the needle-like crystals, which quality did not allow for determining their structure on the base of single crystals x-ray diffractometry. The use of 1:2 EtOH/H_2O mixture in the recrystallization process allowed for the isolation of colorless crystals after five days. Their stability in air and light was higher than pure $Ag(O_2CC_2F_5)$. Analysis of single crystal X-ray diffraction data exhibits the formation of Ag(I) complex of the formula $[Ag_5(O_2CC_2F_5)_5(H_2O)_3]$, which crystallizes in the triclinic system, space group P-1 (Figure 1, Table 1).

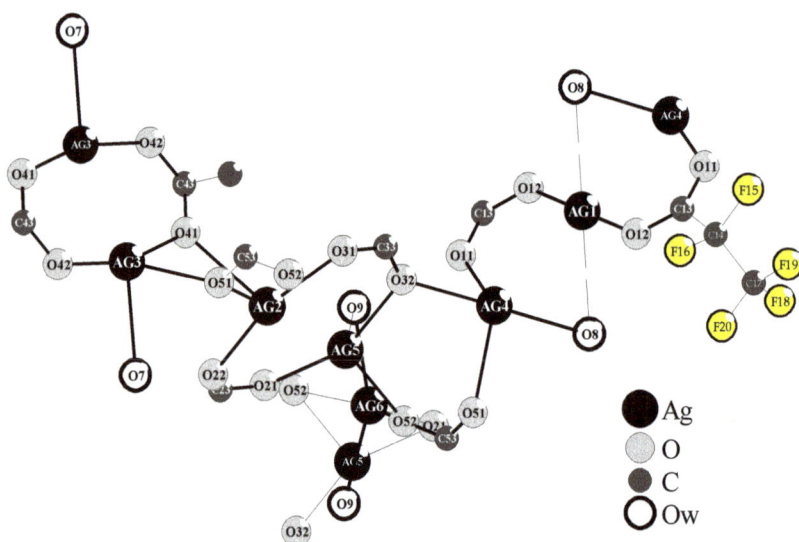

Figure 1. The scheme of the structure of $[Ag_5(O_2CC_2F_5)_5(H_2O)_3]$. For clarity, the position of only one C_2F_5 group has been presented (the full structure image is shown in the supplementary file: pp11a-2000-sch_shape-13b3-checkcif).

The structure of this complex is formed by five differently surrounded Ag(I) atoms, which are linked by carboxylate bridges and water molecules. However, the presence of three water molecules (O_w = O7, O8, O9) in the structure of this Ag(I) compound influences on its novelty and thereby its use as a new silver CVD precursor. Analysis of the single crystal X-ray diffraction data revealed that O7 bridges Ag3 and Ag4 atoms, forming the metal-oxide bonds of lengths 2.547 and 2.426 Å (Table 1). Simultaneously, O12-O7 and O22-O7 of 2.776 and 2.729 Å suggest the formation of middle hydrogen bonds [30]. Whereas, O8 and O9 molecules are strongly bonded by Ag4 (2.577 Å) and Ag6 (2.318 Å) and they are in the field of interactions with Ag1 (2.825 Å) and Ag5 (2.545 Å) (Table 1). The O_w-O and O_w-F distances below 2.8–3.4 Å, which were found between O42-O8-F70 (2.924 and 3.404 Å), O11-O9-F15 (2.777 and 3.051 Å) suggest the formation of middle and weak hydrogen bonds.

Results of the thermal analysis (TG/DTG/DTA) revealed that the thermal decomposition of this compound proceeds in one step and it is an endothermic process (beginning—453 K, max—598 K, and ending—613 K).

Table 1. Selected bonds lengths [Å] and angles [°] for $[Ag_5(O_2CC_2F_5)_5(H_2O)_3]$.

Bond Length					
Ag1-O12	2.138(5)	Ag3-O42	2.205(4)	Ag5-O21	2.234(5)
Ag1-O12 [i]	2.138(5)	Ag3-O41	2.218(4)	Ag5-O32	2.249(4)
Ag1-O8	2.825(6)	Ag3-O7	2.547(4)	Ag5-O52	2.379(5)
Ag1-Ag4 [i]	3.0058(5)	Ag3-O51	2.588(4)	Ag5-O9	2.641(6)
Ag1-Ag4	3.0058(5)	Ag3-Ag3 [iii]	2.8932(9)	Ag5-Ag2 [ii]	2.8951(7)
Ag2-O31 [ii]	2.219(5)	Ag4-O11	2.324(4)	Ag5-Ag6	3.239(2)
Ag2-O22	2.237(5)	Ag4-O7	2.426(4)	Ag6-O9	2.318(7)
Ag2-O51	2.553(4)	Ag4-O51	2.511(4)	Ag6-O9 [iv]	2.412(8)
Ag2-O41	2.610(4)	Ag4-O32	2.540(5)	Ag6-O52 [iv]	2.472(6)
Ag2-Ag5 [ii]	2.8950(7)	Ag4-O8	2.577(6)	Ag6-O52	2.579(6)
Ag2-Ag5	3.3236(8)			Ag6-O21	2.593(5)
Angles					
O12-Ag1-O12 [i]	180.0	O42-Ag3-O41	162.69(16)	O21-Ag5-O32	155.4(2)
O12-Ag1-Ag4 [i]	95.09(12)	O42-Ag3-O7	91.16(14)	O21-Ag5-O52	93.4(2)
O12 [i] -Ag1-Ag4 [i]	84.91(12)	O41-Ag3-O7	104.90(15)	O32-Ag5-O52	108.9(2)
O12-Ag1-Ag4	84.91(12)	O42-Ag3-O51	106.76(16)	O21-Ag5-Ag2 [ii]	82.41(13)
O12 [i] -Ag1-Ag4	95.09(12)	O41-Ag3-O51	84.01(16)	O32-Ag5-Ag2 [ii]	81.11(12)
Ag4 [i] -Ag1-Ag4	180.0	O7-Ag3-O51	75.23(13)	O52-Ag5-Ag2 [ii]	156.0(2)
O31 [i] -Ag2-O22	159.63(18)	O42-Ag3-Ag3 [iii]	82.99(11)	O21-Ag5-Ag6	52.75(13)
O31 [i] -Ag2-O51	91.49(16)	O41-Ag3-Ag3 [iii]	79.70(11)	O32-Ag5-Ag6	136.00(13)
O22-Ag2-O51	108.03(16)	O7-Ag3-Ag3 [iii]	158.57(10)	O52-Ag5-Ag6	51.94(14)
O31 [i] -Ag2-Ag5 [ii]	81.38(12)	O51-Ag3-Ag3 [iii]	126.20(10)	Ag2 [ii] -Ag5-Ag6	133.99(5)
O22-Ag2-Ag5 [ii]	78.26(13)	O11-Ag4-O7	153.99(15)	O21-Ag5-Ag2	134.02(15)
O51-Ag2-Ag5 [ii]	163.39(10)	O11-Ag4-O51	126.55(15)	O32-Ag5-Ag2	61.71(13)
O31 [ii] -Ag2-Ag5	61.33(14)	O7-Ag4-O51	78.78(14)	O52-Ag5-Ag2	82.47(13)
O22-Ag2-Ag5	120.51(13)	O11-Ag4-O32	88.59(16)	Ag2 [ii] -Ag5-Ag2	83.578(19)
O51-Ag2-Ag5	67.06(10)	O7-Ag4-O32	100.33(14)	Ag6-Ag5-Ag2	133.31(6)
Ag5 [ii] -Ag2-Ag5	96.42(2)	O51-Ag4-O32	85.85(15)	O9-Ag6-O9 [iv]	154.41(13)
		O11-Ag4-O8	93.10(17)	O9-Ag6-O52 [iv]	98.8(3)
		O7-Ag4-O8	78.17(17)	O9 [iv] -Ag6-O52 [iv]	78.7(2)
		O51-Ag4-O8	93.07(17)	O9-Ag6-O52	78.3(2)
		O32-Ag4-O8	178.31(16)	O9 [iv] -Ag6-O52	93.6(2)
		O11-Ag4-Ag1	74.81(10)	O52 [iv] -Ag6-O52	156.09(11)
		O7-Ag4-Ag1	79.64(9)	O9-Ag6-O21	79.1(2)
		O51-Ag4-Ag1	148.61(10)	O9 [iv] -Ag6-O21	124.0(2)
		O32-Ag4-Ag1	120.40(11)	O52 [iv] -Ag6-O21	122.1(2)
		O8-Ag4-Ag1	60.22(15)	O52-Ag6-O21	81.03(19)
				O9-Ag6-Ag5	53.70(16)
				O9 [iv] -Ag6-Ag5	134.01(17)
				O52 [iv] -Ag6-Ag5	147.11(18)
				O52-Ag6-Ag5	46.58(12)
				O21-Ag6-Ag5	43.30(11)

[i] -x,-y,-z; [ii] -x,-y-1,-z-1; [iii] -x,-y,-z-1; [iv] -x-1,-y-1,-z-1.

The analysis of the TG curve revealed that during heating of $[Ag_5(O_2CC_2F_5)_5(H_2O)_3]$ between 298 and 773 K under an inert atmosphere (N_2), the weight loss of the studied sample was c.a. 63.3%. It suggested that the metallic silver was a final product of this compound pyrolysis. To assess the volatility of the isolated Ag(I) compound, the variable temperature IR (VTIR) and the mass spectrometry (MS EI) studies have been carried out. The use of the VTIR method allowed for the estimation of the thermal stability of isolated crystals in the temperature range 303-523 K. According to

VTIR data, the dehydration of $[Ag_5(O_2CC_2F_5)_5(H_2O)_3]$ (disappearance of bands at 3436 and 3669 cm^{-1}) and the clear changes in the way of carboxyl groups interaction with Ag(I) ions (splitting of the ν_{as}(COO) band) were found between 303 and 398 K (Figure 2). The further heating of the studied compound of up to 523 K led to the formation of the stable system, which consisted of dimeric species. It was confirmed by the appearance of a single ν_{as}(COO) band at 1690 cm^{-1} [28].

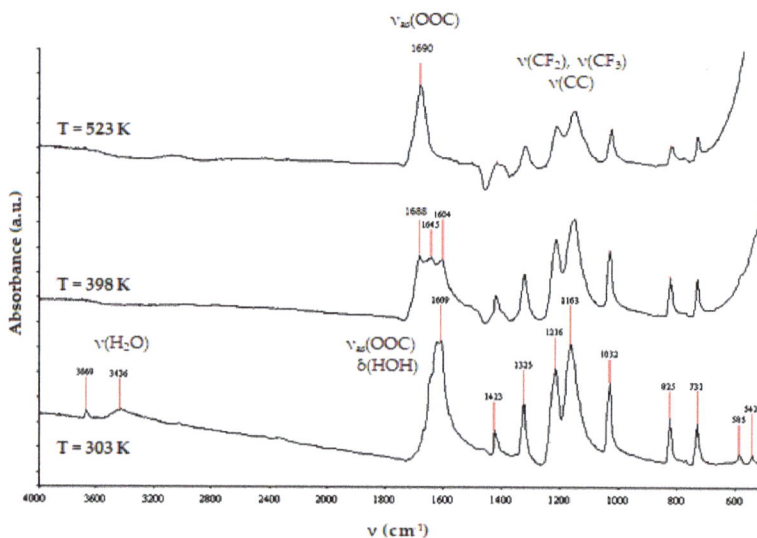

Figure 2. IR spectra of $[Ag_5(O_2CC_2F_5)_5(H_2O)_3]$ registered at 303, 398, and 523 K.

The use of mass spectrometry (MS EI) allowed for the determination of the vapor composition at temperatures 403 and 513 K during the heating of $[Ag_5(O_2CC_2F_5)_5(H_2O)_3]$ (Table 2) [28,29]. Analysis of these data allowed for identifying the following silver(I) containing species: $[Ag(O_2CC_2F_5)(H_2O)]^+$, $[Ag(O_2CC_2F_5)_2(H_2O)]^+$, $[Ag_2(O_2CC_2F_5)_3(OC)(H_2O)]^+$, and $[Ag_2(O_2CC_2F_5)_3(OOC)(H_2O)_2]^+$ in the spectrum registered at 403 K. It can indicate that dehydration process proceeds with the partial decomposition of trihydrate Ag(I) compound. The data presented in Table 2 suggest that the complete decomposition of this compound proceeds above 503 K, and the following Ag(I) containing species will be transported in vapors: $[Ag(C_2F_5)]^+$, $[Ag_2(C_2F_5)]^+$, and $[Ag_2(O_2CC_2F_5)]^+$. Their appearance in vapors suggests that they can be the main source of the metallic silver in CVD processes.

2.1.2. The Enrichment of Ti6Al4V and Ti6Al4V/TNT Substrates by Silver Nanoparticles (AgNPs)

When considering the results of thermal decomposition and volatility studies of $[Ag_5(O_2CC_2F_5)_5 (H_2O)_3]$, the overall conditions for carrying out the CVD processes were established. The optimal conditions have been determined during deposition experiments and the obtained results are listed in Table 3. The use of scanning electron microscopy (Scanning Electron Microscopy with Energy Dispersive Spectroscopy (SEM/EDS) method allowed for confirming that the result of the deposition processes were metallic silver nanoparticles (Figure 3).

Table 2. Silver(I) fragmentation ions present on the mass spectra (MS EI) of [Ag$_5$(O$_2$CC$_2$F$_5$)$_5$(H$_2$O)$_3$] registered at 403 and 513 K.

Fragmentation Ions	m/z	403 K	503 K	523 K
[Ag(CO)]$^+$	136	8	-	-
[Ag(O$_2$C)]$^+$	147	21	11	4
[Ag(O$_2$CF)]$^+$	171	23	>2	-
[Ag(C$_2$F$_5$)]$^+$	209	10	31	12
[Ag(O$_2$CC$_2$F$_5$)(H$_2$O)]$^+$	289	100	-	-
[Ag$_2$(C$_2$F$_5$)]$^+$	335	58	100	38
[Ag$_2$(O$_2$CC$_2$F$_5$)]$^+$	379	-	68	6
[Ag(O$_2$CC$_2$F$_5$)$_2$(H$_2$O)]$^+$	452	10	-	-
[Ag$_2$(O$_2$CC$_2$F$_5$)(C$_2$F$_5$)]$^+$	498	30	5	>2
[Ag$_2$(O$_2$CC$_2$F$_5$)$_2$(CO)]$^+$	586	>2	>1	-
[Ag$_2$(O$_2$CC$_2$F$_5$)$_2$(CO)$_2$]$^+$	598	>1	>1	-
[Ag$_3$(O$_2$CC$_2$F$_5$)(C$_2$F$_5$)(CO)]$^+$	635	>2	-	-
[Ag$_3$(O$_2$CC$_2$F$_5$)$_2$(CO)]$^+$	679	>1	-	-
[Ag$_2$(O$_2$CC$_2$F$_5$)$_3$(OC)(H$_2$O)]$^+$	752	>1	-	-
[Ag$_2$(O$_2$CC$_2$F$_5$)$_3$(OOC)(H$_2$O)$_2$]$^+$	784	>1	-	-

Table 3. Summary of chemical vapor deposition (CVD) conditions for the deposition of silver nanograins.

	[Ag$_5$(O$_2$CC$_2$F$_5$)$_5$(H$_2$O)$_3$]	Ag(O$_2$CC$_2$F$_5$) [29]
Total reactor pressure (p) [hPa]	5×10^{-1}	4
Substrate temperature (T$_D$) [K]	553	563
Vaporization temperature (T$_V$) [K]	508	513
Deposition rate (r_D) [mg·min^{-1}]	2.25–2.57	2.56
Carrier gas	Ar	Ar
Deposition time [min]	30	30
Precursor mass [mg]	100	100

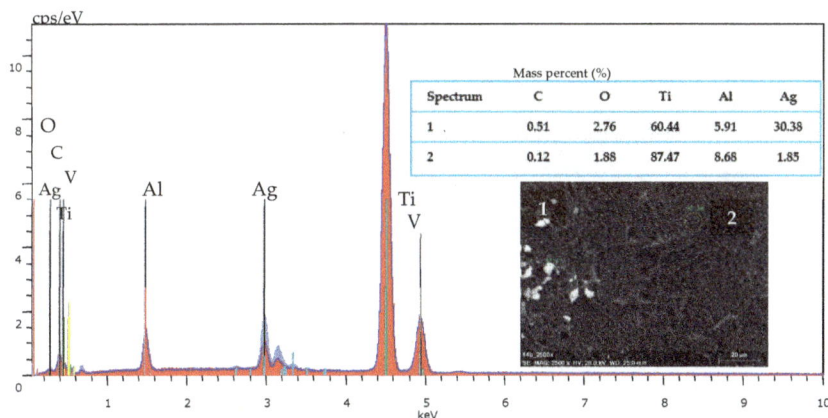

Figure 3. Energy Dispersive Spectroscopy (EDS) spectrum of Ti6Al4V/AgNPs sample and quantitative data.

Analysis of SEM images of the Ti6Al4V implant (Figure 4a) that were enriched by AgNPs revealed that the substrate surface is uniformly covered by Ag spherical grains of diameters from 15 up to 27 nm (r_D = 2.57 mg·min^{-1}; Figure 4b). Ti6Al4V/TNT/AgNPs nanocomposites were produced during the two-step procedure. In the first step, the surface of Ti6Al4V implants (produced by the selective laser sintering (SLS) technique) was modified by titania nanotubes layer using the electrochemical anodization method. The anodization process was carried out using 5, 15, and 20V potentials, and the obtained samples were designated as Ti6Al4V/TNT5, Ti6Al4V/TNT15, and Ti6Al4V/TNT20, respectively. The SEM images of the produced Ti6Al4V/TNT nanocomposites are presented in Figure 4c,e,g. According to these data, the produced TNT layers consisted of nanotubes of diameters ca. 35–45 nm (TNT5), 70–80 nm (TNT15), and 100–120 nm (TNT20). Analysis of Raman and DRIFT (diffuse reflectance infrared Furrier transformation) spectra proved the formation TiO$_2$ amorphous layers.

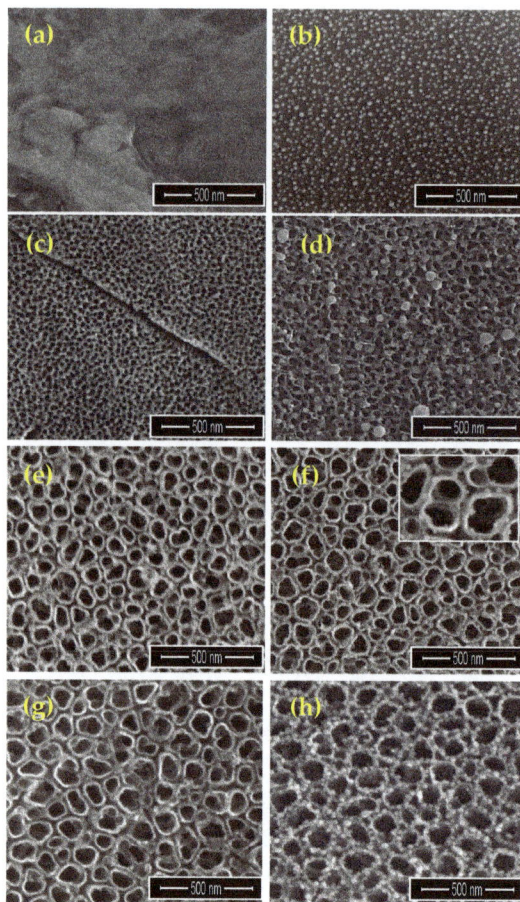

Figure 4. Scanning electron microscopy (SEM) images of Ti6Al4V (**a**), Ti6Al4V/AgNPs (**b**), Ti6Al4V/TNT5 (**c**), Ti6Al4V/TNT5/AgNPs (**d**), Ti6Al4V/TNT15 (**e**), Ti6Al4V/TNT15/AgNPs (**f**), Ti6Al4V/TNT20 (**g**), and Ti6Al4V/TNT20/AgNPs (**h**).

Enrichment of TNT layers by AgNPs using CVD technique was the next step. SEM images of modified titanium alloy implants are presented in Figure 4d,f,h. Dependently to the type of the TNT morphology layer, the differences in the size and distribution of AgNPs were noticed. Mass differences before and after CVD process of Ti6Al4V/TNT/AgNPs samples suggest the formation of coatings containing ca. 1.7 wt% (TNT5), 1.4 wt% (TNT15), and 1.9 wt% (TNT20) of silver grains. On the surface of Ti6Al4V/TNT5 coating, the dispersed nanoparticles of diameters 34–80 nm, were localized mainly on the layer surface (r_D = 2.54 mg·min^{-1}; Figure 4d). In the case of TNT15 coating, which consists of tubes of diameters 70–80 nm (TNT15) the size of AgNPs decreased up to 10–18 nm (r_D = 2.25 mg·min^{-1}; Figure 4f). The deposited metallic grains were localized inside of tubes on their walls. A further increase in the nanotubes diameter (TNT20) was accompanied by an increase of the nanograins size up to 25–35 nm (r_D = 2.42 mg·min^{-1}, Figure 4h). Also, in this case, AgNP were located on the surface of tube walls.

2.2. Measurement of the Contact Angle and Surface Free Energy of Biomaterials

Wettability of studied coatings surface and their surface free energy (SFE) were estimated using two different liquids, i.e., water as a polar liquid and diiodomethane as a dispersive one. In all studied cases, the values of contact angles, which were measured for water, were larger in comparison to adequate value for diiodomethane (Table 4). According to these data, the wettability of the Ti6Al4V/TNT surfaces increases in the row TNT5 < TNT15 < TNT20 (for TNT layers produced using potential of 5, 15, and 20V, respectively) and is better than for pure Ti6Al4V. The enrichment of these materials with AgNPs leads to the wettability decrease (increase of hydrophobic properties) and surface free energy decrease.

Table 4. The results of the contact angle measurement, which was made three times using distilled water and diiodomethane and the results of the surface free energy (SFE) of biomaterials used in Owens-Wendt method.

Biomaterial Sample	Average Contact Angle [°] ± Standard Deviation		SFE [mJ/m^2]
	Measuring Liquid		
	Water	Diiodomethane	
Ti6Al4V	108.3 ± 0.09	37 ± 0.16	45.37 ± 0.05
Ti6Al4V/AgNPs	131.9 ± 0.12	89.6 ± 0.50	15.09 ± 0.09
Ti6Al4V/TNT5	<10	36 ± 6.82	>72.06
Ti6Al4V/TNT15	<10	32.3 ± 2.75	>72.30
Ti6Al4V/TNT20	<10	30.7 ± 2.18	>72.42
Ti6Al4V/TNT5/AgNPs	124.2 ± 0.06	41.9 ± 0.47	51.97 ± 0.15
Ti6Al4V/TNT15/AgNPs	120.5 ± 0.1	67.3 ± 0.96	28.46 ± 0.23
Ti6Al4V/TNT20/AgNPs	110.2 ± 0.5	72.3 ± 0.73	21.7 ± 0.05

2.3. Mechanical Properties of Ti6Al4V/AgNPs, Ti6Al4V/TNT, and Ti6Al4V/TNT/AgNPs

The studies have been carried out for Ti6Al4V/AgNPs, Ti6Al4V/TNT, and Ti6Al4V/TNT/AgNPs systems, where TNT layers were produced using 5V (TNT5) and 15V (TNT15) potentials. The aim was to estimate mechanical property changes of two different types of TNT coatings that were enriched by AgNPs, i.e., the network formed by densely packed TiO$_2$ tubes (TNT5) and the layer composed of separated and ordered nanotubes (TNT15).

2.3.1. Surface Topography

Surface topographies of the obtained coatings and the reference Ti alloy were examined by means of atomic force microscopy (AFM, NaniteAFM, Great Britain) in the 50 × 50 μm area. Surface roughness parameter S$_a$, was determined using software that is an integral part of the device. As demonstrated by the conducted research, electrochemical anodization increases the roughness parameter S$_a$ for both coatings that were produced at 5 V and 15 V. For the coating that was obtained at a voltage of 5V, the increase in the S$_a$ parameter was almost threefold and for the coating obtained with 15V, more than five times. Also, the implantation of silver ions into electrochemically anodized coatings further increases the S$_a$ parameter. In the case of the Ti6Al4V/TNT5/AgNPs coating, the S$_a$ parameter increased by a further 32% and for the Ti6Al4V/TNT15/AgNPs coating by 9.3%. The implantation of silver ions into the Ti6Al4V substrate causes a threefold increase in surface roughness, from S$_a$ = 0.027 μm to S$_a$ = 0.078 μm (Figure 5).

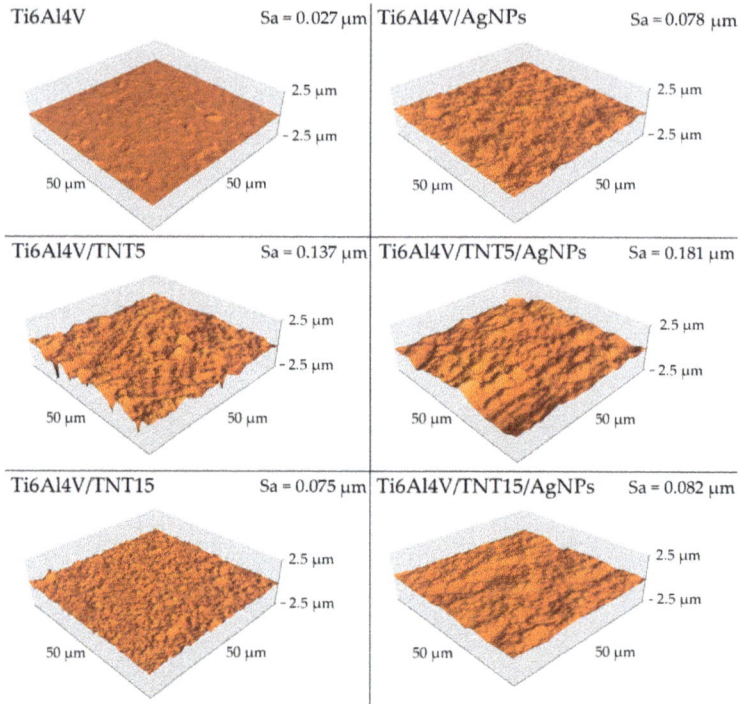

Figure 5. Atomic force microscopy (AFM) surface topography and S_a parameter of reference Ti6Al4V, Ti6Al4V/AgNPs, and Ti6Al4V/TNT/AgNPs composites.

2.3.2. Hardness and Young's Modulus

Hardness tests were carried out using a Berkovich indenter. All of the tested samples were subjected to 25 (5 × 5) measurements of nanoindentation. Individual indentations were spaced 20 μm apart (in both axes). Table 5 presents the mechanical properties measured in nanoindentation tests and Figure 6 exemplary hardness distribution on the area of 50 × 50 μm. The obtained results revealed that the surface implantation of the Ti6Al4V alloy with silver ions causes a slight increase in hardness, from 6.18 GPa to 6.81 GPa. On the other hand, after electrochemical anodization of the titanium alloy surface, the increase in hardness is greater than after surface implantation with silver ions. A particularly large, more than two and a half times, increase in hardness was noted for electrochemically anodized coatings that were obtained at 15 V (Ti6Al4V/TNT15).

Table 5. Mechanical properties (hardness, Young's Modulus and maximum depth of indentation) of reference Ti6Al4V, Ti6Al4V/AgNPs, and Ti6Al4V/TNT/AgNPs composites.

Biomaterial Sample	Hardness [GPa]	Young's Modulus [GPa]	Maximum Depth of Indentation [nm]
Ti6Al4V	6.18 ± 2.88	230.12 ± 21.68	162.18 ± 12.18
Ti6Al4V/AgNPs	6.81 ± 2.55	187.54 ± 54.33	253.09 ± 51.55
Ti6Al4V/TNT5	7.42 ± 3.30	229.71 ± 88.07	302.40 ± 61.85
Ti6Al4V/TNT15	16.23 ± 8.81	350.64 ± 157.57	168.11 ± 46.04
Ti6Al4V/TNT5/AgNPs	9.86 ± 4.61	253.93 ± 87.14	211.53 ± 56.38
Ti6Al4V/TNT15/AgNPs	13.60 ± 7.24	287.03 ± 92.92	184.46 ± 40.60

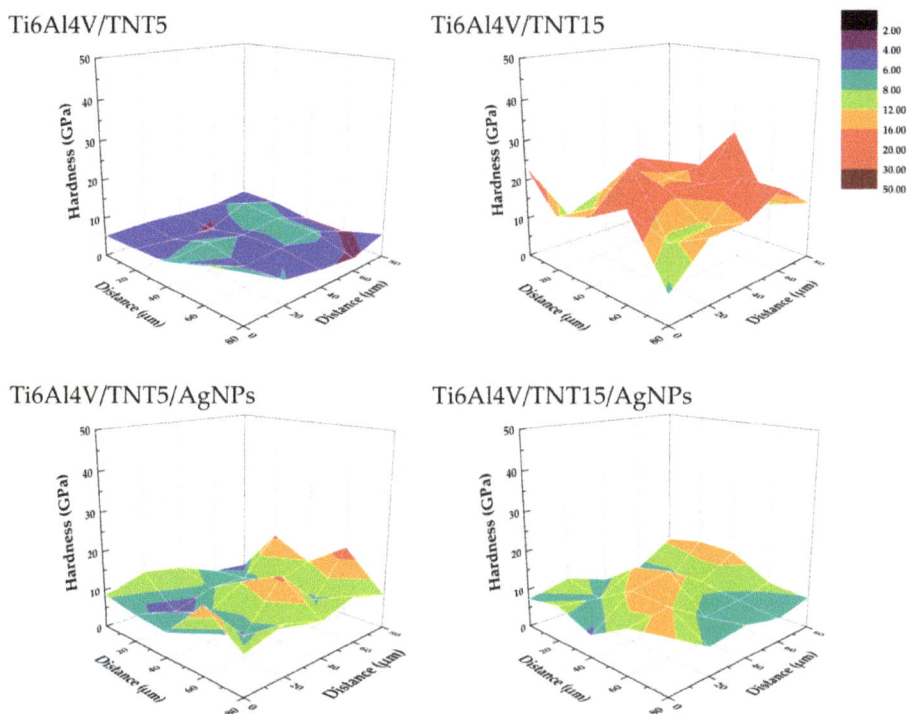

Figure 6. Hardness distribution of Ti6Al4V/TNT5, Ti6Al4V/TNT15, Ti6Al4V/TNT5/AgNPs, and Ti6Al4V/TNT15/AgNPs composites.

For coating obtained at 5 V (Ti6Al4V/TNT5), the increase in hardness was not so significant; the value only increased by 20%. After implantation with silver ions electrochemically anodized coatings, it can be noticed that, depending on the anodizing voltage, the hardness either decreases or increases. In the case of anodized coating that was obtained at 5 V, after implantation with silver ions an increase in hardness by 33% (from 7.42 GPa to 9.46 GPa) is observed. However, for anodized coating that was obtained at 15 V, after implantation with silver ions, a 16% reduction in hardness is observed (from 16.23 GPa to 13.60 GPa). Similar changes after the silver ion implantation of anodized coatings, as in the case of hardness, can also be observed for the measured Young's modulus, i.e., an increase in stiffness for the Ti6Al4V/TNT5/AgNPs composite and a decrease in stiffness for the Ti6Al4V/TNT15/AgNPs composite. In turn, the implantation with silver ions of the surface of the Ti6Al4V alloy results in the reduction of the Young's modulus from 230.12 GPa to 187.54 GPa (18.5%).

2.3.3. Adhesion Tests of Ti6Al4V/TNT and Ti6Al4V/TNT/AgNPs Composites

The coatings were subjected to five scratch tests (individual nanosporks were spaced apart by 250 μm). Table 6 presents aggregate results for all investigated coatings and Figure 7 shows exemplary curves that were obtained in the scratch test. As can be seen from the data presented in Table 6, electrochemical anodization at 15 volts allows for obtaining coatings with greater adhesion to the substrate than anodizing at five volts. The critical force resulting in the loss of adhesion is about 39% higher for the coating obtained at 15 volts than for the coating obtained at five volts.

Table 6. Results of nano scratch-tests of Ti6Al4V/TNT and Ti6Al4V/TNT/Ag composites.

Biomaterial Sample	Nano Scratch—Test Properties	
	Critical Friction [mN]	Critical Load [mN]
Ti6Al4V/TNT5	155.76 ± 69.02	197.713 ± 78.62
Ti6Al4V/TNT15	234.68 ± 21.05	275.03 ± 28.91
Ti6Al4V/TNT5/AgNPs	213.57 ± 49.50	275.11 ± 58.15
Ti6Al4V/TNT15/AgNPs	238.27 ± 53.54	267.74 ± 75.73

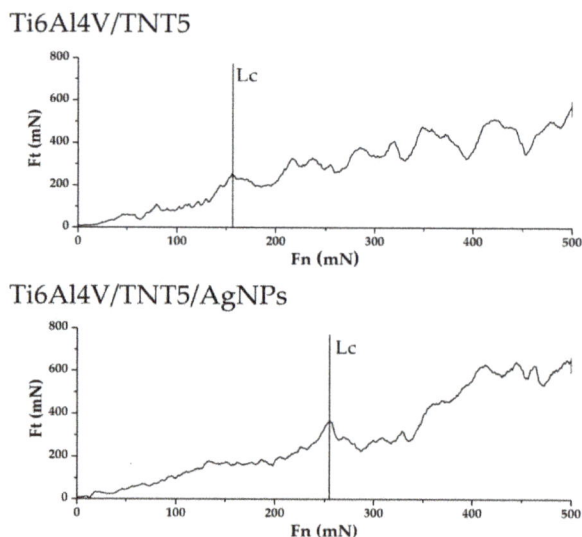

Figure 7. Examples of results obtained in the scratch test for Ti6Al4V/TNT5 coating and for Ti6Al4V/TNT5/AgNPs composite.

In addition, the standard deviation of the average results of the critical force causing the loss of adhesion to the substrate is about three times greater in the case of coatings that were obtained with the voltage of five volts. Implantation of electrochemically anodized coatings with Ag ions contributes to changes in the critical force that causes loss of coating adhesion. Electrochemically anodized coatings that were obtained at 5 V, after their implantation with Ag ions, show an increase in critical force of about 39%, while the implantation with Ag ions of coatings obtained at 15 V causes a slight decrease in adhesion by about 3.6%.

2.4. Evaluation of Stability and Durability of Coating Materials in the Body Fluid Environment

The processes of silver ions releasing from the surface of Ti6Al4V/AgNPs and Ti6Al4V/TNT/AgNPs samples, immersed in phosphate-buffered saline (PBS) solutions at human body temperature (310 K), were studied for five weeks. The results of these investigations are presented in Figure 8.

Analysis of inductively coupled plasma mass spectrometry (ICP-MS) data revealed that 3 h after immersion of Ti6Al4V/AgNPs system in the PBS solution, the concentration of Ag^+ ions was 0.18 ppm, and after 24 and 48 h, it was 0.40 and 0.86 ppm, respectively (Figure 8; extracted graph). Over the next 12 days, a further increase in the concentration of Ag^+ in PBS solution was observed up to 2.52 ppm after 14 days.

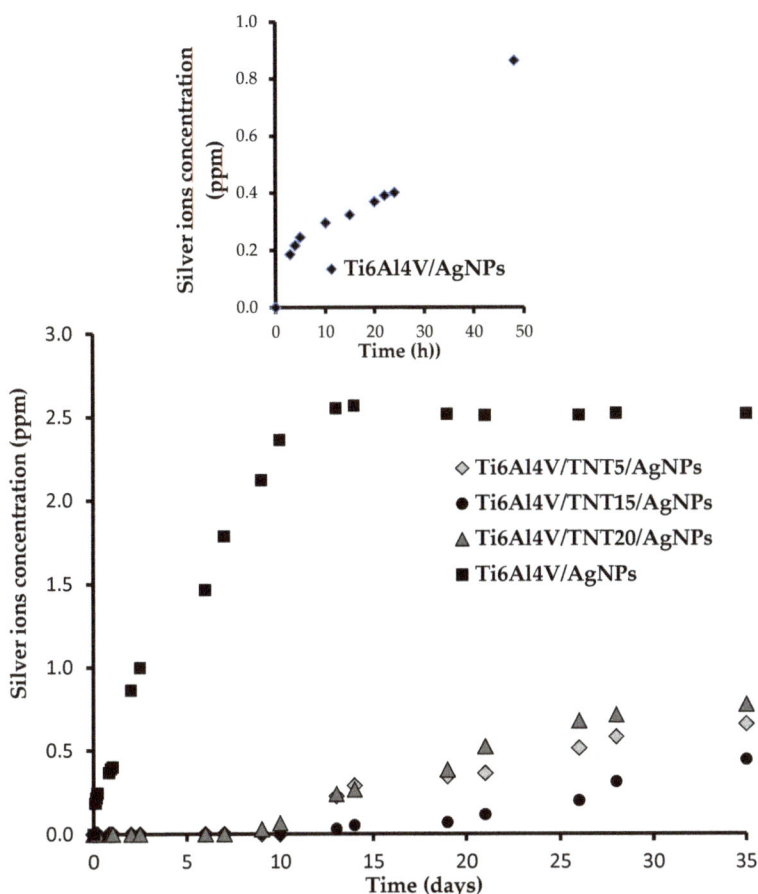

Figure 8. Silver ions released from Ti6Al4V/AgNPs coatings and Ti6Al4V/TNT5/AgNPs; Ti6Al4V/TNT15/AgNPs; and, Ti6Al4V/TNT20/AgNPs coatings immersed in phosphate buffered saline (PBS). The extracted graph shows the concentration changes of silver ions released from the surface of Ti6Al4V/AgNPs in the first 48 h after immersion off the sample in PBS solution.

During the next 14 days, the concentration of Ag^+ ions did not change significantly remaining at 2.51–2.57 ppm (Figure 8). Studies of Ti6Al4V/TNT/AgNPs composites, which were produced by the deposition of silver nanoparticles on the surface of titanium dioxide nanotubes, showed that the Ag^+ releasing processes were preceded in another way (Figure 8). In our experiments, we have used Ti6Al4V/TNT substrates of nanotube diameters ca. 35–45 nm (TNT5), 70–80 nm (TNT15), and 100–120 nm (TNT20). Obtained results revealed that during the first 14 days, the release of silver ions from the surface of all studied Ti6Al4V/TNT/AgNPs composites immersed in PBS solution, was not observed. After this time there was a slow increase of Ag^+ concentration, reaching the highest value ca. 0.77 ppm (Ti6Al4V/TNT20/AgNPs) after 35 days, which was three times lower than in the case of the Ti6Al4V/AgNPs system (Figure 8). Simultaneously, the lowest concentration of Ag^+, which amounted ca. 0.44 ppm (after 35 days) has been noticed for Ti6Al4V/TNT15/AgNPs. The obtained results revealed the clear impact of AgNPs diameter and the manner of their arrangement on the surface of TNT layers on the concentration of released silver ions (Figure 4).

3. Discussion

The implant samples that were fabricated by the selective laser sintering of Ti6Al4V powders (SLS technology) were used as substrates in all of our experiments, and the results are discussed in this paper. The appropriate porosity of substrates was obtained by covering their surface with TiO_2 nanotube coatings (TNT), which were produced using the electrochemical anodization method [9,10,31]. In our works, we have focused on studies on amorphous Ti6Al4V/TNT systems that are produced using potentials: 5 V (Ti6Al4V/TNT5), 15 V (Ti6Al4V/TNT15), and 20 V (Ti6Al4V/TNT20), which showed the promising bioactivity [10]. The conversion of amorphous TiO_2 layers into anatase phase during their heating up to 573 K were not noticed, which was confirmed by the analysis of IR and Raman spectra. This fact was important for our further works that are associated with the use of CVD technique in order to the enrichment of the Ti6Al4V and Ti6Al4V/TNT substrate surfaces with the AgNPs. Our earlier experience with the use of the different Ag(I) precursors in CVD experiments prompted us to choose $Ag(O_2CC_2F_5)$ as a suitable source of dispersed AgNPs [28,29]. However, during the recrystallization of this compound from a 1:2 $EtOH/H_2O$ mixture, the colorless crystals have been isolated after five days. The analysis of single crystals X-ray diffraction data proved the formation of trihydrate species of the general formula $[Ag_5(O_2CC_2F_5)_5(H_2O)_3]$ (Figure 1, Table 1). Three water molecules, which are presented in the structure of this compound (differently coordinated with silver atoms and taking part in possible interactions by hydrogen bonds with oxygen and fluorine atoms), should impact its properties as the CVD precursor. The results of volatility studies (VT IR and MS EI) showed that the heating of this compound in the range 303–503 K proceeded with its dehydration and the releasing of dimeric Ag(I) carboxylate species. Carried out studies exhibited that, in comparison to its anhydrous form, the isolated trihydrate crystals are characterized by the lower vaporization temperature at the pressure 5×10^{-1} hPa. Moreover, CVD experiments proved that the deposition of dispersed AgNPs proceeded with the lower deposition rate $r_D = 2.25$–2.57 mg·min^{-1} at $T_D = 553$ K (Table 3). The SEM images that are presented in Figure 4 indicated that spherical nanoparticles of silver of dimeters ca. 34–80 and 24–35 nm were localized on the surface of Ti6Al4V/TNT5/AgNPs and Ti6Al4V/TNT20/AgNPs coatings, respectively. Simultaneously, in the case of Ti6Al4V/TNT15/AgNPs, most of silver particles were localized on the walls, inside of tubes. The differences that are mentioned above can explain the noticed changes in wettability (hydrophilicity) and in the way of silver ions releasing from Ti6Al4V/TNT/AgNPs composites.

The direct consequence of the TNT layer formation on the surface of Ti6Al4V implant is the surface wettability increase with simultaneous surface free energy growth (Table 4). The obtained results are in good accordance with previous reports [32]. The enrichment of Ti6Al4V and Ti6Al4V/TNT surface by AgNPs lead to a decrease of their wettability and SFE value. The analysis of water contact angle changes revealed that the AgNPs deposition on the surface of hydrophilic surface of TNT coatings (water contact angle < 10 deg) lead to the increase of their hydrophobicity (water contact angle was changed to 110.2–124.2 deg) (Table 4). Simultaneously, it should be noted that hydrophobicity of studied samples decreases in the row: Ti6Al4V/TNT5/AgNPs > Ti6Al4V/TNT15/AgNPs > Ti6Al4V/TNT20/AgNPs. This effect can be related to the increase of nanotubes diameter from 35–45 nm up to 100–120 nm, and thus a higher ability to penetrate the interior of the nanotubes by the liquid. The increase of hydrophobicity of TNT layers (diameter 33 nm) after their decoration by AgNPs (diameter 35 nm) was also noticed by Caihong et al. [33]. The insertion of an AgNPs-enriched implant into an aqueous solution is associated with the oxidation of metal nanoparticles and the releasing of silver ions, which has direct impact on potential antimicrobial properties of the produced coatings [34]. Figure 8 shows that Ag$^+$ ions were released with the high rate in the first 12 days from the surface of Ti6Al4V/AgNPs system. After this time, the release rate changes indicate saturated behavior, and the concentration of the Ag$^+$ ions in PBS was close to 2.5 ppm. The higher concentration of these ions than 10 ppm in the human body can be toxic [35]. Zaho et al. showed a similar way of Ag$^+$ releasing from the surface of Ti/AgNPs substrates, however the concentration of silver ions in PBS solution after seven days stabilized on the level 0.13 ppm [36]. In the case of Ti6Al4V/TNT/AgNPs coatings that are immersed in PBS solution, the different silver ion releasing pathway has been noticed (Figure 8). Independently to the TNT diameters (TNT5—35–45 nm, TNT15—70–80 nm, TNT20—100–120 nm), the silver ions release process was not observed in first seven days. After this time, the Ag^+ ions concentration slowly increases, reaching values 0.44–0.77 ppm after 35 days dependently to the TNT diameter, AgNPs size, and the way of their distribution. Attention is drawn to the fact that the release rate of silver ions from Ti6Al4V/TNT20/AgNPs (lowest value of SFE and water contact angle 110.2 deg) is higher in comparison to Ti6Al4V/TNT15/AgNPs (highest value of SFE and water contact angle 124.2 deg) (Table 4). The earlier studies of Ti/TNT(anatase)/AgNPs composites (tube diameters were 50, 75, and 100 nm) revealed that Ag^+ ions were releasing with the high rate in the first two days and maintaining concentration at the level 0.25–0.28 ppm [36]. Also, in this case, the highest release rate was noticed for AgNPs that were deposited on the surface of TNT layer consisted from tubes of diameter 100 nm. Our earlier studies of Ag^+ ions releasing from Ti/TNT/AgNPs composites revealed that, after 21 days, the concentrations of these ions in PBS solution were close to 0.005–0.008 ppm, and after 28 days they increased to the level 0.15–0.22 ppm [9].

Up till now, the mechanical properties of TiO_2 coatings that are produced by the electrochemical anodization of titanium alloys have been poorly explored. However, Young's modulus, hardness, and adhesion of the coating to the substrate can be the decisive factors in terms of applications, for example, when considering the production of such coatings on implants elements. The strong integration of an implant with the bone tissue is crucial for the safe operation of the implant. The loss of coherence between the bone and the implant due to friction contributes to the implant wear. The abrasive wear of the implant may additionally cause inflammatory reactions in the patient's body. Von Wilmowsky et.al reported in their work [37] that TiO_2 nanotube coatings have shown a "good" qualitative adherence, but other reports [38,39] have qualified such coatings as "not very well adherent". The results of our nano scratch-tests revealed that the adhesion of the Ti6Al4V/TNT15 coating to the substrate is slightly greater than the Ti6Al4V/TNT5 coating (Figure 7). This difference can be result in differences in the way of TNT coatings architecture. The TNT5 layer is composed of dense packed nanotubes of diameter ca. 35–45 nm and wall thicknesses ca. 12 nm, while the TNT15 consists of separated tubes of diameter ca. 70–80 nm and wall thicknesses ca. 20 nm. After the AgNPs deposition on the surface of TNT5 coating, the dispersed nanoparticles with diameters of 34–80 nm were located mainly on the layer surface, which should not impact the adhesion of the coating to the substrate. However, AgNPs of diameters that are similar to tube sizes can be located inside of tubes or close them. This can explain the adhesion increase of TNT5 coating from 197.7 mN up to 275.03 mN after the deposition AgNPs on its surface. In the case of TNT15 coating, which consists of tubes of diameters 70–80 nm (TNT15), the size of AgNPs decreased to 10–18 nm maximum. The metallic grains were located inside the tubes on their walls, which caused a slight decrease in the adhesion of the coating to the substrate (from 275.1 mN to 267.7 mN).

Analysis of data presented in Table 5 revealed a clear increase of nano-mechanical properties (hardness, Young's modulus) of Ti6Al4V substrate surface after the formation Ti6Al4V/TNT system. However, it should be noted that the magnitude of these changes is associated with the morphology of the produced TNT coatings. The results of our measurements indicate that the hardness (7.42 GPa) and Young's modulus (229.71 GPa) of Ti6Al4V/TNT5 coatings are significantly lower than that of Ti6Al4V/TNT15 coatings (hardness 16.23 GPa and Young modulus 350.64 GPa), despite the fact that they have a finer nanotubes structure. It can be explained by the increase of both stiffness and hardness of coatings, which results from an increase of the used voltage in the anodization process. The earlier works revealed that voltage increase (at a constant process time) is accompanied with an increase nanotubes diameter, wall thickness, and also their length [40]. Moreover, the increase in voltage is accompanied by the growth of barrier layer thickness in the lower part of the nanotube, which results in the formation of larger pores and greater distance between them [41]. As a result of these processes, the nanotubes of larger diameter and wall thickness are formed. Bauer et al. [42], revealed that the anodizing of titanium using voltages 10–25 V led to the formation of coatings that are composed of separated and ordered nanotubes, while the use of lower voltages led to the formation of a structure resembling an ordered network. For studied coatings, the above-mentioned morphology changes were noticed for TNT5 and TNT15 respectively. Simultaneously, the values of the surface roughness decreased from $S_a = 0.137$ μm for Ti6Al4V/TNT5 to $S_a = 0.075$ μm for Ti6Al4V/TNT15. The phase structure of nanotubes is the next important factor that can influence the hardness and stiffness of TNT coatings. Analysis of the previous reports showed that anodized TiO_2 nanotubes are amorphous in nature [43–45]. However, the method and parameters of coatings production as well as their heat treatment may affect the occurrence of crystalline TiO_2 phases, such as anatase, rutile, or a mixture of these polycrystalline forms. It was indicated that amorphous TiO_2 nanotubes are softer than mixture of amorphous and crystallized TiO_2 (anatase) nanotubes. Also, the geometry of nanotubes, i.e., their length and wall thickness will affect the hardness measurement result [46–48]. In addition, it is well-known that the radius of the nanoindenter's tip rounding has an effect on the test results. The porosity of the coatings constitutes an additional parameter affecting their nano-mechanical properties. Munirathinam and Neelakantan reported [49] that porosity has a significant influence on the elastic modulus of the nanotubes. In our experiments, the potential of 15 V applied during the anodizing of the Ti6Al4V alloy was low enough for the formation of a structure with low porosity, as evidenced by the high Young's modulus of the Ti6Al4V/TNT15 coating. After enriching produced coatings by AgNPs, the hardness of the Ti6Al4V/TNT5 coating increased from 7.42 GPa to 9.86 GPa, while for the Ti6Al4V/TNT15 coating, the hardness decreased from 16.23 GPa to 13.60 GPa. In both cases, the values of S_a parameters also increased (Figure 5). When considering that during the CVD process, the amorphousness of TNT coatings does not change, the increase of hardness and the roughness (from $S_a = 0.137$ μm up to $S_a = 0.181$ μm) of the Ti6Al4V/TNT5/AgNPs composite can be explained by the deposition of AgNPs on the surface of an ordered network, which consists of the densely packed TiO_2 nanotubes. In the case of Ti6Al4V/TNT15/AgNPs, the increase of tubes diameter (70–80 nm) and the deposition of silver nanoparticles in their interior slightly increase the coating roughness ($S_a = 0.082$ μm), but its hardness decreases in comparison to the layer, which consists of separated and ordered TiO_2 nanotubes.

4. Materials and Methods

4.1. Synthesis of Silver CVD Precursor and Conditions CVD Processes Carry Out

The silver(I) pentafluoropropionate has been synthesized according to previously reported procedure [28,29]. The slow recrystallization of this salt from the 1:2 EtOH/H_2O solution led to the isolation of stable in air and colorless crystals after five days. The light sensitivity of these crystals required their storage in the light protected container.

Yield: 89.2%; anal. calculated for $C_{10}H_6F_{25}O_{13}Ag_5$: C, 8.98%, H, 0.44%; Found: C, 9.09%, H, 0.47%. ^{13}C NMR (75 MHz), δ (ppm): 59.04 (CF_2), 117.49 (CF_3), 167.25 (COO), ^{19}F NMR ($CDCl_3$, 376 MHz), δ (ppm): −82.91 (s, 3F), −118.38 (d, *J* = 21.1 Hz, 2F) Solid NMR spectra were recorded in a Varian Gemini 200 MHz NMR spectrometer in $CDCl_3$ (Varian Inc., Palo Alto, CA, USA). Single crystal X-ray diffraction data were collected from a crystal of dimensions 0.57 × 0.51 × 0.38 nm with an Oxford Diffraction KM4 CCD diffractometer (Oxford Diffraction Ltd., Abingdon, Oxfordshire, UK) (Mo Kα about wavelength λ = 0.71073 Å). The structure was solved by direct methods and it was refined with the full-matrix least squares on F^2 using the software package SHELX-97 [50]. All of the figures were prepared in DIAMOND [51] and ORTEP-3 [52]. CCDC: 1477237; contains the supplementary crystallographic data for $[Ag_5(O_2CC_2F_5)_5(H_2O)_3]$. These data can be obtained free of charge via http://www.ccdc.cam.ac.uk/conts/retrieving.html, or from the Cambridge Crystallographic Data Centre, 12 Union Road, Cambridge CB2 1EZ, UK; fax: (+44)1223-336-033; or e-mail: deposit@ccdc.cam.ac.uk. Crystal data and structure refinement for this compound are given in Table 7.

Table 7. Crystal data and structure refinement for [Ag$_5$(O$_2$CC$_2$F$_5$)$_5$(H$_2$O)$_3$)].

Formula sum	C$_{15}$ H$_6$ Ag$_5$ F$_{25}$ O$_{13}$
Formula weight	1408.55
Crystal system	triclinic
Space group	P-1
Unit cell dimensions	$a = 11.3277(5)$ Å
	$b = 13.0765(5)$ Å
	$c = 13.7547(5)$ Å
	$\alpha = 116.746(4)°$
	$\beta = 100.869(3)°$
	$\gamma = 99.819(3)°$
Cell volume [Å^3]	1709.36(13)
Density (calculated) [Mg/m^3]	2.737
Z	2
Absorption coefficient [mm^{-1}]	3.005
F(000)	1320
Crystal size [mm]	$0.57 \times 0.51 \times 0.38$
Theta range for data collection [deg°]	2.16 to 26.37
Index ranges	$-14 \leq h \leq 14$
	$-16 \leq k \leq 16$
	$-17 \leq l \leq 17$
Reflections collected	18624
Reflections unique/R$_{int}$	6960/0.0424
Completeness to theta = 26.37	99.5%
Transmission Max/Min	0.3947/0.2792
Refinement method	Full-matrix least-squares on F^2
Data/restraints/parameters	6960/20/592
Goodness-of-fit on F^2	1.043
Final R indices [I > 2sigma(I)]	R1 = 0.0474 wR2 = 0.1339
R indices (all data)	R1 = 0.0627 wR2 = 0.1439
Largest diff. peak and hole [e.Å^{-3}]	0.937 and −0.846

$R1 = \Sigma||F_o| - |F_c||/\Sigma|F_o|$; $wR2 = \{\Sigma[w(F_o^2 - F_c^2)^2]/\Sigma[w(F_o^2)^2]\}^{1/2}$.

The enrichment of AgNPs on the surface of Ti6Al4V and Ti6Al4V/TNT substrates were carried out while using CVD method (the horizontal hot-wall reactor) under the conditions that are presented in Table 2 [29].

4.2. The Production of Ti6Al4V/TNT Substrates and Its Characteristics

Titanium dioxide nanotube layers (TNTs) were fabricated on the surface of implants of the radial bone (total area implant—20.53 cm^2, formed by selective laser sintering (SLS) of Ti6Al4V powder) as a result of the electrochemical anodic oxidation, according to the method previously described [9,10,31]. This process was carried out at the following voltages: 5 V (TNT5), 15 V (TNT15), and 20 V (TNT20). The anodizing time was t = 30 min. The morphology of the produced coatings was examined using a Quanta scanning electron microscope with field emission (SEM, Quanta 3D FEG, Huston USA). The structure of the produced TiO$_2$ nanotube layers was studied while using Raman spectroscopy (RamanMicro 200 PerkinElmer (PerkinElmer Inc., Waltham, MA, USA) (λ = 785 nm)) and diffuse reflectance infrared Fourier transform spectroscopy (DRIFT, Spectrum2000, PerkinElmer Inc., Waltham, MA, USA).

4.3. Measurement of the Contact Angle and Surface Free Energy of Biomaterials

Determination of the wettability was carried out by the measuring of the contact angle. The contact angle was measured using a goniometer with drop shape analysis software (DSA 10 Krüss GmbH, Hamburg, Germany). The liquids that were selected for measuring the contact angle were distilled water (H$_2$O) and diiodomethane (CH$_2$I$_2$). In the case of distilled water, the volume of the drop in the contact angle measurement was 3 μL, and in the case of diiodomethane 4 μL. The measurement of the contact angle was carried out immediately after deposition of the drop. In order to determine the surface free energy, mathematical calculations were performed using the Owens-Wendt method. Each measurement was carried out three times.

4.4. Topographies and Mechanical Properties of the Produced Nanocoatings on the Surface of 3D Printed Implants

Surface topographies were examined by means of atomic force microscopy (AFM, NaniteAFM, Nanosurf AG, Liestal, Switzerland) using a contactless module with a force of 55 mN in the 50 × 50 μm area. Hardness tests and Young modulus measurements were carried out using a nanoindenter (NanoTest Vantage, Micro Materials Ltd., Wrexham, UK) using a pyramidal, diamond, three-sided Berkovich indenter, with an apical angle of 124.4°. Hardness tests were performed for the loads of 10 mN. The time of load increase from the zero value to the maximum load 10 mN was 15 s. Indentation involving one cycle with 5 s dwell at maximum load. Hardness values (H), reduced Young's modulus (Er), and Young's modulus were determined using the Oliver-Pharr method using the NanoTest results analysis program. In order to convert the reduced Young's modulus into the Young's modulus, a Poisson coefficient of 0.25 was assumed for the coatings.

Tests of coatings adhesion were made using nanoindenter (NanoTest Vantage, Micro Materials Ltd., Wrexham, UK) and using the Berkovich indenter, as in the case of the nanoindentation tests.

The parameters of scratch tests were as follows: scratch load—0 to 500 mN, loading rate—3.3 mN/s, scan velocity—3 μm/s, and scan length—500 μm. Based on the dependence of the friction force (Ft) on the normal force (Fn) in the program for the analysis of NanoTest results, the values of critical friction force (Lf) and critical force (Lc), which caused the separation of the layer from the substrate, were determined.

4.5. Evaluation of Stability and Durability of Coating Materials in the Body Fluid Environment

The analysis was carried out on Ag-enriched (a) titanium foil (Ti6Al4V, gradation 5, 99.7% purity, STREM) and (b) titanium foil with tiatnium dioxide nanotube modified surface (i.e., arrangements Ti/Ag and Ti/TNT5/Ag; Ti/TNT15/Ag; Ti/TNT20/Ag). Both variants were cut into 7 mm × 7 mm pieces. These composites were additionally protected with polyglycolide (PGA) and were also analyzed. The prepared materials were immersed in 15 mL of buffered saline solution, with the concentration of ions and the osmotic pressure being comparable to that which prevails in human body fluids. This solution was made by dissolving a PBS tablet with the following composition: 140 mM NaCl, 10 mM phosphate buffer, 3 mM KCl in 100 mL distilled water. The samples were kept in an incubator at 310 K for 1, 2, 3, 4, 6, 7, 9, 10, 13, 14, 21, 26, 28, and 35 days. Estimation of silver concentration was performed by mass spectrometry with plasma ionization inductively coupled to a quadrupole analyzer using an ICP-MS 7500 CX spectrometer with Agilent Technologies collision chamber (Agilent Technologies Inc., Tokyo, Japan).

5. Conclusions

The direct result of our works was the fabrication of coatings that are composed of the dispersed AgNPs and/or the TNT/AgNPs nanocomposites on the surface of Ti6Al4V implants, produced in SLS technology. The TNT layers were produced using electrochemical anodization of Ti6Al4V at 5, 15, and 20 V, while the CVD method was used in order to enrich Ti6Al4V and Ti6Al4/TNT substrates by silver nanoparticles. The carried out CVD experiments proved the utility of $[Ag_5(O_2CC_2F_5)_5(H_2O)_3)]$ as a precursor of metallic silver nanoparticles. The enrichment of Ti6Al4V and Ti6Al4V/TNT substrates with AgNPs increased their surface free energy, roughness, hardness, and Young's modulus. It also caused the increase of the hydrophobic properties. Studies of Ti6Al4V/TNT/AgNPs composites that were immersed in PBS solution proved that the concentration of silver ions released form the surface of these materials changes between 0.44 and 0.77 ppm after 35 days. This value is definitely below the critical level, which could have any negative effect on mammalian cells [35].

Supplementary Materials: Files containing crystallographic data were included as Supplementary Materials at http://www.mdpi.com/1422-0067/19/12/3962/s1.

Author Contributions: Conceptualization, P.P. and A.R.; Methodology, P.P., A.R., M.S., M.B.; Formal Analysis, A.R., M.E., M.S., T.M.M., P.P.; Investigation, A.R., T.M.M., M.G., M.E., M.B.; Writing-Original Draft Preparation, P.P.

Funding: This research was funded by the Regional Operational Programme of the Kuyavian-Pomeranian Voivodeship (1.3.1. Support for research and development processes in academic enterprises), within the grant obtained by Nano-implant Ltd.

Conflicts of Interest: The authors declare no conflict of interest.

References

1. Vorndran, E.; Moseke, C.; Gbureck, U. 3D printing of ceramic implants. *MRS Bull.* **2015**, *40*, 127–136. [CrossRef]
2. Imquist, A.; Omar, O.M.; Esposito, M.; Lausmaa, J.; Thomsen, P. Titanium oral implants: Surface characteristics, interface biology and clinical outcome. *J. Roy. Soc. Interface* **2010**, *7*, 515–527. [CrossRef] [PubMed]
3. Streckbein, P.; Streckbein, R.G.; Wilbrand, J.F.; Malik, C.Y.; Schaaf, H.; Howaldt, H.P.; Flach, M. Non-linear 3D Evaluation of Different Oral Implant-Abutment Connections. *J. Dent. Res.* **2012**, *91*, 1184–1189. [CrossRef] [PubMed]
4. Dekker, T.J.; Steele, J.R.; Federer, A.E.; Hamid, K.S.; Adams, S.B. Use of Patient-Specific 3D-Printed Titanium Implants for Complex Foot and Ankle Limb Salvage, Deformity Correction, and Arthrodesis Procedures. *Foot Ankle Int.* **2018**, *36*, 1–6. [CrossRef] [PubMed]
5. Tedesco, J.; Lee, B.E.J.; Lin, A.Y.W.; Binkley, D.M.; Delaney, K.H.; Kwiecien, J.M.; Grandfield, K. Osseointegration of a 3D Printed Stemmed Titanium Dental Implant: A Pilot Study. *Int. J. Dent.* **2017**, *2017*, 1–11. [CrossRef] [PubMed]
6. Shirazi, S.F.S.; Gharehkhani, S.; Mehrali, M.; Yarmand, H.; Metselaar, H.S.C.; Kadri, N.A.; Osman, N.A.A. A review on powder-based additive manufacturing for tissue engineering: Selective laser sintering and inkjet 3D printing. *J. Sci. Technol. Adv. Mater.* **2015**, *16*, 1–20. [CrossRef] [PubMed]
7. Tran, Q.H.; Nguyen, V.Q.; Le, A.T. Silver nanoparticles: Synthesis, properties, toxicology, applications and perspectives. *Adv. Nat. Sci. Nanosci. Nanotechnol.* **2013**, *4*, 1–20. [CrossRef]
8. Kang, C.G.; Park, Y.B.; Choi, H.; Oh, S.; Lee, K.W.; Choi, S.H.; Shim, J.S. Osseointegration of Implants Surface-Treated with Various Diameters of TiO_2 Nanotubes in Rabbit. *J. Nanomater.* **2015**, *2015*, 1–11. [CrossRef]
9. Piszczek, P.; Lewandowska, Ż.; Radtke, A.; Jedrzejewski, T.; Kozak, W.; Sadowska, B.; Szubka, M.; Talik, E.; Fiori, F. Biocompatibility of Titania Nanotube Coatings Enriched with Silver Nanograins by Chemical Vapor Deposition. *Nanomaterials* **2017**, *7*, 1–19. [CrossRef]
10. Radtke, A.; Topolski, A.; Jędrzejewski, T.; Kozak, W.; Sadowska, B.; Więckowska-Szakiel, M.; Szubka, M.; Talik, E.; Nielsen, L.P.; Piszczek, P. The Bioactivity and Photocatalytic Properties of Titania Nanotube Coatings Produced with the Use of the Low-Potential Anodization of Ti6Al4V Alloy Surface. *Nanomaterials* **2017**, *7*, 1–15. [CrossRef]
11. Zhao, L.; Chu, P.K.; Zhang, Y.; Wu, Z. Antibacterial Coatings on Titanium Implants. *J. Biomed. Mater. Res.* **2009**, *91*, 470–480. [CrossRef] [PubMed]
12. Costerton, J.W.; Montanaro, L.; Arciola, C.R. Biofilm in implant infections: Its production and regulation. *Int. J. Artif. Organs* **2005**, *28*, 1062–1068. [CrossRef] [PubMed]
13. Fürst, M.M.; Salvi, G.E.; Lang, N.P.; Persson, G.R. Bacterial colonization immediately after installation on oral titanium implants. *Clin. Oral Implants Res.* **2007**, *18*, 501–508. [CrossRef] [PubMed]
14. Fielding, G.A.; Roy, M.; Bandyopadhyay, A.; Bose, S. Antibacterial and biological characteristics of silver containing and strontium doped plasma sprayed hydroxyapatite coatings. *Acta Biomater.* **2012**, *8*, 3144–3152. [CrossRef] [PubMed]

15. Milić, M.; Leitinger, G.; Pavičić, I.; Avdičević, M.Z.; Dobrović, S.; Goessler, W.; Vrček, I.V. Cellular uptake and toxicity effects of silver nanoparticles in mammalian kidney cells. *J. Appl. Toxicol.* **2015**, *35*, 581–592. [CrossRef] [PubMed]

16. Feng, Q.L.; Wu, J.; Chen, G.Q.; Cui, F.Z.; Kim, T.N.; Kim, J.O. A mechanistic study of the antibacterial effect of silver ions on Escherichia coli and Staphylococcus aureus. *J. Biomed. Mater. Res.* **2000**, *52*, 662–668. [CrossRef]

17. Marambio-Jones, C.; Hoek, E.M.V. A review of the antibacterial effects of silver nanomaterials and potential implications for human health and the environment. *J. Nanopart. Res.* **2010**, *12*, 1531–1551. [CrossRef]

18. Dastjerdi, R.; Montazer, M. A review on the application of inorganic nano-structured materials in the modification of textiles: Focus on antimicrobial properties. *Colloids Surf. B Biointerfaces* **2010**, *79*, 5–18. [CrossRef]

19. Rai, M.K.; Deshmukh, S.D.; Ingle, A.P.; Gade, A.K. Silver nanoparticles: The powerful nanoweapon against multidrug-resistant bacteria. *J. Appl. Microbiol.* **2012**, *112*, 841–852. [CrossRef]

20. Franci, G.; Falanga, A.; Galdiero, S.; Palomba, L.; Rai, M.; Morelli, G.; Galdiero, M. Silver nanoparticles as potential antibacterial agents. *Molecules* **2015**, *20*, 8856–8874. [CrossRef]

21. Santillána, M.J.; Quarantab, N.E.; Boccaccinic, A.R. Titania and titania–silver nanocomposite coatings grown by electrophoretic deposition from aqueous suspensions. *Surf. Coat. Technol.* **2010**, *205*, 2562–2571. [CrossRef]

22. Akhavan, O.; Ghaderi, E. Self-accumulated Ag nanoparticles on mesoporous TiO$_2$ thin film with high bactericidal activities. *Surf. Coat. Technol.* **2010**, *204*, 3676–3683. [CrossRef]

23. Yates, H.M.; Brook, L.A.; Sheel, D.W. Photoactive Thin Silver Films by Atmospheric Pressure CVD. *Int. J. Photoenergy* **2008**, 1–8. [CrossRef]

24. Golrokhi, Z.; Chalker, S.; Sutcliffe, C.J.; Potter, R.J. Self-limiting atomic layer deposition of conformal nanostructured silver films. *Appl. Surf. Sci.* **2016**, *364*, 789–797. [CrossRef]

25. Grodzicki, A.; Łakomska, I.; Piszczek, P.; Szymańska, I.; Szłyk, E. Copper(I), silver(I) and gold(I) carboxylate complexes as precursors in chemical vapour deposition of thin metallic films. *Coord. Chem. Rev.* **2005**, *249*, 2232–2258. [CrossRef]

26. Dryden, N.H.; Vittal, J.J.; Puddephatt, R.J. New precursors for chemical vapor deposition of silver. *Chem. Mater.* **1993**, *5*, 765–766. [CrossRef]

27. Piszczek, P.; Szłyk, E.; Chaberski, M.; Taeschner, C.; Leonhardt, A.; Bała, W.; Bartkiewicz, K. Characterization of Silver Trimethylacetate Complexes with Tertiary Phosphines as CVD Precursors of Thin Silver Films. *Chem. Vap. Depos.* **2005**, *11*, 53–59. [CrossRef]

28. Szłyk, E.; Piszczek, P.; Chaberski, M.; Goliński, A. Studies of thermal decomposition process of Ag(I) perfluorinated carboxylates with temperature variable IR and MS. *Polyhedron* **2001**, *20*, 2853–2861. [CrossRef]

29. Szłyk, E.; Piszczek, P.; Grodzicki, A.; Chaberski, M.; Goliński, A.; Szatkowski, J.; Błaszczyk, T. CVD of AgI Complexes with tertiary Phosphines and Perfluorinated Carboxylates—A New Class of Silver Precursors. *Chem. Vap. Dep.* **2001**, *7*, 1–6.

30. Lutz, H.D. Bonding and structure of water molecules in solid hydrates. Correlation of spectroscopic and structural data. In *Solid Materials; Structure and Bonding*; Springer: Berlin/Heidelberg, Germany, 1988; Volume 69, pp. 97–125.

31. Lewandowska, Ż.; Piszczek, P.; Radtke, A.; Jędrzejewski, T.; Kozak, W.; Sadowska, B. The evaluation of the impact of titania nanotube covers morphology and crystal phase on their biological properties. *J. Mater. Sci.* **2015**, *26*, 163. [CrossRef]

32. Brammer, K.S.; Oh, S.; Cobb, C.J.; Bjursten, L.M.; van der Heyde, H.; Jin, S. Improved bone forming functionality on diameter-controlled TiO$_2$ nanotube surface. *Acta Biomater.* **2009**, *5*, 215–3223. [CrossRef] [PubMed]

33. Caihong, L.; Jiang, W.; Xiaoming, L. A visible-light-controlled platform for prolonged drug release based on Ag-doped TiO$_2$ nanotubes with a hydrophobic layer. *Beilstein J. Nanotechnol.* **2018**, *9*, 1793–1801. [CrossRef]

34. Piszczek, P.; Radtke, A. Silver Nanoparticles Fabricated Using Chemical Vapor Deposition and Atomic Layer Deposition Techniques: Properties. Applications and Perspectives: Review. In *Noble and Precious Metals*; Seehra, M.S., Bristow, A.D., Eds.; IntechOpen: London, UK, 2018; pp. 187–213, ISBN 978-1-78923-292-9.

35. Antoci, V., Jr.; Adams, C.S.; Parvizi, J.; Davidson, H.M.; Composto, R.J.; Freeman, T.A.; Wickstrom, E.; Ducheyne, P.; Jungkind, D.; Shapiro, I.M.; et al. The inhibition of Staphylococcus epidermidis biofilm formation by vancomycin-modified titanium alloy and implications for the treatment of periprosthetic infection. *Biomaterials* **2008**, *29*, 4684–4690. [CrossRef] [PubMed]

36. Zaho, C.; Feng, B.; Li, Y.; Tan, J.; Lu, X.; Weng, J. Preparation and antibacterial activity of titanium nanotubes loaded with Ag Nanoparticles in the dark and under the UV light. *Appl. Surf. Sci.* **2013**, *280*, 8–14. [CrossRef]

37. Von Wilmowsky, C.; Bauer, S.; Lutz, R.; Meisel, M.; Neukam, F.W.; Toyoshima, T.; Schmuki, P.; Nkenke, E.; Schlegel, K.A. In vivo evaluation of anodic TiO$_2$ nanotubes: An experimental study in the pig. *J. Biomed. Mater. Res. Part B Appl. Biomater.* **2008**, *89B*, 165–171. [CrossRef] [PubMed]

38. Kim, D.; Lee, K.; Roy, P.; Birajdar, B.I.; Spiecker, E.; Schmuki, P. Formation of a Non-Thickness-Limited Titanium Dioxide Mesosponge and its Use in Dye-Sensitized Solar Cells. *Angew. Chem.* **2009**, *121*, 9490–9493. [CrossRef]

39. Lee, K.; Kim, D.; Roy, P.; Paramasivam, I.; Birajdar, B.I.; Spiecker, E.; Schmuki, P. Anodic Formation of Thick Anatase TiO$_2$ Mesosponge Layers for High-Efficiency Photocatalysis. *J. Am. Chem. Soc.* **2010**, *132*, 1478–1479. [CrossRef] [PubMed]

40. Regonini, D.; Satka, A.; Jaroenworaluck, A.; Allsopp, D.W.E.; Bowen, C.R.; Stevens, R. Factors influencing surface morphology of anodized TiO$_2$ nanotubes. *Electrochim. Acta* **2012**, *74*, 244–253. [CrossRef]

41. Macak, J.M.; Hildebrand, H.; Marten-Jahns, U.; Schmuki, P. Mechanistic aspects and growth of large diameter self-organized TiO$_2$ nanotubes. *J. Electroanal. Chem.* **2008**, *621*, 254–266. [CrossRef]

42. Bauer, S.; Kleber, S.; Schmuki, P. TiO$_2$ nanotubes: Tailoring the geometry in H3PO4/HF electrolytes. *Electrochem. Commun.* **2006**, *8*, 1321–1325. [CrossRef]

43. Kar, A.; Raja, K.S.; Misra, M. Electrodeposition of hydroxyapatite onto nanotubular TiO$_2$ for implant applications. *Surf. Coat. Technol.* **2006**, *201*, 3723–3731. [CrossRef]

44. Xiong, J.; Wang, Y.; Li, Y.; Hodgson, P. Phase transformation and thermal structure stability of titania nanotube films with different morphologies. *Thin Solid Films* **2012**, *526*, 116–119. [CrossRef]

45. Yang, B.; Ng, C.K.; Fung, M.K.; Ling, C.C.; Djurišić, A.B.; Fung, S. Annealing study of titanium oxide nanotube arrays. *Mater. Chem. Phys.* **2011**, *130*, 1227–1231. [CrossRef]

46. Kaczmarek, D.; Domaradzki, J.; Wojcieszak, D.; Prociow, E.; Mazur, M.; Placido, F.; Lapp, S. Hardness of Nanocrystalline TiO$_2$ Thin Films. *J. Nano Res.* **2012**, *18–19*, 195–200. [CrossRef]

47. Wojcieszak, D.; Mazur, M.; Indyka, J.; Jurkowska, A.; Kalisz, M.; Domanowski, P.; Kaczmarek, D.; Domaradzki, J. Mechanical and structural properties of titanium dioxide deposited by innovative magnetron sputtering process. *Mater. Sci.-Poland* **2015**, *33*, 660–668. [CrossRef]

48. Oh, K.; Lee, K.; Choi, J. Influence of geometry and crystal structures of TiO$_2$ nanotubes on micro Vickers hardness. *Mater. Lett.* **2017**, *192*, 137–141. [CrossRef]

49. Munirathinam, B.; Neelakantan, L. Role of crystallinity on the nanomechanical and electrochemical properties of TiO$_2$ nanotubes. *J. Electroanal. Chem.* **2016**, *770*, 73–83. [CrossRef]

50. Sheldrick, G.M. Crystal structure refinement with SHELXL. *Acta Cryst.* **2015**, *C71*, 3–8. [CrossRef]

51. Brandenburg, K. *DIAMOND, Release 2.1e.*; Crystal Impact GbR: Bonn, Germany, 2001.

52. Farrugia, L.J. WinGX and ORTEP for Windows: An update. *J. Appl. Crystallogr.* **2012**, *45*, 849–854. [CrossRef]

International Journal of
Molecular Sciences

MDPI

Article

Control of Silver Coating on Raman Label Incorporated Gold Nanoparticles Assembled Silica Nanoparticles

Xuan-Hung Pham [1], Eunil Hahm [1], Eunji Kang [1], Byung Sung Son [1], Yuna Ha [1], Hyung-Mo Kim [1], Dae Hong Jeong [2] and Bong-Hyun Jun [1,*]

[1] Department of Bioscience and Biotechnology, Konkuk University, Seoul 143-701, Korea;
    phamricky@gmail.com (X.-H.P.); greenice@konkuk.ac.kr (E.H.); ejkang@konkuk.ac.kr (E.K.);
    imsonbs@konkuk.ac.kr (B.S.S.); wes0510@konkuk.ac.kr (Y.H.); hmkim0109@konkuk.ac.kr (H.-M.K.)
[2] Department of Chemistry Education and Center for Educational Research, Seoul National University,
    Seoul 151-742, Korea; jeongdh@snu.ac.kr
*   Correspondence: bjun@konkuk.ac.kr; Tel.: +82-2-450-0521; Fax: +82-2-3437-1977

Received: 28 January 2019; Accepted: 9 March 2019; Published: 13 March 2019

Abstract: Signal reproducibility in surface-enhanced Raman scattering (SERS) remains a challenge, limiting the scope of the quantitative applications of SERS. This drawback in quantitative SERS sensing can be overcome by incorporating internal standard chemicals between the core and shell structures of metal nanoparticles (NPs). Herein, we prepared a SERS-active core Raman labeling compound (RLC) shell material, based on Au–Ag NPs and assembled silica NPs (SiO_2@Au@RLC@Ag NPs). Three types of RLCs were used as candidates for internal standards, including 4-mercaptobenzoic acid (4-MBA), 4-aminothiophenol (4-ATP) and 4-methylbenzenethiol (4-MBT), and their effects on the deposition of a silver shell were investigated. The formation of the Ag shell was strongly dependent on the concentration of the silver ion. The negative charge of SiO_2@Au@RLCs facilitated the formation of an Ag shell. In various pH solutions, the size of the Ag NPs was larger at a low pH and smaller at a higher pH, due to a decrease in the reduction rate. The results provide a deeper understanding of features in silver deposition, to guide further research and development of a strong and reliable SERS probe based on SiO_2@Au@RLC@Ag NPs.

Keywords: silver shell; silica template; Au–Ag alloy; nanogaps; SERS detection

1. Introduction

Surface-enhanced Raman scattering (SERS) has been widely used for various applications due to its excellent ultrasensitive molecular fingerprinting, and its non-destructive and photostable properties [1–5]. Much effort has been focused on the use of different nanoparticles (NPs) as a substrate for SERS detection, such as silver NPs [6,7], gold NPs [8–11], and metal-embedded graphene oxide [12,13]. Although these nanostructures can enhance the SERS signal, difficulty in controlling the density of hot spots on the surface of a SERS substrate makes them unsuitable for accurate quantitative SERS assays [14].

Internal standards have been used to correct variations in SERS intensity in quantitative SERS assays [14–17]. Internal standard-based quantitative SERS methods can be classified into three categories [14]: (i) internal standard addition detection mode [18,19]; (ii) internal standard tagging detection mode [20–22]; (iii) and ratiometric SERS indicator-based detection mode [14,23]. However, the concurrent presence of target molecules and internal standard compounds on the surface of a SERS-enhancing substrate can lead to the issue of competitive adsorption between the internal standard and the target analytes in both the addition and tagging detection modes. On the other hand, the ratiometric SERS indicator-based detection mode may avoid competition between the internal standard and target molecules, as the target molecules cannot adsorb onto the surface of the SERS substrate. However, difficulties in finding or synthesizing an appropriate SERS probe for a specified target have been a limiting factor in the general application of the ratiometric SERS indicator-based detection mode [14].

Core-shell nanomaterials have attracted attention and have been employed for various applications, such as solar cells [24–26], photocatalysis [27–30], sensors [31,32], biomedical diagnosis [33–35], and imaging [36,37]. This is due to their outstanding features [38], including versatility [39], economy [40], tunability [41,42], stability, dispersibility [43], biocompatibility [44], and controllability [45]. Since their localized surface plasmon resonance (LSPR) can become tunable by controlling the bimetallic component or structure, core-shell nanomaterials have been extensively used as a substrate to enhance Raman signals of probe molecules with exquisite sensitivity. The dynamic exchange between the target molecules and internal standard is bypassed, as the internal standard is embedded between the core and shell layers. However, the unstable sol form of "core-shell" substrates can cause faster agglomeration than solid substrates [46,47]. To overcome this problem, SERS-active core-Raman labeling chemical (RLC)-shell NPs (CRLCS NPs) have been used in SERS application, especially to avoid the competitive adsorption between the internal standard and target molecules, by embedding the internal standard in core-shell NPs as enhancing substrates [15,17,48,49]. Although the presence of RLC between the Au core and the Ag shell enables a strong and reliable SERS probe, to our knowledge the effect of RLC property on the growth of an Ag shell—which can be a critical factor in fabricating the homogeneous structure of core-shell materials—has not been investigated.

Recently, our group reported Au–Raman Labeling Chemical–Ag NP assembled silica NPs (SiO$_2$@Au@RLC@Ag NPs) as strong and reliable SERS probes with an internal standard. SiO$_2$@Au@RLC@Ag NPs were synthesized using an Au seed-mediated Ag growth method on the surface of a silica template, followed by incorporating RLC on their surfaces [50–52]. Herein, we investigated the effect of experimental conditions and RLC properties on the growth of an Ag shell on the surface of SiO$_2$@Au. Three kinds of RLCs with a positive charge (4-aminothiphenol: 4-ATP), a negative charge (4-mercaptobenzoic acid: 4-MBA), and a neutral charge (4-methylbenzenthiol: 4-MBT) were used to investigate the effect of the charge properties of RLC on the growth of the Au shell. In addition, the influence of pH on the formation of the Ag shell was investigated.

2. Results and Discussion

To prepare SiO$_2$@Au@RLC@Ag NPs, silica NPs (ca. 150 nm in diameter) were synthesized using the Stöber method [53] and used as a template for embedding the Au NPs. The surface of silica NPs was first functionalized with amine groups by (3-Aminopropyl) triethoxysilane (APTS) to prepare the aminated silica NPs, as shown in Figure 1. Simultaneously, colloidal Au NPs (7 nm) were prepared by NaBH$_4$, according to the method reported by Martin et al., although with slight modifications [54,55]. Then, the Au NPs were incubated with the aminated silica NPs by gentle shaking to prepare an Au NPs embedded SiO$_2$ (SiO$_2$@Au NPs), since an amine functional group plays a crucial role in attaching the Au NPs through strong electrostatic attraction. Subsequently, three types of RLC with a positive charge (4-aminothiphenol: 4-ATP), a negative charge (4-mercaptobenzoic acid: 4-MBA) and a neutral charge (4-methylbenzenthiol: 4-MBT) were introduced on the surface of SiO$_2$@Au NPs through the strong affinity between thiol groups and Au, to investigate the effect of charge properties of RLCs on the growth of the Au shell. Finally, the Ag shell was deposited on the SiO$_2$@Au@RLC, to enhance the Raman signal of RLCs by reducing a silver precursor (AgNO$_3$) in the presence of ascorbic acid and polyvinyl pyrrolidine (PVP) as a stabilizer and structure-directing agent under mild reducing conditions [51]. In addition, the presence of the Ag shell can prevent the leakage of RLC from the Au surface, and also provide a better chance of generating numerous hot spots on the silica surface to detect target molecules.

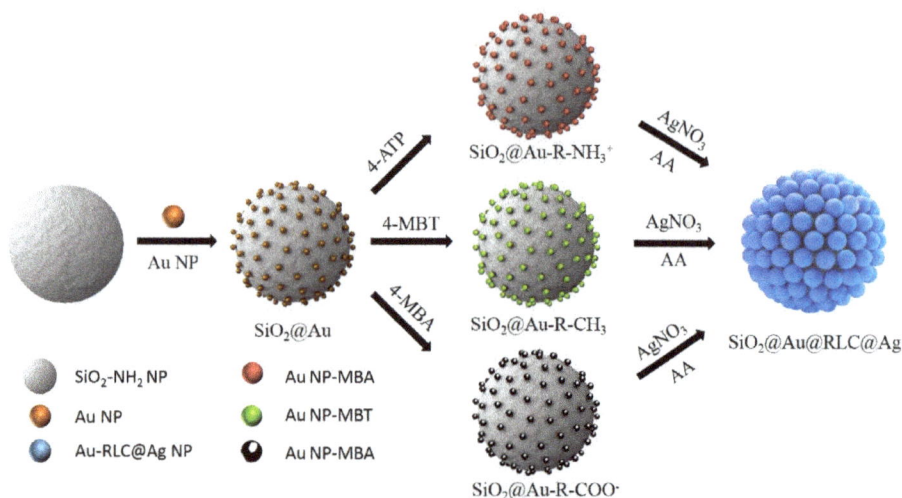

Figure 1. Illustration of a typical preparation of Au@Raman Labeling Compound@Ag embedded silica nanoparticles for a surface-enhanced Raman scattering (SERS) probe. Au NPs embedded silica nanoparticles were incubated with three different Raman labeling compounds, including 4-ATP, 4-MBA, and 4-MBT, and coated with an Ag shell by the reduction of silver nitrate in the presence of ascorbic acid and polyvinyl pyrrolidone.

As expected, the Au NPs exhibited a typical UV peak at ~520 nm, as shown in Figure S1a. After the Au NPs were coated on the surface of SiO_2, the maximum peak of SiO_2@Au was red-shifted to 530 nm. The zeta potential was used to confirm the result, and the SiO_2 NPs had a zeta potential value of -44.6 ± 0.1 mV. When the surface of the SiO_2 NP was incubated with APTS, the zeta potential value of SiO_2@NH_2 was increased to -27.7 ± 0.6 mV, due to the positive property of NH_2 groups. Throughout the entire NH_2 groups, the Au NPs were immobilized on the surface of SiO_2@NH_2 due to electrostatic attraction. Since the surface of the Au NPs was stabilized by BH_4^-, the zeta potential of SiO_2@Au was decreased to -55.4 ± 6.1 mV (Figure S1b).

2.1. Preparation of SiO_2@Au@RLC@Ag

Three types of SiO_2@Au@RLC@Ag nanomaterials with three different RLCs were successfully prepared in our study. The RLCs included 4-aminothiophenol (4-ATP) with a positive $-NH_3^+$ group; 4-MBA with a negative $-COO^-$ group; and 4-methylbenzenethiol (4-MBT) with a neutral $-CH_3$ group. The presence of -SH groups on their structures ensured that the RLCs bound to the surface of SiO_2@Au, and exhibited their functional groups of $-NH_3^+$, $-COO^-$, or $-CH_3$ in the solution. As can be seen in Figure 2a, the structure of SiO_2@Au@RLC@Ag was confirmed by the TEM analysis to show that the Ag shell was well coated on the surface of all RLCs-modified SiO_2@Au.

The UV-Vis spectra of SiO_2@Au@RLC@Ag were consistent with the TEM images (Figure 2b). In general, all solutions of SiO_2@Au@RLC@Ag NPs showed a broad band from 320 to 800 nm, indicating the generation of bumpy structures on the Ag shell and the creation of hot-spot structures on the surface of SiO_2@Au@RLC@Ag NPs [56]. At 300 μm $AgNO_3$, a typical peak of SiO_2@Au@RLCs was around 450 nm, due to the increase in the particle size of Au@RLC@Ag. However, the differences in the size of Au@Ag alloys and the distance of the nanogap between these alloys greatly affected their plasmon properties in the range of 700–800 nm, producing a continuous spectrum of resonant multimode [50,52,56–59]. The zeta potential of SiO_2@Au@RLCs was measured (Figure S2) to explain the formation of the Ag shell on the surface of SiO_2@Au@RLCs. As mentioned previously, the zeta potential of SiO_2@Au was -55.4 ± 6.1 mV. When RLCs were modified on the surface of SiO_2@Au, the zeta potential of all structures increased significantly. RLCs possess the -SH groups, which have a stronger affinity to Au NPs than NH_2 groups on the surface of SiO_2. Thus, RLCs may absorb on the surface of Au NPs, and some of the Au-RLC complex can migrate from the surface of SiO_2@Au NPs, leading the zeta potential of RLCs-modified SiO_2@Au NPs to be less negative. Yet, since the difference exists in functional groups of RLCs, SiO_2@Au@RLC still possess a difference in surface charge of -35.2 ± 0.5 mV (4-ATP), -33.4 ± 1.3 mV (4-MBT) and -44.4 ± 6.9 mV (4-MBA), respectively. Nevertheless, the presence of negative charges on the surface of SiO_2@Au@RLC facilitated the attraction of Ag^+ ions to their surface and reduced them to Ag NPs.

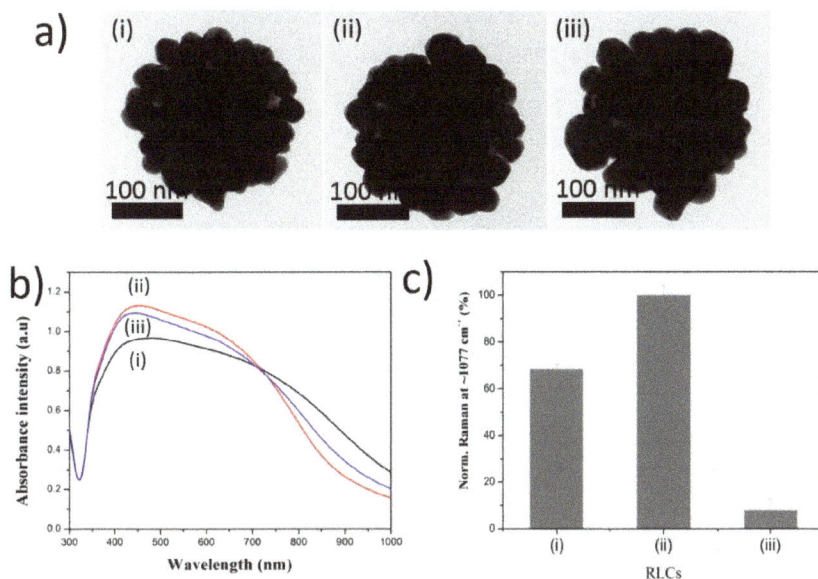

Figure 2. (a) Transmission electron microscopy (TEM) images, (b) UV-Vis absorption spectra of (i) SiO_2@Au@4-ATP@Ag, (ii) SiO_2@Au@4-MBA@Ag and (iii) SiO_2@Au@4-MBT@Ag synthesized in water, and (c) their normalized Raman intensity at 1077 cm^{-1}. All SiO_2@Au was fixed at 200 μg. Concentration of Raman Labeling Chemical was 1 mM and that of $AgNO_3$ was 300 μM.

Raman signals of three SiO_2@Au@RLC@Ag nanomaterials were also measured (Figure 2c). The Raman intensity of SiO_2@Au@4-MBA@Ag at 1075 cm^{-1} was the strongest compared to that of SiO_2@Au@4-ATP@Ag and SiO_2@Au@4-MBT@Ag. Raman signals of SiO_2@Au@4-ATP@Ag and SiO_2@Au@4-MBT@Ag were equal to those of the 68.3% and 7.9% of SiO_2@Au@4-ATP@Ag, respectively.

2.2. Effect of Silver Ion Concentration on Ag Shell Coating on SiO_2@Au@RLCs

To examine the effect of silver ion concentration on a silver shell coating of SiO_2@Au@RLC, 4-MBA, 4-ATP, and 4-MBT were first introduced on the surface SiO_2@Au NPs. The Ag shell was then deposited onto SiO_2@Au@RLCs by the reduction of $AgNO_3$, using ascorbic acid. The TEM analysis was performed to confirm the structure of SiO_2@Au@RLC@Ag, as shown in Figures S3–S5. When the $AgNO_3$ concentration was increased from 50 to 300 μM, the size of Au@RLC@Ag alloy NPs became greater. However, Ag NPs (ca. 50–100 nm) appeared separately at higher concentrations of $AgNO_3$ (>300 μM). This is possibly due to the formation of extra Ag NPs, made by nucleation in the solution during the reduction of high the $AgNO_3$ concentration.

UV-Vis spectroscopies of the solution of SiO_2@Au@RLC@Ag nanomaterials were recorded (Figure 3). The absorbance band of the SiO_2@Au@RLC@Ag prepared with 4-ATP, 4-MBA, and 4-MBT appeared at 430–450 nm at low concentrations of $AgNO_3$ (50 μM). The bands extended from 430 nm to

1000 nm when the AgNO$_3$ concentration was increased to 700 µM. At the same time, their absorbance intensities were increased with a higher AgNO$_3$ concentration. The results indicated that the silver shell was well coated on the surface of SiO$_2$@Au@RLC in deionized water. Indeed, the Raman intensities of the SiO$_2$@Au@RLC@Ag prepared with 4-ATP, 4-MBA, and 4-MBT became greater with an increase in the thickness of the Ag shell when AgNO$_3$ increased from 50 µM to 200 µM. The Raman intensity plateaued when AgNO$_3$ increased up to 300 µM. To compare the exact effects of Ag coating on the Raman signal of SiO$_2$@Au@RLC@Ag without considering the differences in the intrinsic Raman properties of RLCs, we calculated the slopes of SiO$_2$@Au@RLC@Ag in the range of 50 to 200 µM. The slopes of the normalized Raman signal were 0.105, 0.156, and 0.012 unit/µM, which correspond to 4-ATP, 4-MBA, and 4-MBT, respectively. The results indicate that the Ag shell coating significantly affected the Raman signals of these three SiO$_2$@Au@RLC@Ag.

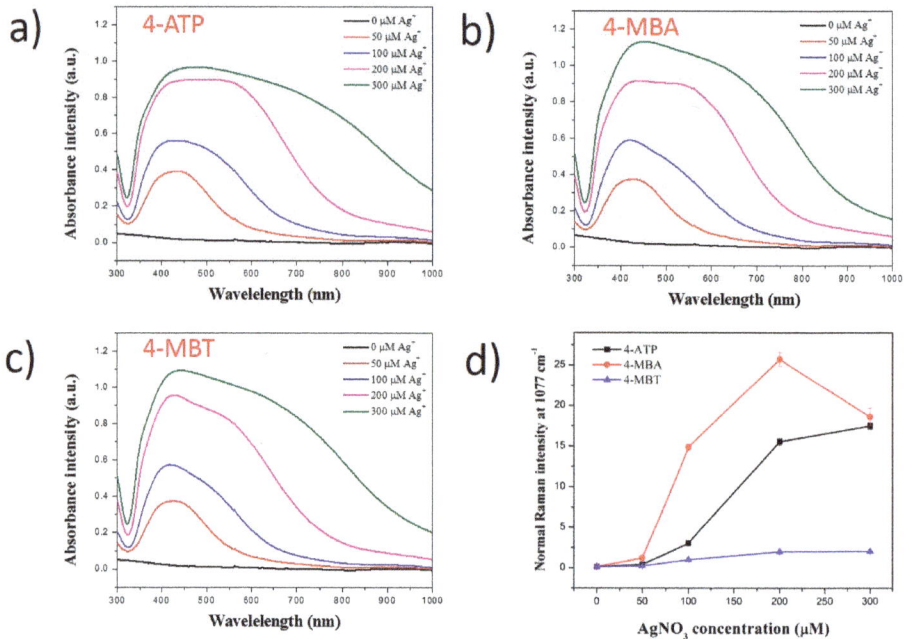

Figure 3. UV-Vis absorption spectra of (**a**) SiO$_2$@Au@4-ATP@Ag, (**b**) SiO$_2$@Au@4-MBA@Ag, (**c**) SiO$_2$@Au@4-MBT@Ag nanoparticles, and (**d**) the normalized Raman spectra of the particles coated with different concentrations of AgNO$_3$ in water. All SiO$_2$@Au was fixed at 200 µg. Concentration of RLCs was 1 mM.

2.3. *Effect of pH Solution on the Ag Shell Coating of SiO$_2$@Au@RLC@Ag NPs*

To confirm the effect of both pH and RLCs characteristics on the Ag shell coating of SiO$_2$@Au@RLCs, we adjusted the pH of the solution during the reduction of Ag$^+$. The coating of the Ag shell on the surface of SiO$_2$@Au@RLCs was strongly dependent on the pH of the solution (Figures 4–6). At a high pH, smaller sized silver nanoparticles were obtained, compared to those obtained at a low pH, due to the low reduction rate of AgNO$_3$ precursors [60]. The coating of the Ag shell on the surface of SiO$_2$@Au@4-MBT was rapid and worked well at a pH of 5.0, but became sluggish and difficult in acidic or basic pH values (Figure 4a and Figure S6). The Raman signals of SiO$_2$@Au@4-MBT@Ag nanomaterials were measured (Figure 4b,c). The Raman signals of SiO$_2$@Au@4-MBT@Ag were too weak and unclear because of small Au@4-MBT@Ag alloys with thin Ag shells. This result was consistent with the TEM images we observed in Figure 4a.

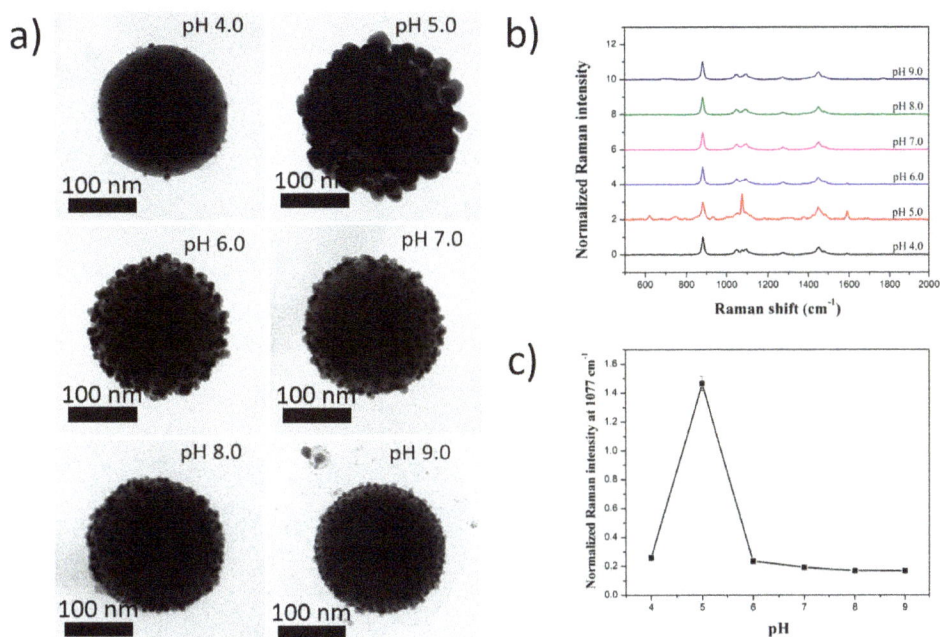

Figure 4. (**a**) TEM images and (**b,c**) Raman spectra of SiO$_2$@Au@4-MBT@Ag synthesized at different pH solutions. All SiO$_2$@Au was fixed at 200 µg. Concentration of RLCs was 1 mM and that of AgNO$_3$ was 300 µM.

When 4-ATP was used as an RLC, the size of SiO$_2$@Au@4-ATP@Ag became smaller when the pH was increased from 4.0 to 9.0 (Figure 5 and Figure S7). The coating of the Ag shell on the surface of SiO$_2$@Au@4-ATP was rapid and worked well from an acidic to a basic pH solution. As a result, the Raman signals of SiO$_2$@Au@4-ATP@Ag were observed clearly (Figure 5b,c). According to previous reports, pK$_a$ values of 4-ATP on a gold surface range from 5.3 to 5.9 [61,62]. At a low pH (pH < 5), NH$_2$ groups of 4-ATP on the surface of Au NPs exist in a protonated form (NH$_3$$^+$), and have a stronger affinity with Ag NPs generated in a bulk solution during the reduction of AgNO$_3$ than with those generated during the deposition of the Ag shell on the surface of the SiO$_2$@Au@4-ATP [63]. This may lead to the formation of large Ag NPs on the surface of SiO$_2$@Au@4-ATP, as can be seen in TEM images (Figure S7), but did not significantly increase the Raman signal of 4-ATP (Figure 5). At a high pH (pH > 6), the deposition of the Ag shell on SiO$_2$@Au@4-ATP dominated more, leading to a greater intensity of Raman signal in 4-ATP (Figure 5a).

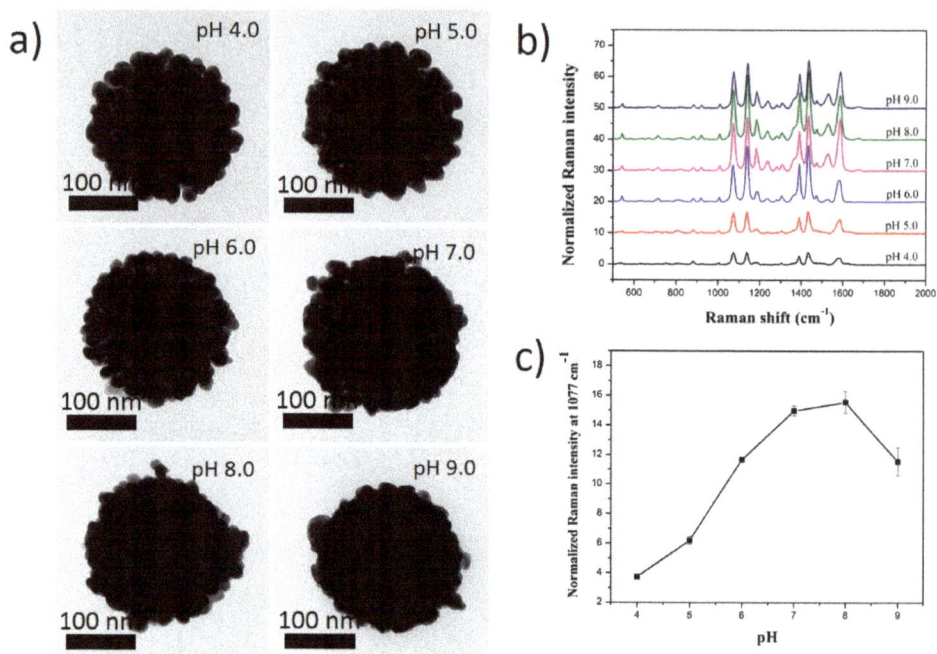

Figure 5. (**a**) TEM images and (**b,c**) Raman spectra of $SiO_2@Au@4\text{-}ATP@Ag$ synthesized at different pH solutions. All $SiO_2@Au$ was fixed at 200 µg. Concentration of RLCs was 1 mM and that of $AgNO_3$ was 300 µM.

Similarly, when 4-MBA was used as an RLC, the size of $SiO_2@Au@4\text{-}MBA@Ag$ became smaller when the pH was increased from 4.0 to 9.0 (Figure 6 and Figure S8). The coating of the Ag shell on the surface of $SiO_2@Au@4\text{-}MBA$ was also well obtained from an acidic to a basic pH solution. The carboxyl groups of 4-MBA existed in a protonated form (-COOH) at a low pH, lower than their pK_a ($pK_a \approx 5$) [64–66]. The presence of -COOH inhibited the coating of the Ag shell on the surface of the $SiO_2@Au@4\text{-}MBA$ (Figure 6) and caused a low signal in 4-MBA (Figure 6). Similarly, the deprotonated form of the carboxylate groups (-COO$^-$) became dominated on the surface of the $SiO_2@Au@4\text{-}MBA$ when the pH of the solution was raised and reached a value higher than the pK_a value of 4-MBA. They also led to an increase of the Raman signal of 4-MBA in the pH range of 5.0 to 6.0. It is known that, as the pH of solution increases continuously, silver oxide or silver chloride is formed [67], which can inhibit the coating of the Ag shell on the surface of $SiO_2@Au@4\text{-}MBA$ (Figure S8), with an obvious decrease in the Raman signal of 4-MBA from a pH of 7.0 to 9.0.

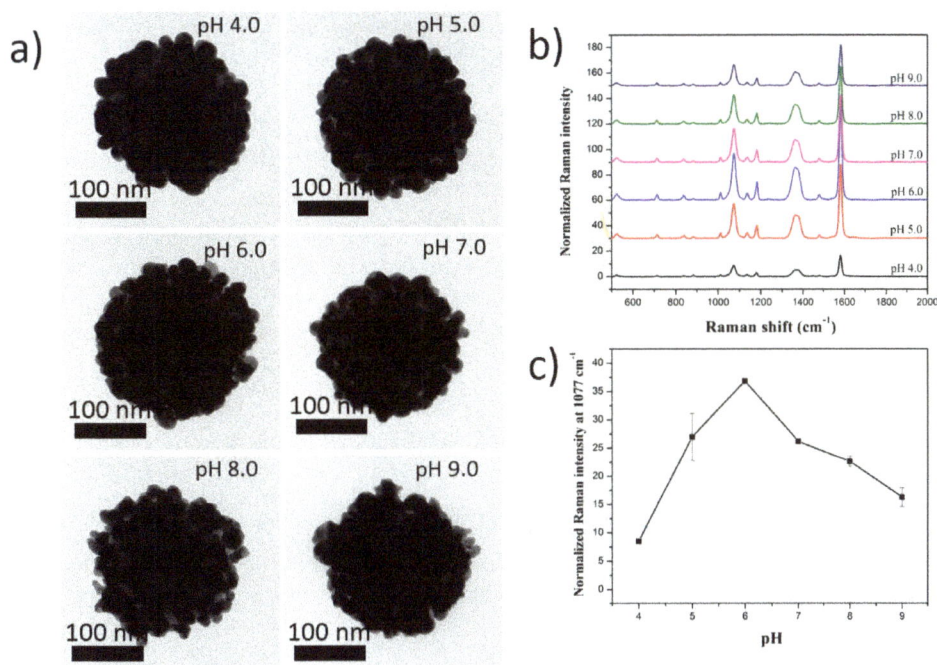

Figure 6. (**a**) TEM images and (**b,c**) Raman spectra of SiO$_2$@Au@4-MBA@Ag synthesized at different pH solutions. All SiO$_2$@Au was fixed at 200 μg. Concentration of RLCs was 1 mM and that of AgNO$_3$ was 300 μM.

3. Experiment

3.1. Materials

Tetraethylorthosilicate (TEOS), 3-aminopropyltriethoxysilane (APTS), silver nitrate (AgNO$_3$), chloroauric acid (HAuCl$_4$), 4-mercaptobenzoic acid (4-MBA), ascorbic acid (AA), polyvinylpyrrolidone (PVP), sodium borohydride (NaBH$_4$), and thiram were purchased from Sigma-Aldrich (St. Louis, MO, USA) and used without further purification. Ethyl alcohol (EtOH) and aqueous ammonium hydroxide (NH$_4$OH, 27%) were purchased from Daejung (Siheung, Korea).

3.2. Preparation of SiO$_2$@Au NP Templates

Silica NPs (~150 nm) were prepared using the Stöber method [53]. The silica NPs (50 mg mL^{-1}, 4 mL) were dispersed in 4 mL of absolute EtOH, and 250 μL of APTS and 40 μL of NH$_4$OH were added to the colloidal solution to aminate the silica NPs. The mixture was stirred vigorously for 6 h at 25 °C, followed by stirring for 1 h at 70 °C. The aminated silica NPs were obtained after centrifugation at 8500 rpm for 15 min, and then washed several times with EtOH to remove excess reagent.

The colloidal Au NPs were prepared by reducing HAuCl$_4$, using NaBH$_4$ as a reducing agent. The reduction of HAuCl$_4$ created small Au NPs (~7 nm) with a net negative surface charge. In order to embed Au NPs into the silica NP surface, the Au NPs (1 mM, 10 mL) and aminated SiO$_2$ solution (1 mg·mL^{-1}, 1 mL) were mixed and sonicated for 30 min and incubated in a shaker overnight [50]. Then, Au NP-embedded silica NPs (SiO$_2$@Au NPs) were obtained by centrifugation at 8500 rpm for 15 min, and washed several times with EtOH to remove unbound Au NPs. The SiO$_2$@Au NPs were re-dispersed in absolute EtOH to obtain a SiO$_2$@Au NP suspension of 1 mg·mL^{-1}.

3.3. Incorporating RLC into SiO₂@Au

RLC solution (1 mL, 10 mM in EtOH) was added to the SiO$_2$@Au (1.0 mg), and the suspension was stirred vigorously for 2 h at 25°C. The colloids were centrifuged and washed several times with EtOH. The NPs were re-dispersed in 1.0 mL of absolute EtOH to obtain 1 mg·mL^{-1} SiO$_2$@Au NPs modified with RLC (SiO$_2$@Au@RLC).

3.4. Preparation of SiO₂@Au@RLC@Ag NPs

Au-Ag core-shell NPs were prepared in an aqueous medium by the reduction and deposition of Ag with ascorbic acid onto the Au NPs in a polyvinylpyrrolidone (PVP) environment. Briefly, 0.2 mg of SiO$_2$@Au@RLC was dispersed in 9.8 mL of water containing 10 mg PVP, and kept still for 30 min. Twenty microliters of 10 mM silver nitrate was added to the solution, followed by the addition of 20 μL of 10 mM ascorbic acid. This solution was incubated for 15 min to reduce the Ag$^+$ ion to Ag. The reduction steps were repeated to obtain the desired AgNO$_3$ concentration. SiO$_2$@Au@4-MBA@Ag NPs were obtained by centrifugation of the solution at 8500 rpm for 15 min, and the NPs were washed several times with EtOH to remove excess reagent. SiO$_2$@Au@4-MBA@Ag NPs were re-dispersed in 0.2 mL of absolute EtOH to obtain 1 mg·mL^{-1} SiO$_2$@Au@4-MBA@Ag NP suspension.

3.5. SERS Measurement of the SiO2@Au@RLC@Ag NPs

SiO$_2$@Au@RLC@Ag NPs were measured in a capillary tube, and SERS signals were measured using a confocal micro-Raman system (LabRam 300, JY-Horiba, Tokyo, Japan) equipped with an optical microscope (BX41, Olympus, Tokyo, Japan). The SERS signals were collected in a back-scattering geometry using a ×10 objective lens (0.90 NA, Olympus) and a spectrometer equipped with a thermoelectric cooled Charge-Coupled Device (CCD) detector. A 532 nm diode-pumped solid-state laser (CL532-100-S; Crystalaser, US) was used as a photo-excitation source, exerting 10 mW laser power at the sample. The strong Rayleigh scattered light was rejected using a long-pass filter. Selected sites were measured at random, and all SERS spectra were integrated for 5 s. The size of the laser beam spot was about 2 μm.

3.6. Transmission Electron Microscopy (TEM) Measurements

Our material was dispersed in EtOH to obtain a final concentration of 1 mg mL^{-1}, and 10 μL of the dispersed solution was dropped onto a 400 mesh Cu grid (Pelco, Fresno, CA, USA) and dried in air. Field energy transmission electron microscopy (Libra 120, Carl Zeiss, Germany) was used to analyze our materials. The acceleration voltage was 120 kV.

4. Conclusions

In summary, we have prepared three types of SiO$_2$@Au@RLC@Ag materials with three different RLCs, including 4-MBA, 4-ATP, and 4-MBT. The effect of RLCs on the deposition of the silver shell was also investigated. The formation of the Ag shell was strongly dependent on the negative charge of SiO$_2$@Au@RLCs, the concentration of the silver ion, and the pH solution. In general, the size of Ag NPs was greater at a lower pH and became smaller at a higher pH due to the decrease in reduction rate. Especially, the pH of the solution played an important role in the formation of the Ag shell on the surface of SiO$_2$@Au@RLCs, by affecting the local surface charge of the RLCs. For the neutral group of -CH$_3$, the Ag shell was coated with difficulty on RLC-modified SiO$_2$@Au, whereas the presence of the positive charge of -NH$_3^+$ on the surface of SiO$_2$@Au facilitated the coating of the Ag shell, leading to a greater intensity of Raman signal in 4-ATP. The negative charge of -COO$^-$ led to a well coated Ag shell, and increased the Raman signal of 4-MBA in the pH range of 5.0 to 6.0. However, it inhibited the coating of the Ag shell on the surface of SiO$_2$@Au@4-MBA, with an obvious decrease in the Raman signal of 4-MBA from a pH of 7.0 to 9.0 due to the formation of silver oxide or silver chloride.

Int. J. Mol. Sci. **2019**, *20*, 1258

This study provides a thorough understanding of silver deposition, to support further research and the development of strong and reliable SERS probes based on SiO_2@Au@RLC@Ag NPs.

Supplementary Materials: Supplementary materials can be found at http://www.mdpi.com/1422-0067/20/6/1258/s1.

Author Contributions: Conceptualization, X.-H.P. and B.-H.J.; Data curation, X.-H.P., E.H., E.K., B.S.S., Y.H. and H.-M.K.; Formal analysis, X.-H.P.; Investigation, X.-H.P.; Methodology, E.H.; Supervision, D.H.J. and B.-H.J.; Writing–original draft, X.-H.P.; Writing–review & editing, D.H.J. and B.-H.J.

Funding: This work was supported by the KU Research Professor Program of Konkuk University and funded by basic Science Research Program through the NRF funded by the Ministry of Education (NRF-2018R1D1A1B07045708) and Science, ICT & Future Planning (NRF 2016M3A9B6918892) and funded by the Korean Health Technology R&D Project, Ministry of Health & Welfare (HI17C1264).

Conflicts of Interest: The authors declare no conflict of interest.

References

1. Schlücker, S. Surface-enhanced raman spectroscopy: Concepts and chemical applications. *Angew. Chem. Int. Ed.* **2014**, *53*, 4756–4795. [CrossRef] [PubMed]

2. Wang, Y.; Yan, B.; Chen, L. Sers tags: Novel optical nanoprobes for bioanalysis. *Chem. Rev.* **2013**, *113*, 1391–1428. [CrossRef] [PubMed]

3. Culha, M.; Cullum, B.; Lavrik, N.; Klutse, C.K. Surface-enhanced raman scattering as an emerging characterization and detection technique. *J. Nanotechnol.* **2012**, *2012*, 15. [CrossRef]

4. Jun, B.-H.; Kim, G.; Jeong, S.; Noh, M.S.; Pham, X.-H.; Kang, H.; Cho, M.-H.; Kim, J.-H.; Lee, Y.-S.; Jeong, D.H. Silica core-based surface-enhanced raman scattering (sers) tag: Advances in multifunctional sers nanoprobes for bioimaging and targeting of biomarkers#. *Bull. Korean Chem. Soc.* **2015**, *36*, 963–978.

5. Goodacre, R.; Graham, D.; Faulds, K. Recent developments in quantitative sers: Moving towards absolute quantification. *TrAC Trends Anal. Chem.* **2018**, *102*, 359–368. [CrossRef]

6. Zhao, J.; Zhang, Z.; Yang, S.; Zheng, H.; Li, Y. Facile synthesis of mos2 nanosheet-silver nanoparticles composite for surface enhanced raman scattering and electrochemical activity. *J. Alloys Compd.* **2013**, *559*, 87–91. [CrossRef]

7. Zhu, C.; Meng, G.; Zheng, P.; Huang, Q.; Li, Z.; Hu, X.; Wang, X.; Huang, Z.; Li, F.; Wu, N. A hierarchically ordered array of silver-nanorod bundles for surface-enhanced raman scattering detection of phenolic pollutants. *Adv. Mater.* **2016**, *28*, 4871–4876. [CrossRef]

8. Du, Y.; Wei, W.; Zhang, X.; Li, Y. Tuning metamaterials nanostructure of janus gold nanoparticle film for surface-enhanced raman scattering. *J. Phys. Chem. C* **2018**, *122*, 7997–8002. [CrossRef]

9. Kasera, S.; Biedermann, F.; Baumberg, J.J.; Scherman, O.A.; Mahajan, S. Quantitative sers using the sequestration of small molecules inside precise plasmonic nanoconstructs. *Nano Lett.* **2012**, *12*, 5924–5928. [CrossRef]

10. Lim, D.-K.; Jeon, K.-S.; Hwang, J.-H.; Kim, H.; Kwon, S.; Suh, Y.D.; Nam, J.-M. Highly uniform and reproducible surface-enhanced raman scattering from DNA-tailorable nanoparticles with 1-nm interior gap. *Nat. Nano* **2011**, *6*, 452–460. [CrossRef]

11. Li, C.; Wang, L.; Luo, Y.; Liang, A.; Wen, G.; Jiang, Z. A sensitive gold nanoplasmonic sers quantitative analysis method for sulfate in serum using fullerene as catalyst. *Nanomaterials* **2018**, *8*, 277. [CrossRef] [PubMed]

12. Liang, A.; Li, X.; Zhang, X.; Wen, G.; Jiang, Z. A sensitive sers quantitative analysis method for ni2+ by the dimethylglyoxime reaction regulating a graphene oxide nanoribbon catalytic gold nanoreaction. *Luminescence* **2018**, *33*, 1033–1039. [CrossRef]

13. Liang, A.; Wang, H.; Yao, D.; Jiang, Z. A simple and sensitive sers quantitative analysis method for urea using the dimethylglyoxime product as molecular probes in nanosilver sol substrate. *Food Chem.* **2019**, *271*, 39–46. [CrossRef] [PubMed]

14. Zhang, X.-Q.; Li, S.-X.; Chen, Z.-P.; Chen, Y.; Yu, R.-Q. Quantitative sers analysis based on multiple-internal-standard embedded core-shell nanoparticles and spectral shape deformation quantitative theory. *Chemometr. Intell. Lab. Syst.* **2018**, *177*, 47–54. [CrossRef]

15. Shen, W.; Lin, X.; Jiang, C.; Li, C.; Lin, H.; Huang, J.; Wang, S.; Liu, G.; Yan, X.; Zhong, Q.; et al. Reliable quantitative sers analysis facilitated by core–shell nanoparticles with embedded internal standards. *Angew. Chem. Int. Ed.* **2015**, *54*, 7308–7312. [CrossRef]

16. Kammer, E.; Olschewski, K.; Bocklitz, T.; Rosch, P.; Weber, K.; Cialla, D.; Popp, J. A new calibration concept for a reproducible quantitative detection based on sers measurements in a microfluidic device demonstrated on the model analyte adenine. *Phys. Chem. Chem. Phys.* **2014**, *16*, 9056–9063. [CrossRef]

17. Zhou, Y.; Ding, R.; Joshi, P.; Zhang, P. Quantitative surface-enhanced raman measurements with embedded internal reference. *Anal. Chim. Acta* **2015**, *874*, 49–53. [CrossRef]

18. Zhang, L.; Li, Q.; Tao, W.; Yu, B.; Du, Y. Quantitative analysis of thymine with surface-enhanced raman spectroscopy and partial least squares (pls) regression. *Anal. Bioanal. Chem.* **2010**, *398*, 1827–1832. [CrossRef]

19. Chen, Y.; Chen, Z.-P.; Zuo, Q.; Shi, C.-X.; Yu, R.-Q. Surface-enhanced raman spectroscopy based on conical holed enhancing substrates. *Anal. Chim. Acta* **2015**, *887*, 45–50. [CrossRef]

20. Lorén, A.; Engelbrektsson, J.; Eliasson, C.; Josefson, M.; Abrahamsson, J.; Johansson, M.; Abrahamsson, K. Internal standard in surface-enhanced raman spectroscopy. *Anal. Chem.* **2004**, *76*, 7391–7395. [CrossRef]

21. Chen, Y.; Chen, Z.-P.; Jin, J.-W.; Yu, R.-Q. Quantitative determination of ametryn in river water using surface-enhanced raman spectroscopy coupled with an advanced chemometric model. *Chemometr. Intell. Lab. Syst.* **2015**, *142*, 166–171. [CrossRef]

22. Xia, T.-H.; Chen, Z.-P.; Chen, Y.; Jin, J.-W.; Yu, R.-Q. Improving the quantitative accuracy of surface-enhanced raman spectroscopy by the combination of microfluidics with a multiplicative effects model. *Anal. Methods* **2014**, *6*, 2363–2370. [CrossRef]

23. Chen, Y.; Chen, Z.-P.; Long, S.-Y.; Yu, R.-Q. Generalized ratiometric indicator based surface-enhanced raman spectroscopy for the detection of cd2+ in environmental water samples. *Anal. Chem.* **2014**, *86*, 12236–12242. [CrossRef]

24. Hammond, P.T. Form and function in multilayer assembly: New applications at the nanoscale. *Adv. Mater.* **2004**, *16*, 1271–1293. [CrossRef]

25. Wang, F.; Deng, R.; Wang, J.; Wang, Q.; Han, Y.; Zhu, H.; Chen, X.; Liu, X. Tuning upconversion through energy migration in core–shell nanoparticles. *Nat. Mater.* **2011**, *10*, 968. [CrossRef]

26. Huang, X.; Han, S.; Huang, W.; Liu, X. Enhancing solar cell efficiency: The search for luminescent materials as spectral converters. *Chem. Soc. Rev.* **2013**, *42*, 173–201. [CrossRef]

27. Maeda, K.; Domen, K. Photocatalytic water splitting: Recent progress and future challenges. *J. Phys. Chem. Lett.* **2010**, *1*, 2655–2661. [CrossRef]

28. Zhang, N.; Liu, S.; Fu, X.; Xu, Y.-J. Synthesis of m@tio2 (m = au, pd, pt) core–shell nanocomposites with tunable photoreactivity. *J. Phys. Chem. C* **2011**, *115*, 9136–9145. [CrossRef]

29. Zhang, N.; Liu, S.; Xu, Y.-J. Recent progress on metal core@semiconductor shell nanocomposites as a promising type of photocatalyst. *Nanoscale* **2012**, *4*, 2227–2238. [CrossRef]

30. Pelaez, M.; Nolan, N.T.; Pillai, S.C.; Seery, M.K.; Falaras, P.; Kontos, A.G.; Dunlop, P.S.M.; Hamilton, J.W.J.; Byrne, J.A.; O'Shea, K.; et al. A review on the visible light active titanium dioxide photocatalysts for environmental applications. *Appl. Catal. B Environ.* **2012**, *125*, 331–349. [CrossRef]

31. Strobbia, P.; Languirand, E.R.; Cullum, B.M. Recent Advances in Plasmonic Nanostructures for Sensing: A Review. *Opt. Eng.* **2015**, *54*, 100902. [CrossRef]

32. Loo, C.; Lin, A.; Hirsch, L.; Lee, M.-H.; Barton, J.; Halas, N.; West, J.; Drezek, R. Nanoshell-enabled photonics-based imaging and therapy of cancer. *Technol. Cancer Res. Treat.* **2004**, *3*, 33–40. [CrossRef]

33. Janib, S.M.; Moses, A.S.; MacKay, J.A. Imaging and drug delivery using theranostic nanoparticles. *Adv. Drug Deliv. Rev.* **2010**, *62*, 1052–1063. [CrossRef] [PubMed]

34. Chen, G.; Roy, I.; Yang, C.; Prasad, P.N. Nanochemistry and nanomedicine for nanoparticle-based diagnostics and therapy. *Chem. Rev.* **2016**, *116*, 2826–2885. [CrossRef] [PubMed]

35. Jain, P.K.; El-Sayed, I.H.; El-Sayed, M.A. Au nanoparticles target cancer. *Nano Today* **2007**, *2*, 18–29. [CrossRef]

36. Gobin, A.M.; Lee, M.H.; Halas, N.J.; James, W.D.; Drezek, R.A.; West, J.L. Near-infrared resonant nanoshells for combined optical imaging and photothermal cancer therapy. *Nano Lett.* **2007**, *7*, 1929–1934. [CrossRef]

37. Loo, C.; Lowery, A.; Halas, N.; West, J.; Drezek, R. Immunotargeted nanoshells for integrated cancer imaging and therapy. *Nano Lett.* **2005**, *5*, 709–711. [CrossRef]

38. Ghosh Chaudhuri, R.; Paria, S. Core/shell nanoparticles: Classes, properties, synthesis mechanisms, characterization, and applications. *Chem. Rev.* **2012**, *112*, 2373–2433. [CrossRef]

39. Pandikumar, A.; Lim, S.-P.; Jayabal, S.; Huang, N.M.; Lim, H.N.; Ramaraj, R. Titania@gold plasmonic nanoarchitectures: An ideal photoanode for dye-sensitized solar cells. *Renew. Sustain. Energy Rev.* **2016**, *60*, 408–420. [CrossRef]

40. Jiang, H.-L.; Akita, T.; Xu, Q. A one-pot protocol for synthesis of non-noble metal-based core–shell nanoparticles under ambient conditions: Toward highly active and cost-effective catalysts for hydrolytic dehydrogenation of nh3bh3. *Chem. Commun.* **2011**, *47*, 10999–11001. [CrossRef]

41. Caruso, F.; Spasova, M.; Salgueiriño-Maceira, V.; Liz-Marzán, L.M. Multilayer assemblies of silica-encapsulated gold nanoparticles on decomposable colloid templates. *Adv. Mater.* **2001**, *13*, 1090–1094. [CrossRef]

42. Oldenburg, S.J.; Averitt, R.D.; Westcott, S.L.; Halas, N.J. Nanoengineering of optical resonances. *Chem. Phys. Lett.* **1998**, *288*, 243–247. [CrossRef]

43. Li, J.-F.; Zhang, Y.-J.; Ding, S.-Y.; Panneerselvam, R.; Tian, Z.-Q. Core–shell nanoparticle-enhanced raman spectroscopy. *Chem. Rev.* **2017**, *117*, 5002–5069. [CrossRef]

44. Anker, J.N.; Hall, W.P.; Lyandres, O.; Shah, N.C.; Zhao, J.; Van Duyne, R.P. Biosensing with plasmonic nanosensors. *Nat. Mater.* **2008**, *7*, 442. [CrossRef]

45. Raemdonck, K.; Demeester, J.; De Smedt, S. Advanced nanogel engineering for drug delivery. *Soft Matter* **2009**, *5*, 707–715. [CrossRef]

46. Gao, J.; Zhao, C.; Zhang, Z.; Li, G. An intrinsic internal standard substrate of au@ps-b-p4vp for rapid quantification by surface enhanced raman scattering. *Analyst* **2017**, *142*, 2936–2944. [CrossRef]

47. Hahm, E.; Cha, M.G.; Kang, E.J.; Pham, X.-H.; Lee, S.H.; Kim, H.-M.; Kim, D.-E.; Lee, Y.-S.; Jeong, D.H.; Jun, B.-H. Multi-layer ag-embedded silica nanostructure as sers-based chemical sensor with dual-function internal standards. *ACS Appl. Mater. Interfaces* **2018**, *10*, 40748–40755. [CrossRef]

48. Feng, Y.; Wang, Y.; Wang, H.; Chen, T.; Tay, Y.Y.; Yao, L.; Yan, Q.; Li, S.; Chen, H. Engineering "hot" nanoparticles for surface-enhanced raman scattering by embedding reporter molecules in metal layers. *Small* **2012**, *8*, 246–251. [CrossRef]

49. Gandra, N.; Singamaneni, S. Bilayered raman-intense gold nanostructures with hidden tags (brights) for high-resolution bioimaging. *Adv. Mater.* **2013**, *25*, 1022–1027. [CrossRef]

50. Pham, X.-H.; Lee, M.; Shim, S.; Jeong, S.; Kim, H.-M.; Hahm, E.; Lee, S.H.; Lee, Y.-S.; Jeong, D.H.; Jun, B.-H. Highly sensitive and reliable sers probes based on nanogap control of a au-ag alloy on silica nanoparticles. *RSC Adv.* **2017**, *7*, 7015–7021. [CrossRef]

51. Shim, S.; Pham, X.-H.; Cha, M.G.; Lee, Y.-S.; Jeong, D.H.; Jun, B.-H. Size effect of gold on ag-coated au nanoparticle-embedded silica nanospheres. *RSC Adv.* **2016**, *6*, 48644–48650. [CrossRef]

52. Pham, X.-H.; Hahm, E.; Kang, E.; Ha, Y.N.; Lee, S.H.; Rho, W.-Y.; Lee, Y.-S.; Jeong, D.H.; Jun, B.-H. Gold-silver bimetallic nanoparticles with a raman labeling chemical assembled on silica nanoparticles as an internal-standard-containing nanoprobe. *J. Alloys Compd.* **2019**, *779*, 360–366. [CrossRef]

53. Stöber, W.; Fink, A.; Bohn, E. Controlled growth of monodisperse silica spheres in the micron size range. *J. Colloid Interface Sci.* **1968**, *26*, 62–69. [CrossRef]

54. Martin, M.N.; Basham, J.I.; Chando, P.; Eah, S.-K. Charged gold nanoparticles in non-polar solvents: 10-min synthesis and 2d self-assembly. *Langmuir* **2010**, *26*, 7410–7417. [CrossRef]

55. Martin, M.N.; Li, D.; Dass, A.; Eah, S.-K. Ultrafast, 2 min synthesis of monolayer-protected gold nanoclusters (d < 2 nm). *Nanoscale* **2012**, *4*, 4091–4094.

56. Bastús, N.G.; Merkoçi, F.; Piella, J.; Puntes, V. Synthesis of highly monodisperse citrate-stabilized silver nanoparticles of up to 200 nm: Kinetic control and catalytic properties. *Chem. Mater.* **2014**, *26*, 2836–2846. [CrossRef]

57. Genov, D.A.; Sarychev, A.K.; Shalaev, V.M. Metal-dielectric composite filters with controlled spectral windows of transparency. *J. Nonlinear Opt. Phys. Mater.* **2003**, *12*, 419–440. [CrossRef]

58. Biswas, A.; Eilers, H.; Hidden, F.; Aktas, O.C.; Kiran, C.V.S. Large broadband visible to infrared plasmonic absorption from ag nanoparticles with a fractal structure embedded in a teflon af® matrix. *Appl. Phys. Lett.* **2006**, *88*, 013103. [CrossRef]

59. Yang, J.-K.; Kang, H.; Lee, H.; Jo, A.; Jeong, S.; Jeon, S.-J.; Kim, H.-I.; Lee, H.-Y.; Jeong, D.H.; Kim, J.-H.; et al. Single-step and rapid growth of silver nanoshells as sers-active nanostructures for label-free detection of pesticides. *ACS Appl. Mater. Interfaces* **2014**, *6*, 12541–12549. [CrossRef] [PubMed]

60. Alqadi, M.K.; Abo Noqtah, O.A.; Alzoubi, F.Y.; Alzouby, J.; Aljarrah, K. Ph effect on the aggregation of silver nanoparticles synthesized by chemical reduction. *Mater. Sci. Poland* **2014**, *32*, 107–111. [CrossRef]
61. Bryant, M.A.; Crooks, R.M. Determination of surface pka values of surface-confined molecules derivatized with ph-sensitive pendant groups. *Langmuir* **1993**, *9*, 385–387. [CrossRef]
62. Zhang, H.; He, H.-X.; Mu, T.; Liu, Z.-F. Force titration of amino group-terminated self-assembled monolayers of 4-aminothiophenol on gold using chemical force microscopy. *Thin Solid Films* **1998**, *327–329*, 778–780. [CrossRef]
63. Bayram, S.; Zahr, O.K.; Blum, A.S. Short ligands offer long-term water stability and plasmon tunability for silver nanoparticles. *RSC Adv.* **2015**, *5*, 6553–6559. [CrossRef]
64. Koivisto, J.; Chen, X.; Donnini, S.; Lahtinen, T.; Häkkinen, H.; Groenhof, G.; Pettersson, M. Acid–base properties and surface charge distribution of the water-soluble au102(pmba)44 nanocluster. *J. Phys. Chem. C* **2016**, *120*, 10041–10050. [CrossRef]
65. Clark, R.A.; Trout, C.J.; Ritchey, L.E.; Marciniak, A.N.; Weinzierl, M.; Schirra, C.N.; Christopher Kurtz, D. Electrochemical titration of carboxylic acid terminated sams on evaporated gold: Understanding the ferricyanide electrochemistry at the electrode surface. *J. Electroanal. Chem.* **2013**, *689*, 284–290. [CrossRef]
66. Zhang, H.; Zhang, H.-L.; He, H.-X.; Zhu, T.; Liu, Z.-F. Study on the surface dissociation properties of 6-(10-mercaptodecaoxyl)quinoline self-assembled monolayer on gold by chemical force titration. *Mate. Sci. Eng. C* **1999**, *8*, 191–194. [CrossRef]
67. Ma, Y.; Li, W.; Cho, E.C.; Li, Z.; Yu, T.; Zeng, J.; Xie, Z.; Xia, Y. Au@ag core−shell nanocubes with finely tuned and well-controlled sizes, shell thicknesses, and optical properties. *ACS Nano* **2010**, *4*, 6725–6734. [CrossRef] [PubMed]

International Journal of
Molecular Sciences

MDPI

Review

Bactericidal and Cytotoxic Properties of Silver Nanoparticles

Chengzhu Liao [1,*], Yuchao Li [2] and Sie Chin Tjong [3,*]

1   Department of Materials Science and Engineering, Southern University of Science and Technology, Shenzhen 518055, China
2   Department of Materials Science and Engineering, Liaocheng University, Liaocheng 252000, China; liyuchao@lcu.edu.cn
3   Department of Physics, City University of Hong Kong, Tat Chee Avenue, Kowloon, Hong Kong, China
*   Correspondence: liaocz@sustc.edu.cn (C.L.); aptjong@gmail.com (S.C.T.)

Received: 7 December 2018; Accepted: 17 January 2019; Published: 21 January 2019

Abstract: Silver nanoparticles (AgNPs) can be synthesized from a variety of techniques including physical, chemical and biological routes. They have been widely used as nanomaterials for manufacturing cosmetic and healthcare products, antimicrobial textiles, wound dressings, antitumor drug carriers, etc. due to their excellent antimicrobial properties. Accordingly, AgNPs have gained access into our daily life, and the inevitable human exposure to these nanoparticles has raised concerns about their potential hazards to the environment, health, and safety in recent years. From in vitro cell cultivation tests, AgNPs have been reported to be toxic to several human cell lines including human bronchial epithelial cells, human umbilical vein endothelial cells, red blood cells, human peripheral blood mononuclear cells, immortal human keratinocytes, liver cells, etc. AgNPs induce a dose-, size- and time-dependent cytotoxicity, particularly for those with sizes ≤10 nm. Furthermore, AgNPs can cross the brain blood barrier of mice through the circulation system on the basis of in vivo animal tests. AgNPs tend to accumulate in mice organs such as liver, spleen, kidney and brain following intravenous, intraperitoneal, and intratracheal routes of administration. In this respect, AgNPs are considered a double-edged sword that can eliminate microorganisms but induce cytotoxicity in mammalian cells. This article provides a state-of-the-art review on the synthesis of AgNPs, and their applications in antimicrobial textile fabrics, food packaging films, and wound dressings. Particular attention is paid to the bactericidal activity and cytotoxic effect in mammalian cells.

Keywords: silver ion; bacteria; cytotoxicity; cell culture; membrane; reactive oxygen species; polymer nanocomposite; food packaging; wound dressing; administration route

1. Introduction

Nanotechnology is a multidisciplinary field that brings together researchers in diverse scientific disciplines such as biology, chemistry, material science and physics for developing advanced functional materials at the nanoscale level. The physicochemical and biological properties of nanomaterials differ significantly from the corresponding bulk materials due to their extremely large surface area to volume ratio. Recent advances in nanotechnology allow the synthesis of various types of novel nanomaterials for industrial and biomedical applications [1–10]. Among these, metal nanoparticles with unique optical properties have gained much attention in the field of nanomedicine, for drug delivery, imaging, and sensing purposes [11–13]. In particular, silver nanoparticles (AgNPs) exhibit several attractive properties, including excellent electrical conductivity, chemical stability, antifungal, and bactericidal properties. As such, AgNPs find attractive applications in textiles, healthcare products, cosmetics, cancer therapies, wound dressings, catalysts, food packaging films, water treatments, electronic devices, corneal replacements, etc. (Figure 1) [14–29]. For example, clothing textiles that are in close contact with human skin create a warm and humid environment for microorganisms. Sweaty fabrics are perfect breeding grounds for bacteria. Thus, antimicrobial fabrics are coated with AgNPs to inhibit bacterial adhesion and growth [30,31]. Biomedical products with AgNPs are typically used to prevent bacterial infections by accelerating wound healing [21,27–29], and treating tumor through a rapid breakdown of infected cells [24,25]. AgNPs with doxorubicin and alendronate serve as effective antitumor drug carriers for the HeLa cell line [13].

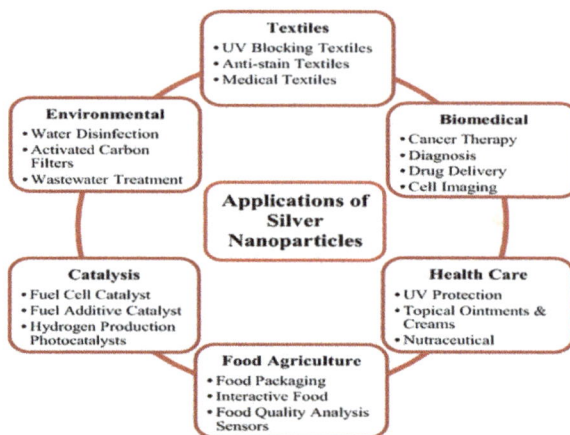

Figure 1. Applications of AgNPs. Reproduced from [22], Springer Open.

Conventional antibiotics such as tetracycline and streptomycin have been widely used to inhibit bacterial infections. However, these antibacterial agents are ineffective to inhibit multidrug-resistant bacterial strains. This is because such bacteria are getting more resistant to biocidal action of antibiotic molecules. Therefore, it deems necessary to develop biomaterials that target drug-resistant bacteria. In this respect, AgNPs with large surface areas can provide a better contact for interacting with bacteria compared to conventional silver microparticles. Nanosilver in the form of colloidal silver and silver nitrate have been used for more than 100 years as a biocidal agent in the United States [32]. AgNPs have been reported to be effective in killing both gram-negative and gram-positive bacteria strains [23,33–35]. However, AgNPs are more effective in destroying gram-negative bacterial strains than gram-positive because gram-positive bacteria have one cytoplasmic membrane and a relatively thick cell wall comprising of several peptidoglycan layers (20–80 nm), while gram-negative bacteria have an external layer of lipopolysaccharide (LPS) followed by a thin layer of peptidoglycan and an innermost plasma membrane (Figure 2) [36,37]. AgNPs can also eliminate multidrug resistant (MDR) bacteria by interfering with their defense mechanisms. They can be used alone or in combination with antibiotics [15,38–40]. Gurunathan et al. indicated that biosynthesized AgNPs are very effective in killing MDR bacteria *Prevotella melaninogenica* and *Arcanobacterium pyogenes* [38]. Katva et al. reported that AgNPs combined with gentamicin and chloramphenicol exhibit a better antibacterial effect in *Enterococcus faecalis* than both antibiotics alone. *Enterococcus faecalis* is a MDR bacteria which is resistant to a wide range of antibiotics [40]. The antibacterial activity of AgNPs is known to be shape-, size-, charge-, and dose-dependent [15,41–43]. Xia et al. reported that a series of Ag nanocrystals with controlled shapes and sizes can be synthesized from silver salts by using different combinations of seeds and capping agents [44]. Recently, Hosseinidoust et al. reported a one-pot green synthesis of colloidally stable AgNPs having triangular, hexagonal and dendritic shapes without using toxic chemicals and seeds [45].

In general, AgNPs act like a double-edged sword with beneficial and harmful effects, i.e., they can eliminate bacteria but also induce cytotoxicity. Due to the versatility of AgNPs in many consumer and health products, there is growing public concern about the risk of using those products because AgNPs may pose potential health hazards. Furthermore, extensive application and production of AgNPs would increase their release into aquatic environments such as rivers and lakes. For instance, AgNPs can be released from antimicrobial fabrics into water during washing, thereby polluting groundwater environment [23,46,47]. Once AgNPs enter freshwater environment, they usually oxidize into Ag^+ ions that are toxic to aquatic organisms. Moreover, ionic silver is immobilized to a large extent as a sparingly soluble salt like AgCl or Ag_2S [23]. By accumulating in aquatic organisms, AgNPs can enter the human body through the food chain. However, little is known about the long-term safety and toxic effects of AgNPs in the aquatic environment. Humans can be exposed to AgNPs via several routes including inhalation, oral ingestion, intravenous injection, and dermal contact. AgNPs then enter human cells either by endosomal uptake or by diffusion (Figure 2) [36]. The American Conference of Governmental Industrial Hygienists (ACGIH) has established threshold limit values for metallic silver (0.1 mg/m^3) and soluble compounds of silver (0.01 mg/m^3). As recognized, extended exposure to Ag through oral and inhalation can lead to Argyria or Argyrosis, i.e., chronic disorders of skin microvessels and eyes in humans [23,48]. In vitro cell culture studies have indicated toxic effects of AgNPs in immortal human skin keratinocytes (HaCaT), human erythrocytes, human neuroblastoma cells, human embryonic kidney cells (HEK293T), human liver cells (HepG2), and human colon cells (Caco2) [49–55]. In vivo animal studies have revealed toxic effects of AgNPs in rodents by accumulating in their liver, spleen, and lung [56,57]. Similarly, AgNPs-mediated cytotoxicity in mammalian cells [55,58–62] depends greatly on the nanoparticle size, shape, surface charge, dosage, oxidation state, and agglomeration condition as well as the cell type. This article provides a state-of-the-art review on the recent development in the synthesis of AgNPs, their antibacterial activity, and cytotoxic effects in mammalian cells, especially in the past five years. Proper understanding of the interactions between AgNPs and mammalian cells is essential for the safe use of these nanoparticles. This knowledge enables scientists to develop functional AgNPs with improved biocompatibility to mammalian cells for combating MDR bacteria.

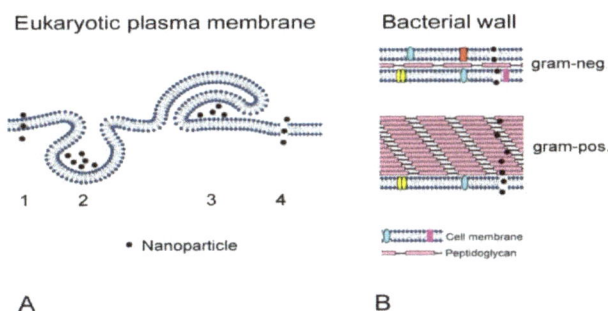

Figure 2. Uptake of AgNPs by mammalian cells (**A**) and by bacteria (**B**). (**A**) AgNPs can cross the plasma membrane by diffusion (**1**), endocytotic uptake (**2,3**), and disruption of membrane integrity (**4**). (**B**) AgNPs permeate the cell walls of gram-negative and gram-positive bacteria. Reproduced from [36], MDPI under the Creative Commons Attribution License.

2. Synthesis of AgNPs and Their Polymer Nanocomposites

AgNPs can be prepared from physical, chemical, and biological routes [63]. The physical route has a distinct advantage in forming high-purity AgNPs. However, it is a low yield process, which often requires high temperature and power consumption, thereby limiting their application in the industrial sector. In contrast, AgNPs can be produced in large quantities through chemical and biological routes due to their cost-effectiveness. In this respect, we focus mainly on nanoparticle fabrication using chemical and biological routes in this section. The physical synthesis process generally employs thermal, laser, or arc discharge power to form AgNPs with nearly narrow size distribution. In this context, evaporation condensation and laser ablation strategies are commonly adopted to form AgNPs. The former approach is carried out by simply placing a source metal, i.e., silver target inside the tube furnace, and then vaporized silver into a carrier gas under atmospheric pressure at high temperatures. The vapor is then condensed to form nanoparticles. The drawbacks of this approach include high power consumption and long heating time to reach thermal stability. Jung et al. modified this process by using a small ceramic heater with a local heating area for vaporizing silver metal [64]. The temperature gradient in the vicinity of the heater surface is very steep, such that the vapor can cool at a rapid rate, thereby condensing into spherical AgNPs with diameters of 6.2–21.5 nm without agglomeration. Alternatively, laser ablation can be used to synthesize AgNPs from a silver target placed in a solution under laser beam irradiation. The nanoparticle size of colloids depends on the laser wavelength, ablation time, and duration of laser pulses [65–67]. The limitation of this process is the high cost of laser facility.

2.1. Wet Chemical Route

The wet chemical route is by far the most economical and commonly used process for preparing nanosilver colloidal dispersions in water or organic solvents. This route includes chemical reduction, microwave-assisted synthesis, microemulsion, photoreduction, electrochemical approach, etc. [41–43,62,68–72]. Among these, chemical reduction is a relatively simple, high yield and cost-effective process through the chemical reduction of silver salt in water or an organic solvent to form a colloidal suspension. This strategy requires silver salt, reducing agent and stabilizing/capping agent. Silver nitrate is the most widely used silver salt precursor. Sodium borohydride ($NaBH_4$), ascorbic acid, glucose, hydrazine, sodium citrate, and ethylene glycol (EG) are typical reductants for reducing silver ions [71–74]. Sodium borohydride is a strong reductant for forming fined and monodispersed AgNPs. Weak reducing agents such as sodium citrate, ascorbic acid, and glucose lead to the formation of relatively large AgNPs, having a wider size distribution; sodium citrate plays the role of a reducing and a stabilizing agent [73]. As recognized, colloidal AgNPs tend to contact or link with each other to form aggregates as a result of the attractive Van der Waals forces. The agglomeration of colloidal AgNPs can be prevented by the use of stabilizing agents. The stabilizing effect of polymer-based capping agents or non-ionic surfactants of AgNPs suspensions is based on the charge repulsion or steric hindrance to counteract the van der Waals attraction between colloidal nanoparticles. As such, electrostatic or steric stabilization of colloidal AgNPs is achieved through adsorption of macromolecules or organic compounds to the surfaces of the nanoparticles. Citrate-capped AgNPs exhibit negative surface charge due to the carboxylic moiety of citrate. This leads to electrostatic repulsion between AgNPs, thereby preventing agglomeration of nanoparticles [71,75]. Typical polymer-based capping agents are polyvinyl alcohol (PVA), polyvinylpyrrolidone (PVP), polyethylene glycol (PEG) and polysaccharides, while non-ionic surfactants including Brij, Tween, and Triton X-100 are employed to stabilize AgNPs during the formation process [76,77]. These stabilizers also play important roles in controlling the size and shape of AgNPs. The properties and bactericidal activities of AgNPs are greatly influenced by their size and shape. Thus, researchers have spent much effort on the size- and shape-controlled synthesis of AgNPs by using different capping agents. Organic solvents have some advantages over aqueous solutions such as high yield and narrow particle size distribution. In certain cases, solvent itself can also serve as a reducing agent [78]. As an example, the polyol process utilizing EG as both solvent and reductant in the presence of PVP at 160 °C is known to be effective to reduce silver nitrate to yield AgNPs [79,80]. In addition, microwave irradiation is more energy efficient than thermal heating in the polyol process. Microwave irradiation offers fast and homogeneous heating of the reaction medium, typically in a time period of few seconds. As such, it provides uniform nucleation and growth conditions for AgNPs [81].

The formation of colloidal solutions via silver salt reduction generally proceeds through two stages: nucleation and subsequent growth. The nucleation stage involves the reduction of Ag^+ to Ag^0 atoms, and subsequent aggregation of Ag^0 atoms to form clusters, $(Ag^0)_n$, according to the reactions:

$nAg^+ \rightarrow nAg^0 \rightarrow (Ag^0)_n$, or via a stepwise reduction mechanism: $Ag^+ \rightarrow Ag^0 \rightarrow Ag_2^+ \rightarrow Ag_4^+ \rightarrow \cdots \rightarrow$ $(Ag^0)_n$. When the critical size is reached, nanoparticle nuclei will grow accordingly [74]. These stages can be manipulated by monitoring reaction conditions such as pH, temperature, precursors, reductants and stabilizers. Thus, the shape, size and size distribution of AgNPs can be controlled by varying the reaction parameters. As mentioned above, sodium citrate is a weak reductant leading to the formation of polydispersed AgNPs. Agnihotri et al. demonstrated that the size of AgNPs can be controlled by using a co-reduction approach, i.e., $NaBH_4$ acts as a primary reductant, while citrate ions act as both a secondary reductant and a capping agent (Figure 3) [41]. Recently, Ajitha et al. fabricated PVA-capped AgNPs by reducing silver nitrate with $NaBH_4$ in ethanol solvent in the presence of PVA [82]. The pH of the suspensions was further regulated by adding sodium hydroxide. They reported that the size of AgNPs decreases with increasing pH of the solutions, as revealed by transmission electron microscope (TEM) images (Figure 4a,b). The lattice spacing (i.e., 0.23) of AgNP (111) plane in high-resolution TEM image (Figure 4c), and the presence of rings in selected area electron diffraction pattern (Figure 4d) reveal the formation of nanocrystalline silver.

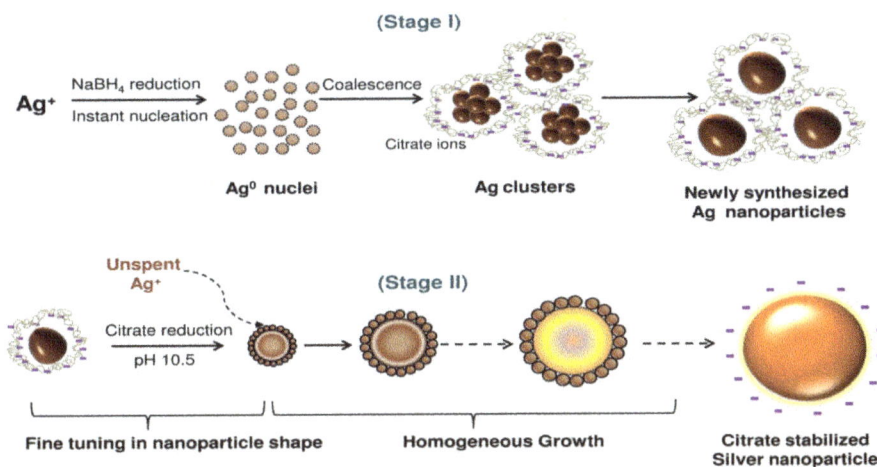

Figure 3. Schematic representation of size-controlled AgNPs synthesis employing the co-reduction strategy. Reproduced from [41], the Royal Society of Chemistry.

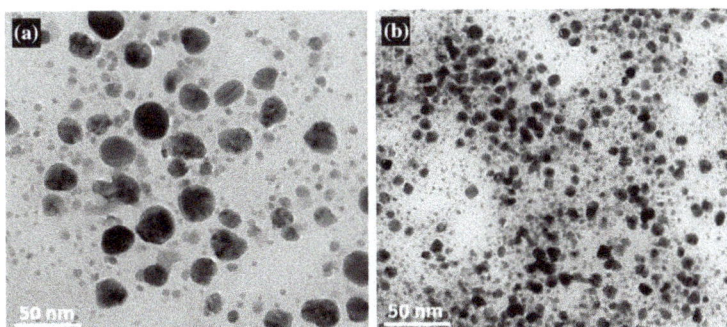

Figure 4. *Cont.*

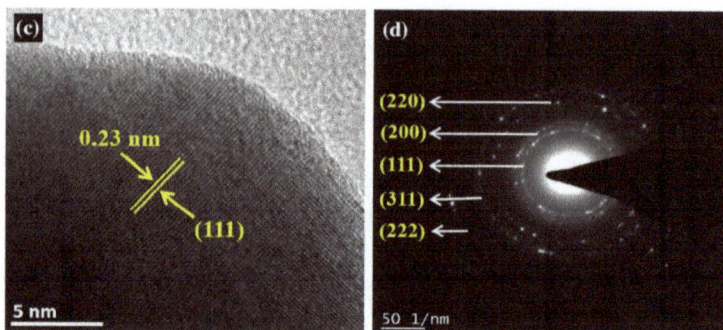

Figure 4. TEM images of AgNPs formed at (**a**) pH 6, and (**b**) pH 12. (**c**) High-resolution TEM image and (**d**) selected area electron diffraction pattern of AgNP. Reproduced from [82] with permission of Elsevier.

The size, geometry, morphology and homogeneity of AgNPs can also be manipulated by means of microemulsion technique. Several surfactants can be used to form microemulsion, including cationic cetyltrimethylammonium bromide (CTAB), anionic sodium dodecyl benzene sulfonate (SDBS) and sodium dodecyl sulfate (SDS), and nonionic Triton X-100 [83,84]. Water-in-oil (W/O) microemulsions or reverse micelles are often employed to prepare metallic nanoparticles [85–87]. They consist of nanosized water droplets suspended in a continuous oil phase and stabilized by surfactant molecules located at the oil/water interface. Thus, reactants can be introduced into water droplets acting as microreactors, leading to the formation of nanoparticles with uniform size distributions. For preparing metallic nanoparticles, two W/O microemulsions containing respective metal salt and reductant are mixed together to produce nanoparticles. As a result, an exchange of the reactants between micelles takes place during the collisions of water droplets, leading to coalescence, fusion, and mixing of the reactants. The size and shape of nanoparticles depend on the size and shape of water droplets, and the type of surfactant employed [88]. For instance, silver nitrate solubilized in the water core of one microemulsion can act as a source of silver ions, hydrazine hydrate dispersed in the water core of another microemulsion as a reducing agent, cyclohexane as the continuous phase, SDS as the surfactant, and isoamylalcohol as the cosurfactant [85]. It is noted that microemulsion process needs large quantities of surfactants and organic solvents, thus increasing the cost of production and polluting the environment. As the surfactants and solvents are mostly toxic [83], recent attention is paid to the use of natural plant extracts as the oil phase and reductant, or microorganisms as biosurfactants to produce AgNPs [86,87].

Similarly, organic solvents and reductants used in the chemical reduction process, such as *N*,*N*-dimethylformamide (DMF), dimethyl sulfoxide (DMSO), hydrazine, and sodium borohydride are toxic chemicals. This toxicity poses a threat to the environmental and living organisms. In this respect, photoreduction of silver salt through ultraviolet (UV) light without using toxic reductants and solvents can be used to form AgNPs. Thus, this is a simple and ecofriendly strategy to produce AgNPs [89–91]. For example, AgNPs can be synthesized from $[Ag(NH_3)_2]^+$ aqueous solution under UV irradiation using PVP as both reducing and stabilizing agents The resultant particles exhibit nano-size (4–6 nm), monodisperse and uniform size distribution [90]. Lu et al. prepared AgNPs with and without PVP as a surface capping agent by employing photochemical synthesis [89].

2.2. Biological Route

In recent years, green synthesis has opened up a new direction for forming AgNPs with different sizes and shapes without using toxic reductants and stabilizers [92]. The advantages of green synthesis over chemical and physical routes including ecofriendly, cost-effective, natural abundance and easy to scale-up for mass production of nanoparticles. In this respect, bacteria, plant extracts, fungi, polysaccharides and their derivatives can be used as the reducing agents and stabilizers. Biosynthesis of AgNPs using certain bacterial strains shows little industrial and medical applications because it may pose health risk to humans. Therefore, plant extracts such as leaves, stems, fruits and seeds are attractive reagent materials to form green AgNPs. Natural plants generally contain carbohydrates, fats, proteins, nucleic acids and pigments that can act as effective reducing agents and stabilizers for silver ions. In particular, polysaccharides possess many functionalities such as hydroxyl groups and a hemiacetal reducing end that play crucial roles in both the reduction and the stabilization of metallic nanoparticles [93]. Polysaccharides and their derivatives include chitosan, cellulose, starch, hyaluronic acid and heparin. In addition, whole leaf extracts are rich in polyphenols such as flavonoids, which are effective reductants for fabricating AgNPs [93]. Furthermore, parameters like nature of plant extract, pH, and reaction time greatly affect the size, shape, and morphology of green AgNPs. Extensive studies have been conducted by researchers on the biosynthesis of AgNPs, especially using plant leaves [94–102].

One main drawback of biosynthesis of AgNPs is the long reaction time. Microwave-assisted synthesis has attracted a great interest in biosynthesis because it can increase the reaction rate and product yield compared to conventional thermal heating [103]. Several successful reports relating microwave-assisted biosynthesis of AgNPs can be found elsewhere [104–110]. Peng et al. synthesized spherical AgNPs with sizes of 8.3–14.8 nm in an aqueous medium using bamboo hemicelluloses as a stabilizer and glucose as a reductant under microwave irradiation [104]. Ali et al. biosynthesized AgNPs with Eucalyptus globulus leaf extract (ELE) and $AgNO_3$ with and without microwave irradiation (Figure 5) [106]. The solution mixture was heated with microwave radiation at 2450 MHz for 30 s. The size of biosynthesized AgNPs ranged from 1.9–4.3 and 5–25 nm, with and without microwave treatment, respectively. The size of microwave-treated AgNPs (scheme-II) was smaller than that formed by conventional process at room temperature (scheme-I), because the extent of nucleation and capping was faster with microwave heating than synthesis at 37 °C. By heating with microwave radiation for 60 s, the reaction rate increased such that more nucleation sites were formed due to the availability of −OH ions at high temperatures. Organic constituents such as flavanoids and terpenes in ELE were reported to be surface active molecules to stabilize AgNPs [106].

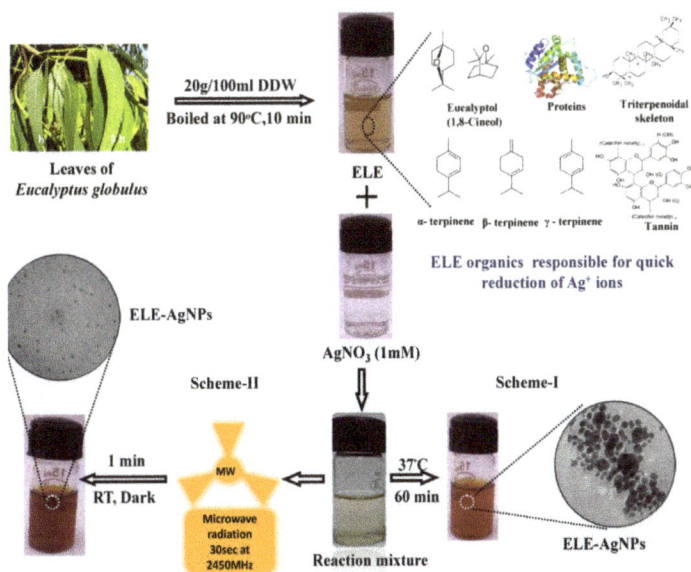

Figure 5. Graphical representation of AgNPs synthesis with Eucalyptus globulus leaf extract (ELE) and silver nitrate depicting scheme-I (without microwave treatment) and scheme-II with microwave irradiation. Reproduced from [106] with permission of Public Library of Science.

2.3. AgNP-Polymer Nanocomposites

As aforementioned, AgNPs find extensive applications in wound dressings, food packaging films or containers, antimicrobial fabrics, clinical scaffolds, etc. For those applications, AgNPs are typically embedded in the polymer matrix to form polymer nanocomposites. In this respect, polymers with high flexibility are ideal materials for protecting nanoparticles from mechanical damage [111,112]. A typical example is Acticoat™ Flex 3 dressing for burn care, i.e., a flexible polyethylene cloth coated with AgNPs at a concentration between 0.69 and 1.64 mg/cm^2. This dressing can release a sustained amount of nanocrystalline silver and silver ions to the wound area [27]. Moreover, immobilization of AgNPs onto textile fibers can also impart colors to the fabrics in addition to antimicrobial features due to the surface plasmon resonance effect of silver. This avoids the use of toxic agents to fix colorants to the textiles [30,31,113]. In this respect, AgNPs function as a simultaneous colorant and antimicrobial agent for fabrics [113]. Traditional polymer microcomposites are usually employed as components for structural engineering applications due to their light weight and low cost [114–118]. However, polymer microcomposites are reinforced with fillers of micrometer scale at loadings $\geq$ 30 wt% to achieve desired mechanical performance. The additions of microfillers with high loading levels lead to poor processability and low ductility of resultant composites. On the contrary, polymer nanocomposites only require low nano-filler loadings (say 0.1–1 wt%) for electronic, medical, and structural engineering applications. These nanocomposites have been prepared by means of solution mixing, electrospinning, extrusion or injection molding [119–122].

The fabrication techniques for AgNP-polymer nanocomposites vary from one to another depending on their specific intended applications. Water-soluble polymers such as PVA, PEG, and polyacrylic acid (PAA) are commonly used as hydrogels in tissue engineering. Hydrogels are crosslinked polymer networks that swell in water. Polymeric hydrogels can be simply prepared by freeze-thawing cyclic processing without the utilization of chemical crosslinkers [123,124]. The freezing step at low temperatures induces a liquid–liquid phase separation due to the formation of ice crystals that expel polymer chains. This creates polymer-poor and polymer-rich phases accordingly. Upon thawing, the ice crystals melt, leaving behind pores between cross-linked polymer regions. Repeated freeze-thawing cycles of the polymer solution lead to the formation of crystallites that act as cross-linking sites, so a hydrogel with a high swelling capacity can be produced [125]. AgNP-polymer composite hydrogels can be prepared by introducing pre-formed AgNPs into a water-soluble polymer matrix, or the formation of AgNPs in-situ through chemical reduction in aqueous polymer solution. In the latter case, metal salt is dissolved in aqueous polymer solution, followed by in situ reduction with sodium borohydride and freeze-thawing cycles [126,127].

In the case of antimicrobial fabrics, the dip-coating method is commonly adopted in which the fabrics are immersed in a silver salt solution followed by chemical reduction [128–133]. In some cases, UV- or microwave-radiation is used to speed up the reaction rate and to control the size of AgNPs on fabric fibers [134,135]. Babaahmadi and Montazer reported one-step in situ synthesis of Ag NPs on polyamide (nylon) fabrics through the reduction of silver nitrate with stannous chloride (SnCl$_2$) using CTAB as a stabilizer [129]. Montazer et al. then prepared AgNPs using [Ag(NH$_3$)$_2$]$^+$ complex with PVP as a reducing/stabilizing agent under UV irradiation. These nanoparticles were finally deposited onto nylon fabric using a simple dip-pad technique [134]. In a later study, they introduced AgNPs within the polymeric chains of polyamide-6 fabric by using [Ag(NH$_3$)$_2$]$^+$ complex [133]. The silver complex was reduced directly by the functional groups of polyamide chains without using any reductants. Moreover, nitrogen atoms of polyamide chains can stabilize AgNPs through coordination between the amide groups and silver.

As recognized, AgNPs can be deposited on fabrics more effectively by using plasma treatment through the creation of active groups on fabric fibers [136]. Recently, Zille et al. carried out dielectric barrier discharge (DBD) plasma treatment on polyamide 6,6 (PA 6,6) fabrics, followed by coating with colloidal AgNPs of different sizes [137]. Plasma pre-treatment promoted formation of oxygen species on fabric fibers, facilitating both ionic and covalent interactions between the oxygen species and AgNPs on the fibers. This led to the deposition of fined AgNPs on PA 6,6 fibers. Ilic et al. employed radio frequency (RF) plasma to etch the fibers of polyester fabrics in order to enhance binding efficiency of colloidal AgNPs onto polyester fibers [138]. They found that plasma treatment positively affected the loading of AgNPs on the fibers and antibacterial activity of polyester nanocomposite materials.

3. Antibacterial Activity

AgNPs are well known for their remarkable antimicrobial properties against various pathogens including bacteria, fungi, and viruses. However, the mechanisms responsible for the bactericidal effect of AgNPs remain unclear. There is an ongoing debate over whether AgNPs or silver ions exert a cytotoxic effect on microorganisms. The killing effect of AgNPs have been proposed to be associated with a direct contact of nanoparticles to the bacterial cell wall, followed by penetrating into cytoplasm. Direct contact of AgNPs with large surface areas on a bacterial cell wall could lead to membrane damage, resulting in the leakage of cellular contents and eventual cell death [139–142]. In particular, AgNPs with sizes below 10 nm are more toxic towards bacteria [141]. By penetrating into the cytoplam, AgNPs may interact with biomolecules such as proteins, lipids, and DNA. In certain cases, AgNPs can interact with the respiratory enzyme system, thereby generating reactive oxygen species (ROS) such as hydrogen peroxide (H_2O_2), hydroxyl (OH^-) and superoxide ($O_2{}^-$) radicals that induce oxidative stress and damage to proteins and nucleic acids [38]. Herein, we present literature reports that support this mechanism. Earlier work by Sondi and Salopek-Sondi indicated that the accumulation of AgNPs (12 nm) on the cell wall of *Escherichia coli* (*E. coli*) leads to the formation of pits [139]. Those pits cause a loss of outer membrane integrity, resulting in the release of LPS molecules and membrane proteins, and causing eventual cell death. Morones et al. reported that AgNPs (1–10 nm) anchor to the cell wall of *E. coli* and disturb its normal function such as permeability and respiration. The nanoparticles also penetrate into cytoplasm and interact with protein and DNA leading to cellular death [143]. Moreover, AgNPs can also release silver ions, resulting in further cell damage. Recently, Gahlawat et al. also demonstrated that AgNPs (10 nm) attach to the cell wall of cholera, thereby interrupting permeability and metabolic pathways of the cell and causing cell death [140].

On the contrary, the cytotoxic effect of AgNPs against bacteria may result from the oxidative dissolution of Ag^+ ions from AgNPs. As is known, metals can be chemically oxidized in aqueous solutions to give metallic ions [144–146]. In this respect, AgNPs can be oxidized in aerated aqueous solutions to yield Ag^+ ions [23]. Xiu et al. demonstrated that silver ions released from AgNPs in aerobic conditions are fully responsible for antibacterial activity (Figure 6) [147]. Small AgNPs (ca 5 nm) can release more Ag^+ ions than large AgNPs (11 nm) under aerobic conditions due to their higher surface-to-volume ratio. In an anaerobic environment, very little Ag^+ ions are released, so AgNPs themselves are non-toxic to bacteria. The antimicrobial action of Ag^+ ions is closely related to their interaction with thiol (sulfhydryl) groups [148]. Thus, Ag^+ ions can react with the -SH groups of cell wall-bound enzymes and proteins, interfering with the respiratory chain of bacteria and disrupting bacterial cell wall. Moreover, those ions released from AgNPs can penetrate the cell wall into the cytoplasm, thereby degrading chromosomal DNA [149], or reacting with thiol groups of the proteins in cytoplasm. Consequently, DNA loses its replication ability and the proteins essential to the ATP production becomes inactivated. In general, smaller silver particles can enter the cytoplasm more easily than larger particles [150]. AgNPs that have penetrated inside the cell can also release Ag^+ ions, thereby generating free radicals and oxidative stress accordingly. Bondarenko et al. reported that there exists a synergistic effect between these two mechanisms for antibacterial activity. Direct cell−nanoparticle contact promotes the release of silver ions from AgNPs, thereby enhancing the amount of cellular uptake of particle-associated Ag^+ ions [151]. Ivask et al. demonstrated that positively charged ions released from AgNPs tend to interfere with the normal function of the bacterial electron transport chain of *E. coli*, thereby facilitating the formation of reactive oxygen species (ROS) [152]. ROS generation is mostly responsible for the bacterial death because it facilitates lipid peroxidation, but inhibits ATP production and DNA replication. Elevated ROS levels in bacterial cells can result in oxidative stress [153]. Figure 7 shows a schematic representation of bactericidal effects due to AgNPs- induced membrane damage and silver ion release from the nanoparticles, or the combination of these two effects [154]. These mechanistic effects can be summarized into an initial attachment of AgNPs or Ag$^+$ ions to the bacterial cell wall, their subsequent penetration inside the cell, followed by ROS and free radical generation, DNA damage and protein denaturation.

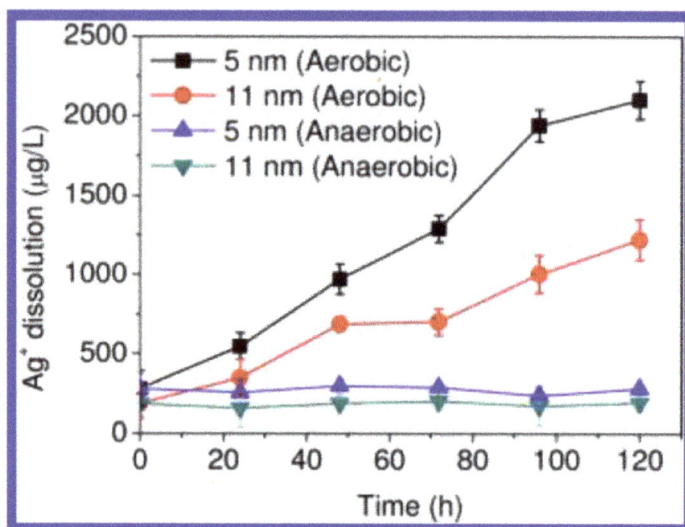

Figure 6. Dissolved Ag$^+$ concentration vs. air exposure time for PEG-AgNPs with sizes of 5 and 11 nm under aerobic conditions. No Ag$^+$ ions can be detected (<1 µg/L) under anaerobic conditions. Reproduced from [147] with permission of the American Chemical Society.

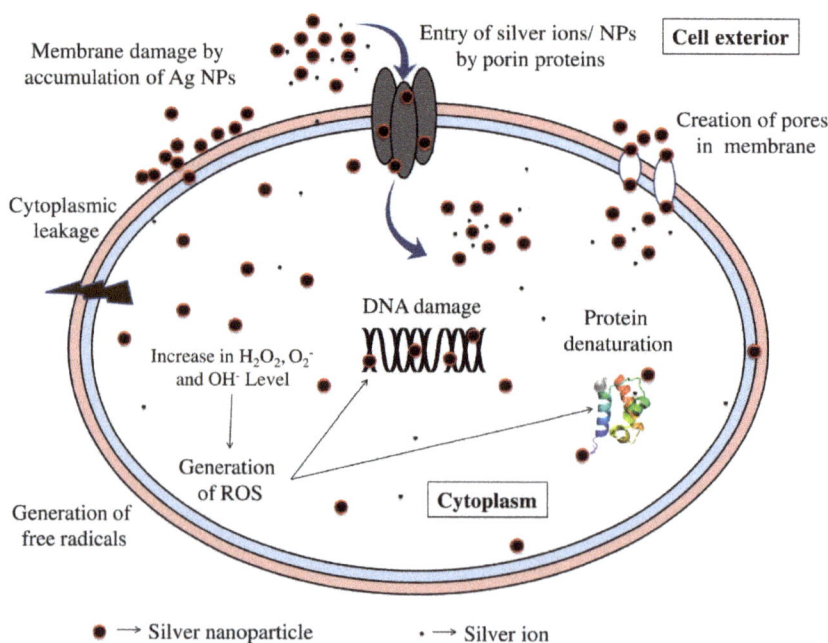

Figure 7. Bactericidal mechanisms of AgNPs due to their direct contact with the bacterial cell wall and the release of silver ions. Reproduced from [154] with permission of Elsevier.

The antibacterial efficacy of AgNPs relates to the kinds of pathogenic bacteria. Gram-negative bacteria are generally more prone to Ag$^+$ invasion than gram-positive bacteria due to the difference in their cell wall structures (Figure 2). As aforementioned, gram-positive bacteria possess a very thick cell wall containing many peptidoglycan layers, thereby serving as a barrier for Ag$^+$ ions penetration into the cytoplasm. However, gram-negative bacteria only have a single peptidoglycan layer, thus Ag$^+$ ions can easily damage the cell wall. The bactericidal effects of AgNPs also depend on the nanoparticle characteristics, including the size, shape, surface charge, dose and particle dispersion state [41–43,133–161]. Generally, well-dispersed AgNPs in physiological solutions exhibit greater bactericidal efficacy than agglomerated AgNPs. Moreover, the killing effect of AgNPs against gram-negative and gram-positive bacteria increases with decreasing particle size. Lu et al. prepared AgNPs of different sizes (~5, 15 and 55 nm) using a simple reduction method and found that AgNPs (5 nm) exhibited the highest antibacterial activity against oral bacteria [157]. Agnihotri et al. synthesized AgNPs of various sizes, i.e., 5, 7, 10, 15, 20, 30, 50, 63, 85 and 100 nm, and reported that AgNPs with sizes below 10 nm exhibited the best antibacterial activity for *E. coli* than larger nanoparticles (Figure 8) [41].

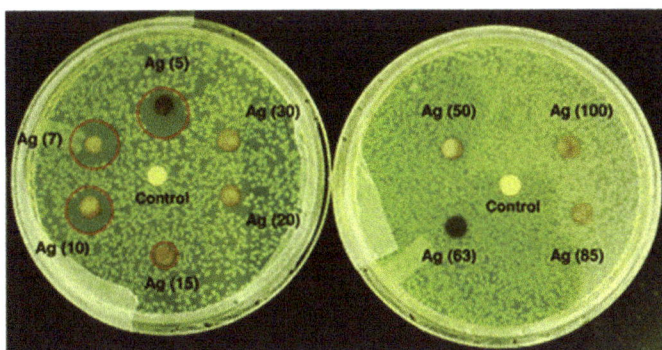

Figure 8. Disk diffusion assay results for AgNPs of various sizes against *E. coli*. The zone of inhibition is highlighted with a dashed circle indicating a noticeable antibacterial effect. Reproduced from [41], the Royal Society of Chemistry.

AgNPs can be produced in different shapes depending upon synthesized conditions [44,45,72]. Accordingly, AgNPs exhibit shape-dependent efficacy of bactericidal activities. Pal et al. reported that AgNPs with the same surface areas but different shapes exhibit dissimilar antibacterial activity. They found that truncated triangular silver nanoplates with a {111} lattice plane as the basal plane exhibit the strongest biocidal effect, compared with spherical and rod-shaped nanoparticles [160]. This is because the reactivity of silver is favored by a high-atom-density {111} plane. From the disk-diffusion tests, the bactericidal efficacy against *E. coli* of 10^7 CFU/mL takes the following order: truncated triangular > spherical > rod-shaped AgNPs. Very recently, Archaya et al. investigated the bactericidal effects of spherical and rod-shaped AgNPs against gram positive (*S. aureus, B. subtilis*) and gram negative (*E. coli, K. pneumoniae AWD5, P. aeruginosa*) bacterial strains [161]. Among these strains, *Klebsiella pneumoniae* can cause pneumonia, bloodstream infection and wound infection. *Klebsiella* bacteria show resistance to antibiotics and pose serious threat to human health, as outlined by the World Health Organization [162]. Their results indicated that the bactericidal activity of both spherical and rod-shaped AgNPs is dose- and time- dependent. Spherical AgNPs are more effective than rod-shaped AgNPs in killing *Klebsiella* bacteria (Figure 9).

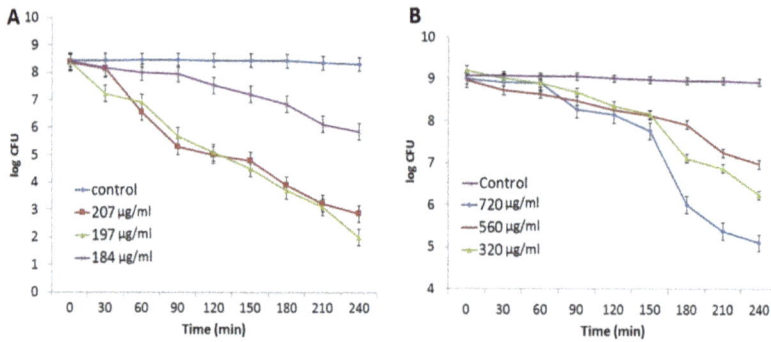

Figure 9. Killing kinetics of *K. pneumoniae* AWD5 exposed to (**A**) spherical AgNPs at concentrations of 184–207 µg/mL and (**B**) rod-shaped AgNPs at 320–720 µg/mL. Results were expressed as means ± SD; *n* = 3. *p* < 0.05 was considered statistically significant. Reproduced from [161], Nature Company under the Creative Commons Attribution License.

From the literature, the positive charge of the Ag$^+$ ions is critical for their bactericidal activity through electrostatic attractions between the negatively charged cell wall of the bacteria and positively charged Ag$^+$ ions [37]. The carboxyl, phosphate, hydroxyl, and amine groups associated with the thick peptidoglycan layer of the cellular wall of gram-positive bacteria render them with a negative charge. Similarly, those functional groups associated with LPS in the outer membrane confer an overall negative charge to the gram-negative cell wall [163]. As stated, the bactericidal effect of AgNPs is influenced by their surface charges. Thus, the capping agent and stabilizers used to prevent the aggregation of colloidal nanoparticles inevitably exert an influence on their surface charges [156,158,164]. Badawy et al. investigated the effect of surface charge of AgNPs capped with PVP, citrate (CT) and branched polyethyleneimine (BPEI) on bactericidal activity against the bacillus species. The BPEI-AgNPs are electrosterically stabilized through adsorption of the BPEI polyelectrolyte containing amine groups, which ionize in the solution to create charged polymers [164]. Electrosteric stabilization derives from both electrostatic repulsion and steric stabilization. Zeta potential of CT-AgNPs has a very negative value of −38 mV due to the carboxylic moiety of citrate. Consequently, there exists an electrostatic repulsion between the negatively charged CT-AgNPs and bacterial cell wall, thereby forming an electrostatic barrier that restrains the cell-particle interactions and thus reducing toxicity. The zeta potential of PVP-AgNPs is less negative, i.e., −10 mV, thus promoting cell-particle interactions and resulting in a higher toxicity than CT-AgNPs. The electrostatic repulsion changes to attraction by exposing bacteria to positively charged BPEI-AgNPs (+40 mV). In this case, BPEI-AgNPs interact strongly with the negatively charged moieties in the bacteria membrane (e.g., proteins) and induce changes in structural integrity of the bacteria cell wall, leading to the leakage of cytoplasmic contents and eventual cell death. These results reveal surface charge-dependent toxicity of AgNPs capped with different stabilizing agents on the *bacillus* species [164]. Thus, electrosterically coated BPEI-AgNPs exhibit a higher toxicity than electrostatically capped CT-AgNPs and sterically stabilized PVP-AgNPs. Similarly, Lee et al. demonstrated that PEI-AgNPs exhibit a positive zeta potential of +49 mV. PEI is a cationic polymer in which the amino groups provide AgNPs with a positive charge and stability against agglomeration [158]. Moreover, PEI-AgNPs show excellent bactericidal activity against *S. aureus* and *K. pneumoniae*. Recently, Abbaszadegan et al. studied the effect of surface charge of AgNPs on antimicrobial activity against gram-positive (*S. aureus*, *Streptococcus mutants*, *and Streptococcus pyogenes*) and gram-negative bacteria (*E. coli* and *Proteus vulgaris*) [156]. They indicated that positively-charged AgNPs exhibit the highest bactericidal activity against all bacterial strains. The negatively charged AgNPs have the least, while neutral AgNPs show intermediate antibacterial activity.

3.1. Biosynthesized AgNPs

Ali et al. employed Eucalyptus globulus leaf extract (ELE) to stabilize colloidal AgNPs during synthesis. They reported that ELE-AgNPs are effective antibacterial and antibiofilm agents for gram-negative *P. aeruginosa* and *E. coli*, gram positive methylene resistant *S. aureus* (MRSA) and methylene sensitive *S. aureus* (MSSA) [106]. Figure 10 shows the disk diffusion assay results for these bacterial strains exposed to ELE-AgNPs with concentrations ranging from 25–100 µL. Gurunathan and coworkers carried out a comprehensive study on antibacterial activity of biosynthesized AgNPs prepared from quercetin against *S. aureus* and *P. aeruginosa* [107]. Several bioassays for detecting colony-forming unit (CFU), lactose dehydrogenase (LDH), ROS generation, malondialdehyde (MDA), glutathione (GSH), etc. were employed to assess antibacterial activity. The AgNPs exhibited a spherical feature with an average size of 11 nm. The minimum inhibitory concentrations (MICs) of AgNPs against *P. aeruginosa* and *S. aureus* were 1 and 2 µg/mL, respectively. The bactericidal effect of AgNPs on bacteria came from the generation of ROS and MDA, and the leakage of proteins and sugars in bacterial cells. Figure 11 shows the dose- and time-dependent bactericidal activities of AgNPs for both bacterial strains. Complete growth inhibition concentration and time were determined to be 1 µg/mL AgNPs and 20 h for *P. aeruginosa*, while those for *S. aureus* were 2 µg/mL and 24 h, respectively. Figure 12 shows the ROS and MDA levels of both bacterial strains treated with AgNPs. Apparently, high ROS and MDA levels were observed in both strains, leading to abnormal cell metabolism and function, and eventual cell death. MDA was a product of lipid peroxidation, thus serving as an indicator of oxidation stress. On the basis of these results, Gurunathan and coworkers demonstrated that biosynthesized AgNPs is an effective therapeutic agent for treating mastitis-infected goats in husbandry.

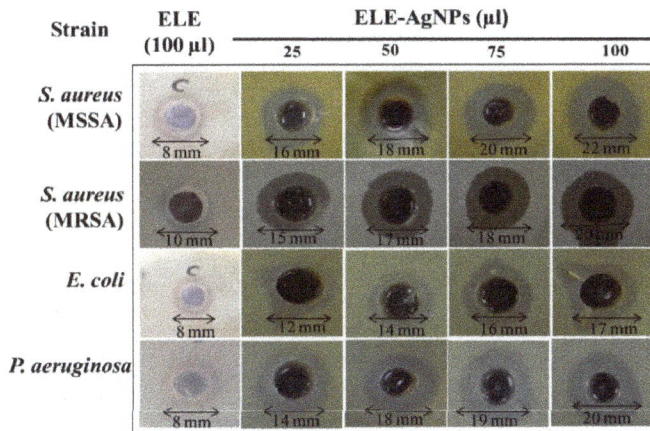

Figure 10. Assessment of antibacterial activity of ELE and ELE-AgNPs by disk diffusion assay. Reproduced from [106] with permission of Public Library of Science.

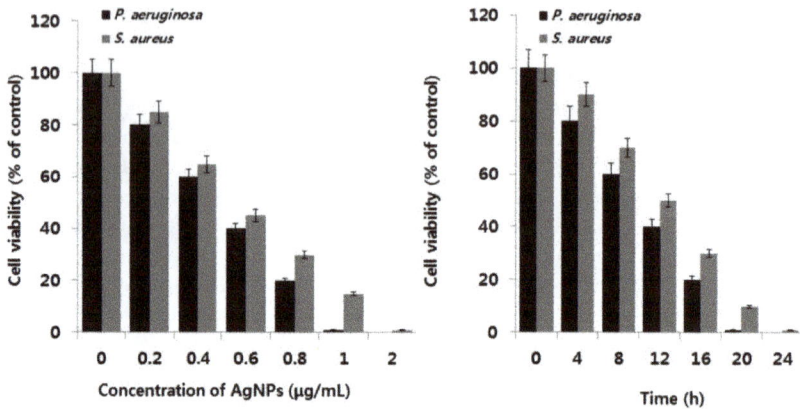

Figure 11. (**Left**): Effect of AgNPs concentration on bacterial cell viability. Bacterial survival was determined at 24 h based on a colony-forming unit (CFU) count assay. (**Right**): Time-dependent bactericidal effect of AgNPs on *P. aeruginosa* and *S. aureus*. Results were expressed as means ± SD; *n* = 3. $p < 0.05$ was considered statistically significant. Reproduced from [107], MDPI under the Creative Commons Attribution License.

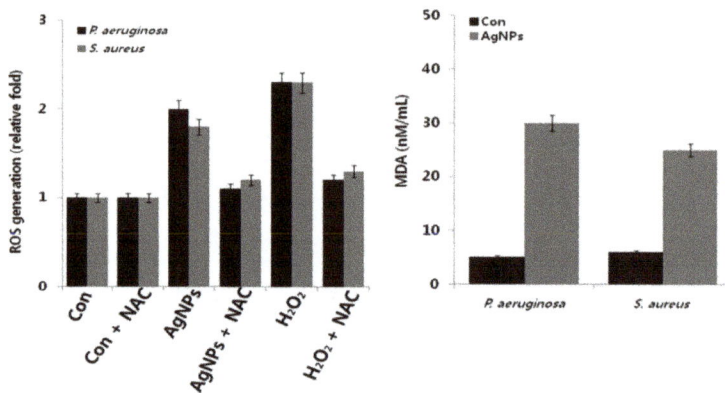

Figure 12. Effects of AgNPs on ROS (left panel) and MDA (right panel) levels. Results were expressed as means ± SD of *n* = 3; $p < 0.05$ was considered statistically significant as compared to control (con) groups. Reproduced from [107], MDPI under the Creative Commons Attribution License.

With the increasing use of antibiotics in animal husbandry, bacteria have developed resistance to antibiotics that pose serious threats to human health. The creation of multi-drug resistant bacteria is increasing at an alarming rate. In this respect, AgNPs appear to be a promising therapeutic agent against microbial pathogens in husbandry, especially for bacteria with antibiotic resistance. Very recently, Gurunathan et al. synthesized AgNPs (10 nm) using apigenin as a reducing and stabilizing agent [38]. They reported that as-synthesized AgNPs were very effective in eliminating multidrug resistant bacteria *Prevotella melaninogenica* and *Arcanobacterium pyogenes*. From the cell viability assay, antibacterial activity of AgNPs was dose-dependent, and the minimum inhibitory concentration (MIC) values of AgNPs against *P. melaninogenica* and *A. pyogenes* were determined to be 0.8 and 1.0 µg/mL, respectively. The minimum bactericidal concentration (MBC) values of AgNPs against *P. melaninogenica* and *A. pyogenes* were 1.0 and 1.5 µg/mL, respectively. The antibacterial activity of AgNPs was derived from the ROS generation, LDH leakage and DNA damage in bacterial cells. Figure 13 depicts a typical anti-biofilm activity of AgNPs on *P. melaninogenica* and *A. pyogenes*.

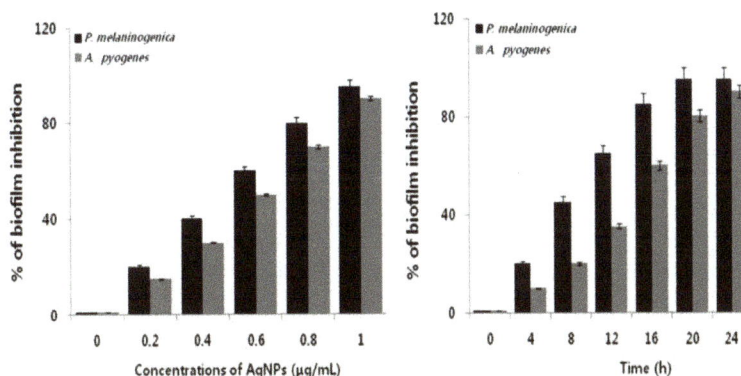

Figure 13. Anti-biofilm behavior of AgNPs on *P. melaninogenica* and *A. pyogenes*. (**Left**): Bacterial strains were treated with AgNPs of different concentrations. (**Right**): Bacterial strains were incubated with 0.8 and 1.0 µg/mL of AgNPs, respectively, for 24 h. $p < 0.05$ was considered statistically significant as compared to control groups. Reproduced from [38], MDPI under the Creative Commons Attribution License.

3.2. Polymer-AgNPs Nanocomposites

3.2.1. Nanocomposite Fabrics

As recognized, direct contact of AgNPs with human body inevitably leads to cytotoxicity and genotoxicity. Accordingly, it is necessary to immobilize AgNPs into polymeric materials to isolate them from the human body, and to control the release of Ag^+ ions. In recent years, considerable attention has been paid to produce antimicrobial composite fabrics due to their attractive applications in healthcare and medical sectors [165]. However, the poor laundering durability of nanocomposite fabrics limits their applications as a result of weak bonding between the polymer fabrics and nanoparticles [166,167]. In particular, hospital textiles are laundered at elevated temperatures for many cycles to minimize the risk of contaminated linens and to prevent the spread of various diseases. Mechanical vibration coupled by high temperature conditions in washing machines can detach AgNPs from the fabrics. Some efforts have been taken by researchers to improve the adherence of AgNPs to fabric fibers including plasma deposition, choice of fabric materials, graft polymerization, etc. [134,168,169]. For instance, El-Rafie et al. applied 50 and 100 ppm biosynthesized AgNPs to cotton fabrics, and reported that the reduction rate of bacterial colonies was higher than 90% against *S. aureus* and *E. coli* before washing [167]. The antibacterial activity of the composite fabrics againsts both bacterial strains reduced by more that 40% after 20 washing cycles. The absence of chemical interactions between the AgNPs and cotton fibers led to poor binding of AgNPs to cotton fabrics. Consequently, some AgNPs were removed from the fabrics during washing cycles. Montazer et al. employed UV radiation to synthesize AgNPs using $[Ag(NH_3)_2]^+$ and PVP. The as-sythesized PVP-AgNPs were deposited onto nylon fabrics using a dip-pad technique [134]. In the process, PVP-AgNPs with respective concentrations of 100 and 200 ppm were deposited onto nylon fabrics, and then exposed to *E. coli* and *S. aureus*. The bacterial reduction levels of unwashed and washed nanocomposite fabrics (10, 20, and 30 washes) were evaluated; the results are listed in Table 1. Apparently, nanocomposite fabrics exhibited good antibacterial property by eliminating *E. coli* up to 99% after 30 washes. This is because AgNPs coated on the fabric fibers were resistant to *E. coli* after repeated laundries. Moreover, capped PVP of the AgNPs can establish chemical linkages with polyamide chains of nylon, leading to a strong adherence of AgNPs to nylon. However, bactericidal activity of the fabric with 100 ppm AgNPs decreased slightly after repeated washing. The bacterial reduction percentage of *S. aureus* decreased slightly from 99.99% to 86.92% after 30 washes. This rate was acceptable for antimicrobial fabrics after several washing cycles.

Table 1. Bacterial reduction percentages of PVP-AgNP/nylon nanocomposite fabrics against *E. coli* and *S. aureus*. Reproduced from [134] with permission of Elsevier.

Bacterium	AgNP Content (ppm) in Fabrics	Percentage of Bacterial Reduction		Number of Washing	
		0	10	20	30
E. coli	100	99.99	99.99	99.46	99.20
E. coli	200	99.99	99.99	99.99	99.55
S. aureus	100	99.99	99.86	99.27	86.92
S. aureus	200	99.99	99.57	99.27	91.03

As mentioned above, Zille et al. pretreated PA6,6 fabrics with DBD-plasma, followed by immersion in commercial colloidal AgNPs dispersions (10, 20, 40, 60 and 100 nm particle size) containing sodium citrate as a stabilizer to form PA6,6/AgNP composite fabrics [137]. Figure 14 shows the percentage of bacterial reduction vs. the size of AgNPs deposited on plasma-treated PA6,6 fabrics upon exposure to *E. coli* and *S. aureus* for 1 day and 30 days It can be seen that bacterial growth inhibition for S. aureus is size-dependent at day 1. The inhibition effect against *S. aureus* increases with decreasing nanoparticle size. The value decreases from 95% for the 10 nm-AgNPs, down to 19% for the 100 nm-AgNPs. The inhibition effect is associated with the release of Ag$^+$ ions from the AgNPs in the solution during antimicrobial tests. Thus, more Ag$^+$ ions are released from smaller AgNPs than larger nanoparticles due to their larger surface area-to-volume ratio. Upon exposure to *E. coli* at day 1, AgNPs with sizes of 10, 20 and 40 nm exhibit full bacteria inhibition, while AgNPs with sizes $\geq$60 nm show partial killing. Thus, AgNPs are more effective in eliminating *E. coli* due to their thin wall structure, as mentioned previously. At day 30, considerable amounts of Ag$^+$ ions are released from the nanocomposite fabrics, and the growth of both *S. aureus* and *E. coli* are completely inhibited with the exception of 100 nm-AgNPs [137].

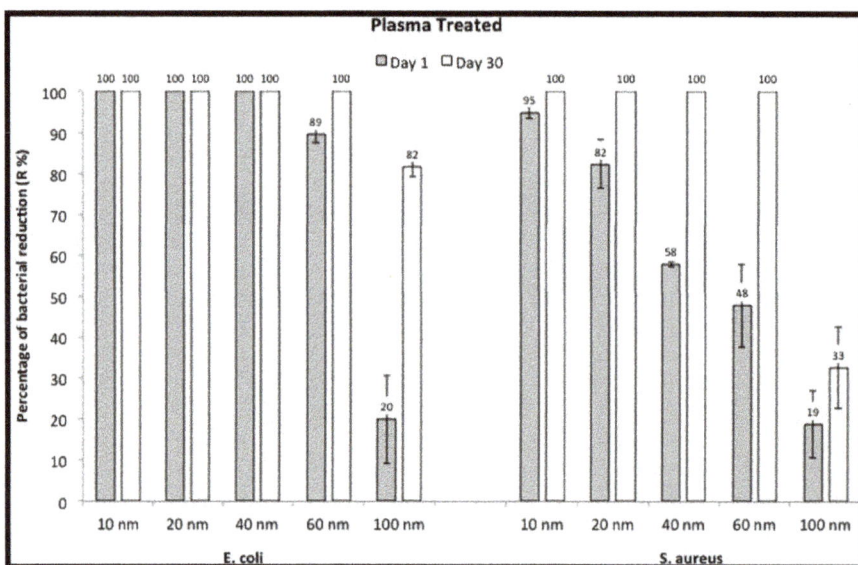

Figure 14. Percentage of bacterial reduction (*E. coli* and *S. aureus*) as a function of the size of AgNPs after exposure of 1 day and 30 days. Data are presented as mean values $\pm$ SD ($n = 3$). Reproduced from [137] with permission of the American Chemical Society.

3.2.2. Food Packaging Nanocomposite Films

In recent years, there has been a growing demand in food industries to develop antimicrobial food packaging films, bottles and containers to avoid microbial food spoilage and to extend or preserve shelf life. Food packaging is employed to protect foods, vegetables and fruits from environmental and bacterial contaminations to ensure their quality and food safety. Oxidation, microbial invasion, and metabolism are the main factors causing deterioration of foods and fruits during production, transportation, and storage [170]. Nowadays, AgNPs, silver nitrate and nanoclay are commonly used in the food packaging industry to resist microbial contamination and to improve barrier properties, thus prolonging shelf life and freshness of packaged foods and drinks [171–174]. As mentioned previously, colloidal nanosilver and silver nitrate have been used for more than 100 years in the United States [32]. Martinez-Abad et al. incorporated silver nitrate (0.1–10%) into ethylene-vinyl alcohol copolymer (EVOH) films and studied their antimicrobial behavior against *listeria monocytogenes* and *salmonella spp.* [175]. They employed the bacterial challenge test [176] to assess antimicrobial resistance of EVOH composite films against low protein food samples (lettuces, apple peels, and eggshells) and high protein food samples (chicken, marinated pork loin, and cheese) contaminated with those bacterial strains. Figure 15 shows representative viable bacterial counts on apple peels with *listeria monocytogenes*, and then treated with EVOH composite films containing 0.1, 1 and 10 wt% AgNO$_3$. Composite film with 10% AgNO$_3$ and control (silver nitrate solution) show a 4–5 log reduction in microbial population, while films with 0.1 and 1 wt% AgNO$_3$ display little antimicrobial effect, i.e., a decrease of about 2 log bacterial counts after 24 h exposure. These results indicate that antimicrobial resistance of the composite films with 0.1 and 1 wt% AgNO$_3$ on food samples are somewhat poorer than aqueous silver nitrate solution. This is due to the confinement of AgNO$_3$ in the polymer matrix, thereby restricting the release of sufficient amounts of Ag$^+$ ions to combat microorganisms. Thus, only composite film with high filler loading level, i.e., 10 wt% AgNO$_3$ can achieve a similar antibacterial effect as silver nitrate solution.

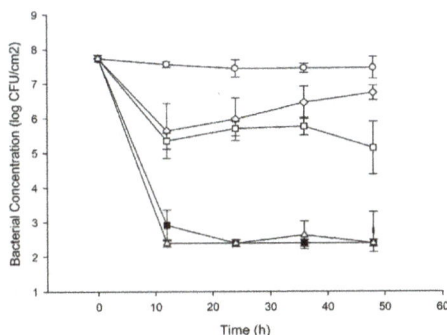

Figure 15. Viable counts in the challenge test on apple peels with *L. monocytogenes* versus silver nitrate aqueous solution (black square), EVOH (circle), and EVOH composite films with 0.1 wt% Ag$^+$ (diamond), 1 wt% Ag$^+$ (square), and 10 wt% Ag$^+$(triangle). Reproduced from [175] with permission of the American Chemical Society.

In general, AgNPs exhibit a beneficial effect over silver nitrate salt in food packaging films because AgNPs allow a sustained release of Ag$^+$ ions due to the size-related Ag$^+$/Ag^0 ratio on their surfaces [177]. In this respect, low AgNPs loadings are added in polymeric films to release sufficient Ag$^+$ ions to ensure effective bactericidal activity [178]. More recently, Tavakoli et al. fabricated polyethylene (PE) films with 1, 2 and 3 wt% AgNPs using the extrusion process [173]. They reported that PE/AgNP packaging films decrease mold and coliform attack on walnuts, hazelnuts, almonds and pistachios for extended periods, thereby increasing shelf life and preserving the quality of nuts. The widespread use of polymer/AgNPs packaging films and containers in the food industry has resulted in increased concerns over the migration of AgNPs from the films or containers into foods. In this context, Huang et al. exposed commercial PE/nanosilver film bags to four kinds of food-simulating solutions, representing water, acid, alcohol and fatty foods, at 25–50 °C for 3 to 15 days, respectively. They found the migration of Ag^0 from commercial PE/nanosilver films into food-stimulants on the basis of atomic absorption spectroscopic measurements [179]. It is considered that Ag^+ ions are also released from nanocomposite films upon exposure to food-simulating solutions. Moreover, Ag^+ ions are easily reduced to Ag^0 in the presence of acid environments. Indeed, Echegoyen and Nerin reported the presence of both elemental Ag^0 and Ag^+ ions in commercial polyolefin packaging films and containers with nanosilver. Furthermore, microwave oven heating accelerates the migration of these species into stimulant solutions due to the structural modification of the polymer matrix [180].

3.2.3. Nanocomposite Wound Dressings

Hydrogels have been developed and used in the medical sector to enhance wound healing. They find attractive clinical applications due to their biocompatibility, high water content, and good absorption of wound exudates. By incorporating AgNPs or silver nitrate into hydrogels, their antimicrobial resistance can be improved through reduction in infections. In this respect, antimicrobial polymer/AgNPs and polymer/AgNO₃ hydrogels for wound dressing applications have attracted considerable attention in recent years [126,127,181–184]. As an example, Oliveira et al. fabricated PVA/AgNO₃ hydrogels loaded with 0.25% and 0.5% AgNO₃ [184]. The nanocomposite hydrogels exhibited significant inhibition against both gram-positive and gram-negative bacteria due to the Ag^+ ions released from silver nitrate (Figure 16a). Culturing mouse fibroblasts with nanocomposite hydrogels revealed good cell membrane integrity and cell viability (Figure 16b), indicating that nanocomposite hydrogels are non-toxic.

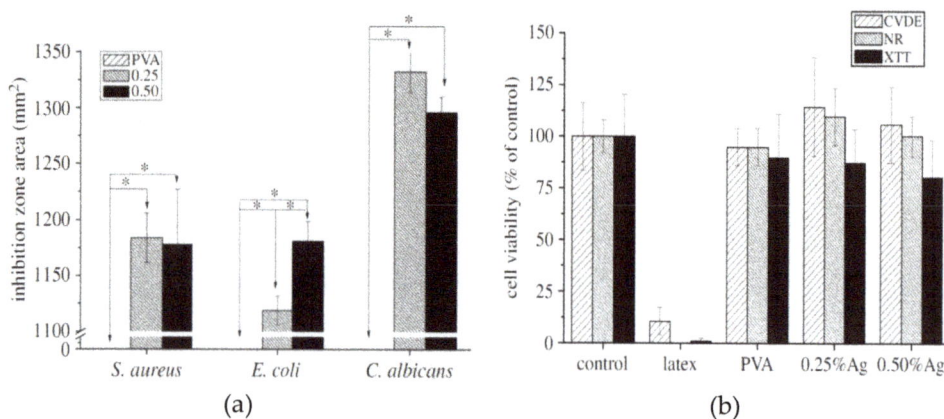

Figure 16. (a) Inhibition zones of all samples exposed to S. aureus, E. coli and C. albicans. There is a significant difference between the levels indicated by arrows, * p < 0.05. (b) Cell viability of mouse fibroblasts after 24 h incubation with nanocomposite hydrogels. CVDE (cell density), NR (membrane integrity assay) and XTT (mitochondrial activity). 'Control' is the negative control, whereas 'latex' is the positive control. Reproduced from [184] with permission of the Royal Society Publishing.

4. In Vitro Cell Cultivation

As aforementioned, AgNPs have been widely used for antibacterial and therapeutic applications, including fabrics, food packaging materials, wound dressings, and cancer therapy [22,24,25,27,30,31,113,171–174,184]. These routes can lead to increasing exposure of AgNPs to human cells [185]. Cellular uptake of AgNPs takes place either via diffusion (translocation), endocytosis or phagocytosis [186]. Upon entering the cytoplasm, AgNPs themselves or Ag^+ ions can generate ROS, leading to DNA damage, protein denaturation, and apoptosis [23,187]. AgNPs of different sizes and shapes tend to accumulate in the mitochondria, thereby inducing mitochondrial dysfunction, i.e., a reduction in mitochondrial membrane potential (MMP), and promoting ROS creation. This leads to the damage of intracellular proteins and nucleic acids (Figure 17a) [52–55,185,188–194]. Grzelak et al. and AshaRani et al. have demonstrated that the disruption of mitochondrial respiratory chain by AgNPs would increase ROS generation and interrupt ATP synthesis, thereby resulting in DNA damage [189,190]. The ROS generation can also cause cell membrane damage through the release of lactate dehydrogenase (LDH). Furthermore, AgNPs can interact with the membrane proteins and activate signaling pathways, leading to the inhibition of cell proliferation. On the other hand, Ag^+ ions released from AgNPs can also induce ROS generation [192,195], especially for cellular uptake through endocytosis [49,51]. In this context, AgNPs confined in an acidic lysosomal environment dissolute into Ag^+ ions. These ions initiate cascades or series of events that lead to intracellular toxicity, termed as the "lysosome-enhanced Trojan horse effect" [51]. Furthermore, some AgNPs, which translocate into cytoplasm through diffusion or channel proteins, are oxidized by cytoplasmic enzymes, thereby releasing Ag^+ ions. Those ions interact with thiol groups of mitochondrial membrane proteins, causing mitochondrial dysfunction and generating ROS accordingly (Figure 17b). In the case of bacterial cells, several factors such as nanoparticle size, shape, surface area, surface charge, surface functionalization, and particle dispersion state also affect cytoxicity in mammalian cells [188,191]. Therefore, AgNPs tend to induce size-, dose- and time-dependent toxicity by creating ROS, oxidative stress, and DNA damage [189,193,194]. Figure 17a,b summarize possible mechanisms of AgNPs-, or Ag^+-induced toxicity in mammalian cells [190,192].

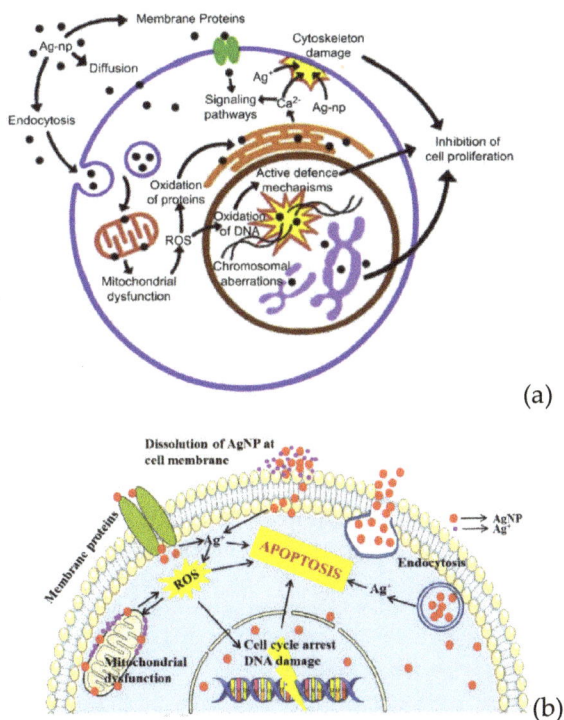

(a)

(b)

Figure 17. Proposed mechanisms of (**a**) AgNPs- and (**b**) silver ion-induced cytotoxicity. Reproduced from [190] and [192] with permission of BioMed Central Ltd and Elsevier, respectively.

Industrial activities involved in manufacturing AgNPs and their associated products have raised concerns over their release into the environment via several processes, including particle synthesis during manufacturing and incorporation into products, recycling, and disposal [196]. During industrial fabrication and laboratory synthesis, AgNPs in the form of powder or liquid may enter the human body through inhalation and dermal contact [197]. Inhaled nanoparticles can reach the lung alveoli, which is the deepest region of the respiratory system. AgNPs have several adverse health effects upon entering the pulmonary alveoli, because of their prolonged interaction with the lung cells. Fordbjerg et al. studied the effect of gene expression profiling in adenocarcinomic human alveolar basal epithelial cells (A549) by treating them with AgNPs [198]. They reported that AgNPs at 12.1 µg/mL modified the regulation of more than 1000 genes of A549 cells. The upregulated genes included members of the metallothionein, heat shock protein, and histone families. Moreover, ROS was also generated but did not cause apoptosis at 12.1 µg/mL AgNPs. Han et al. prepared AgNPs using both green and chemical reduction methods [193]. They demonstrated that the toxicity of AgNPs in A549 cells was dose-dependent, resulting from ROS generation and oxidative stress. In addition, green AgNPs were more toxic at lower concentrations than chem-AgNPs. The IC_{50} values for bio-AgNPs and chem-AgNPs were 25 µg/mL and 70 µg/mL, respectively. IC_{50} was the concentration of AgNPs with a 50% reduction in cell viability. Very recently, Gurunathan et al. synthesized green AgNPs from silver nitrate using biomolecule quercetin, and treated A549 with a combination of AgNPs and antitumor drug, i.e., MS-275 derived from histone deacetylases (HDACs) [199]. The as-synthesized AgNPs exhibited dose- and size-dependent toxicity against A549 cells. Combined AgNPs and MS-275 markedly induced apoptosis as a result of ROS accumulation, LDH leakage, mitochondria dysfunction, activation of caspase 9/3, up and down regulation of pro-apoptotic genes and anti-apoptotic genes, respectively. For human bronchial epithelium (BEAS-2B) cells [200–202], Gliga et al. employed pristine AgNPs (50 nm), PVP-AgNPs (10 nm,) and CT-AgNPs (10, 40 and 75 nm) to interact with BEAS-2B [201]. They reported that only AgNPs with a size of 10 nm induce cytotoxicity regardless of the surface coating. The cytotoxicity was associated with DNA damage and the release of intracellular Ag. Similarly, Kim et al. demonstrated that Ag-NPs induced a significant increase in the ROS level and oxidative DNA damage in the BEAS-2B cells [202].

The widespread use of antimicrobial textiles, wound dressings, and cosmetics containing AgNPs has increased human dermal exposure to those nanoparticles. Sapkota et al. demonstrated that biosynthesized AgNPs exhibit dose-dependent toxicity towards human keratinocytes (CRL-2310) [203]. At 10 µg/mL, cell viability was 98.76%, but the viability further decreased to 74.5% at 100 µg/mL. Carrola et al. indicated that Ag^+ ions released intracellularly from CT-AgNPs caused a dose-dependent ROS generation in human skin keratinocytes (HaCaT) [49]. Further, CT-AgNPs (10 nm) agglomerated considerably in culture medium compared to CT-AgNPs (30 nm). As such, agglomerated CT-AgNPs (10 nm) became less cytotoxic than CT-AgNPs (30 nm). Avalos et al. studied genotoxic effects of PEI/PVP-coated AgNPs (4.7 nm) and uncoated AgNPs (42 nm) on normal human dermal fibroblasts (NHDFs) and human pulmonary fibroblasts (HPFs) [204]. In vitro exposure of NHDFs and HPFs to coated (0.1–1.6 µg/mL) and uncoated AgNPs (0.1–6.7 µg/mL) for 24 h triggered DNA strand fragmentation in a dose- and size-dependent manner. Furthermore, smaller PEI/PVP-AgNPs were more genotoxic than larger AgNPs. In another study, they also found that smaller AgNPs (4.7 nm) were more toxic than pristine AgNPs (42 nm) in NHDFs on the basis of MTT and LDH measurements. The oxidative stress parameters showed a dramatic increase of ROS but a depletion in glutathione levels [205].

Hou et al. studied toxicity of AgNPs (20 nm) in three human cell lines, i.e., human bronchial epithelial cells (16HBE), human umbilical vein endothelial cells (HUVECs), and human hepatocellular liver carcinoma cells (HepG2) (Figure 18A–D) [206]. HUVECs are commonly used in vitro model for assessing toxicity of nanoparticles to endothelium [207–209]. 16HBE cells originate from human airway epithelial cells, thus representing potential toxicity due to inhalation, while HUVECs and HepG2 cells are the target cells for AgNPs upon entering blood circulation. Human blood vessels are composed of a thin layer of endothelial cells known as the endothelium. Capillary endothelium differs in structure depending upon the tissue type in which it is located. Continuous endothelium is closely packed together and linked with tight junctions, anchored to a basement membrane. It is found in the blood vessel, skin, lung, and nervous tissues. Fenestrated endothelium is found in the capillaries of kidney and endocrine glands, while discontinuous endothelium is found in the liver. As recognized, nanoparticles come into first contact with vascular endothelium once they enter the circulation system. Vascular endothelium in different tissues has its own distinctive properties including surface receptors and intercellular junctions [209]. From Figure 18A–C, a dose- and time-dependent manner of cell viability can be readily seen, especially for 16HBE. Thus, the toxicity of AgNPs on these cell lines takes the order: 16HBE > HepG2 > HUVECs. The toxicity of 16HBE arises from the activation of endoplasmic reticulum (ER) stress signaling pathway. ER stress response is markedly induced in the 16HBE cells, but not in HUVECs and HepG2 cells [206]. Shi et al. also reported a dose-dependent toxicity of AgNPs on HUVECS [54]. In their study, AgNPs induce intracellular ROS formation, reduce cell proliferation, and cause cell membrane damage, leading to cell dysfunction and eventual apoptosis. These adverse effects are attributed to the activation of IKK/NF-κB pathways as a result of the oxidative stress. Guo et al. investigated the cytotoxicity of citrate-coated AgNPs (10, 75, and 110 nm) towards HUVECs [210]. AgNPs can be readily taken up by vascular endothelial cells, resulting in cell leakiness via altering inter-endothelial junctions.

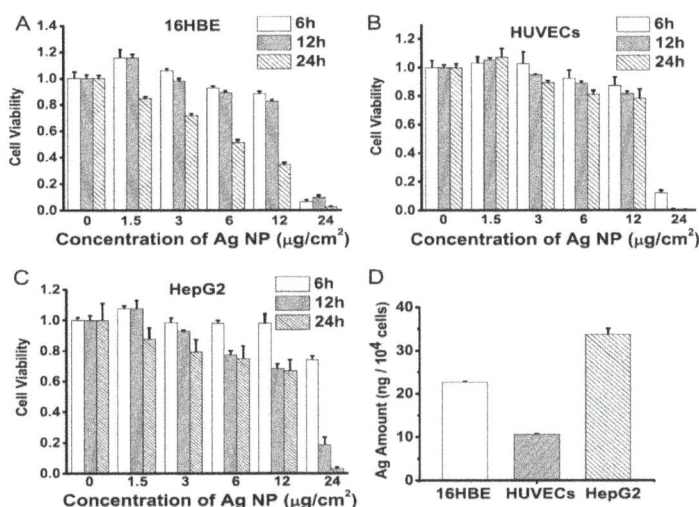

Figure 18. (**A–C**) Cell viability vs AgNP concentration for 16HBE, HUVECs and HepG2 cells determined from CCK-8 assay at different time points. (**D**) Inductively coupled plasma mass spectrometry results showing cellular uptake of AgNPs upon exposure at a dose of 2 mg/cm^2 AgNPs for 24 h. Data are expressed as means ± SD, $n = 5$. Reproduced from [206] with permission of Elsevier.

From the in vitro model in the literature, AgNPs triggered pro-inflammatory cytokines in brain endothelial cells, thereby causing an increased permeability of the cell layer. Trickler et al. studied inflammatory responses of rat brain microvessel endothelial cells (rBMECs) exposed to AgNPs of different sizes (25, 40 and 80 nm) and concentrations [58]. They reported that exposure of AgNPs to BMECs induce pro-inflammatory cytokines such as interleukin IL-1β, tumor necrosis factor (TNF-α), and prostaglandin E$_2$ (PGE$_2$). The pro-inflammatory response followed a size- and time-dependent manner, with IL-1β preceding both TNF-α and PGE$_2$ for AgNPs (25 nm). The interactions of the Ag-NPs with endothelial cells also induced cellular damage in the form of perforations in rBMEC monolayers. The secretion of pro-inflammatory cytokines together with an increase of vascular permeability of rBMECs allowed the entry of substances into the brain tissues, inducing neuronal cell death. Very recently, Sokolowska et al. studied toxic effects induced by AgNPs on three kinds of endothelial cell lines, i.e., HUVEC, human brain endothelial cell (HBEC5i) and human endothelial cell line for blood vessel (EA.hy926) [59]. The viability of these three cell lines decreased with increasing AgNPs concentration. HBEC5i cells were much less vulnerable to AgNPs induced toxicity than EA.hy926 and HUVEC cells (Figure 19A). These three cell lines also exhibited a dose-dependent membrane damage, in which HBEC5i cells were less susceptible to the damage compared to EA.hy926 and HUVEC cells (Figure 19B). They attributed the higher cell viability against AgNPs to the presence of specialized cellular components of the brain barrier.

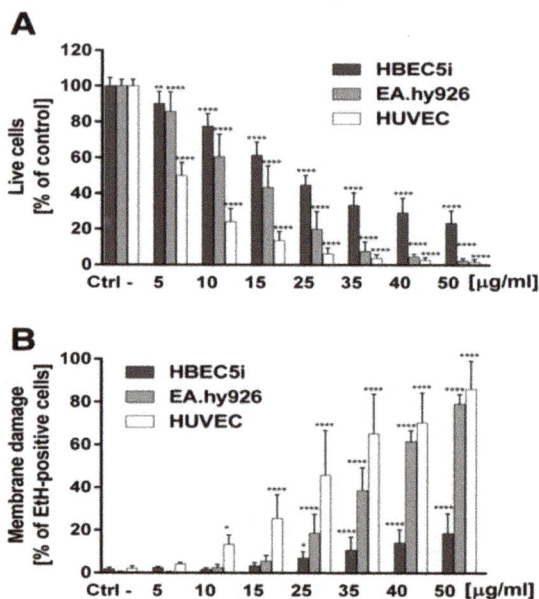

Figure 19. (**A**) Cell viability and (**B**) membrane damage of HBEC5i, HUVEC and EA.hy926 cells vs AgNPs concentrations after 24 h exposure to nanoparticles. Data are presented as means ± SD. * $p <$ 0.05; ** $p < 0.01$; **** $p < 0.0001$. Reproduced from [59] with permission of Elsevier.

The liver is one of the target organs once AgNPs enter the bloodstream [194,211,212]. Xue et al. demonstrated that AgNPs (15 nm) induce toxicity in HepG2 cells under a dose- and time-dependent manner. They also assessed the effect of solvents (deionized water and culture medium) for dispersing AgNPs on cytotoxicity [194]. The toxic effects were attributed to ROS generation, mitochondrial injury, and oxidative stress, leading to cell apoptosis (Figures 20 and 21). AgNPs-induced cytotoxicity was more severe in water than culture medium because of the dissolution of AgNPs into Ag$^+$ ions in water. Singh et al. biosynthesized AgNPs from silver nitrate using leaf extract of Morus alba as a reductant [212]. They then exposed green AgNPs to HepG2, and observed a dose-dependent cytotoxicity with an IC$_{50}$ value of 20 µg/mL. The cytotoxic effect of green AgNPs was compared with the standard anticancer drug 5-Fluorouracil (5-FU) and pure Morus alba extract. The IC$_{50}$ values of 5-FU and M. alba were recorded, respectively, as 30 and 80 µg/mL. Apparently, AgNPs showed nearly a same trend in destroying cancer cells as that of standard drug, showing potential application for hepatocellular therapy.

Red blood cells (RBCs) or erythrocytes contain no nucleus and organelles such as mitochondria; thus, they have limited repair capability following injury. Direct interaction of nanoparticles with RBCs can damage their membranes, leading to membrane rupture or hemolysis. Kim and Shin studied hemolysis, deformability, and morphological change of human RBCs exposed to AgNPs (30 and 100 nm) and silver nanowires (AgNWs) for 2 h [50]. They reported that hemolysis of RBCs is size- and dose-dependent in which small AgNPs induce higher hemolysis than large AgNPs. The shape of silver nanomaterials had little influence on hemolysis. They attributed cytotoxicity to the direct interaction of AgNPs with the RBCs, leading to the generation of oxidative stress, membrane injury, and eventual hemolysis. Chen et al. also reported a size- dependent hemolysis effect for murine RBCs (Figure 22a) [213]. Serious hemolysis was found at AgNPs (15 nm) contents $\geq$ 10 µg/mL (Figure 22b).

Figure 22c showed the TEM image of RBCs prior to AgNPs exposure. Figure 22d showed the internalized AgNPs in RBCs, leading to membrane injury, lipid peroxidation, and eventual hemolysis. Very recently, Ferdous et al. studied the interactions of PVP and citrate coated AgNPs (10 nm) of various concentrations (2.5, 10, 40 μg/mL) with murine RBCs [214]. AgNPs induced significant dose-dependent hemolysis, resulting from cellular uptake of AgNPs and oxidative stress generation.

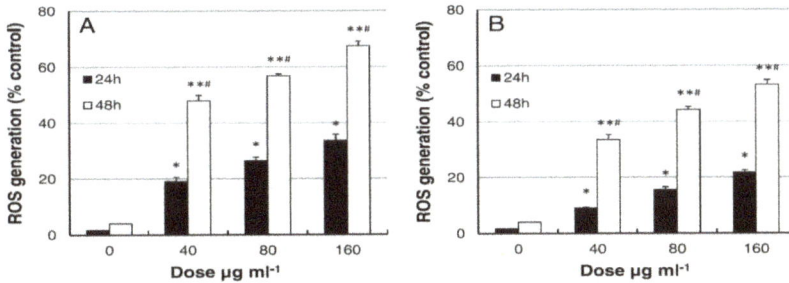

Figure 20. Dose-and time-dependent ROS generation in HepG2 cells exposed to AgNPs in: (**A**) deionized water and (**B**) cell culture medium. Data are expressed as means ± SD. There was significant difference between the treated and control groups (* $p < 0.05$; ** $p < 0.01$), and between the 24- and 48-h groups (# $p < 0.05$). Reproduced from [194] with permission of Wiley.

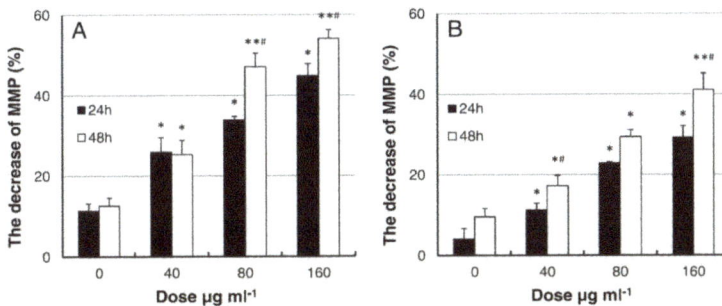

Figure 21. Dose-and time-dependent MMP reduction of HepG2 cells exposed to AgNPs in (**A**) deionized water and (**B**) cell culture medium. Data are expressed as means ± SD. There was significant difference between the treated and control groups (* $p < 0.05$; ** $p < 0.01$), and between the 24- and 48-h groups (# $p < 0.05$). Reproduced from [194] with permission of Wiley.

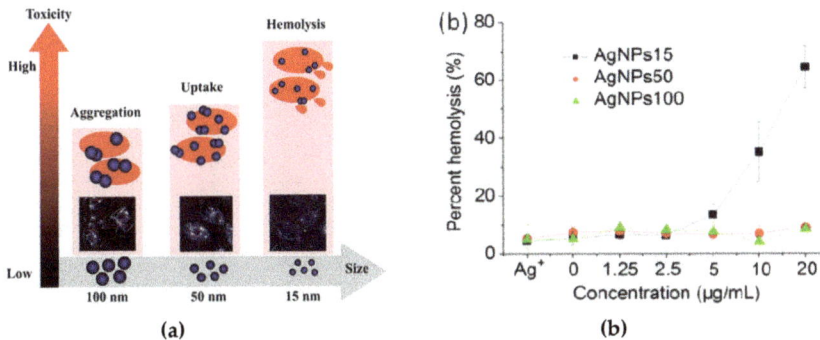

Figure 22. *Cont.*

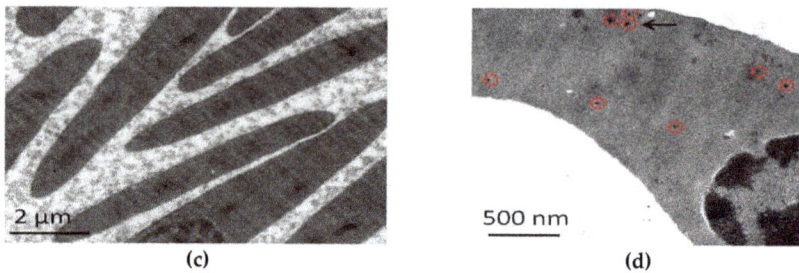

(c)   **(d)**

Figure 22. (**a**) Schematic representation showing size-dependent hemolysis of RBCs due to AgNPs. (**b**) Percentage hemolysis vs AgNPs concentrations. TEM images of RBCs (**c**) without and (**d**) with AgNPs (15 nm) treatment. Individual AgNP in (**d**) is outlined with a red circle, while AgNPs are aggregate using black arrows. Reproduced from [213] with permission of the American Chemical Society.

Macrophages are well known phagocytic cells of the innate immune system, acting as a first line of defense against pathogens. They exist in nearly all mammalian tissues and are involved in bacteria killing, wound healing, restoring tissue homeostasis, and regulating immune response. AgNPs also exhibit a toxic effect on macrophages, especially those with the smallest particle sizes. Figure 23 shows typical size- and dose- dependent toxicity in murine alveolar macrophages induced by AgNPs of different sizes [215]. The IC_{50} values of AgNPs (15 nm), AgNPs (30 nm) and Ag (55 nm) were recorded as 27.87 ± 12.23, 33.38 ± 11.48, and >75 μg/mL respectively. Apparently, AgNPs (15 nm) showed the highest cytotoxicity as expected. Yang et al. studied the cytotoxic and immunological effect of AgNPs (5 nm, 28 nm and 100 nm) on innate immunity using human peripheral blood mononuclear cells (PBMCs) (Figure 24a,b) [216]. They reported a dose-dependent toxicity of AgNPs on PBMCs in which AgNPs (5 nm) were the most toxic nanoparticles. Furthermore, AgNPs with sizes of 5 nm and 28 nm induced inflammasomes to generate IL-1β and subsequent caspase-1 activation. Inflammasomes formation was derived from the leakage of cathepsins due to the disruption of lysosomal membranes, and the K^+ efflux via cell membrane pores triggered by AgNPs. In addition, AgNPs (5 nm) and AgNPs (28 nm) increased the production of mitochondrial superoxide. At the same concentration, AgNPs (5 nm) induced more production of hydrogen peroxide that was toxic to cells [216]. Martinez-Gutierrez et al. treated the human monocytic cell line (THP-1) with AgNPs (24 nm), and reported that monocytes secrete inflammatory cytokines IL-6 and TNF-α at AgNPs contents ≥ 10 μg/mL [217]. Butler et al. examined the genotoxic effects of AgNPs (10, 20, 50 and 100 nm) on THP-1 cells, and indicated that AgNPs (10 and 20 nm) induce micronucleus nucleation and DNA strand breaks [60]. Micronucleus formation only required very low AgNPs dosages, i.e., 15 μg/mL for AgNPs (10 nm), and 20 μg/mL for AgNPs (20 nm). Silver ions released from AgNPs endocytosed by THP-1 were mainly responsible for the DNA damages.

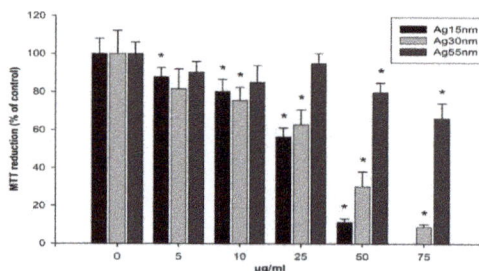

Figure 23. Effect of AgNPs concentration on mitochondrial metabolism (MTT assay) in murine alveolar macrophages treated with AgNPs for 24 h. The data were expressed as means ± SD ($n = 3$). $p < 0.05$ was considered significant. Reproduced from [215] with permission of the American Chemical Society.

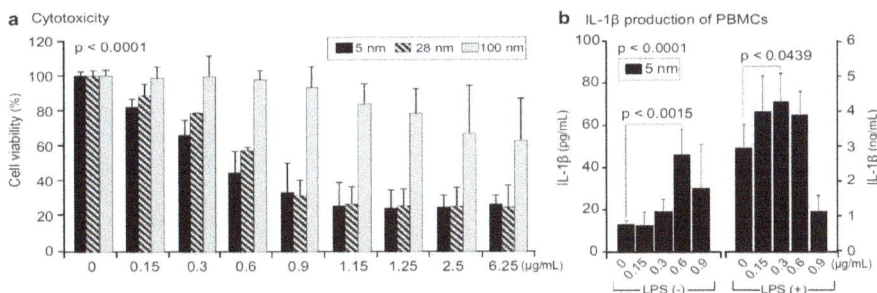

Figure 24. Cytotoxicity and IL-1β generation in PBMCs. (**a**) PBMCs were treated with AgNPs for 6 h and cell viability was determined with CCK-8 assay. (**b**) PBMCs were treated with AgNPs (5 nm) for 6 h and supernatant levels of IL-1β were assessed by ELISA. LPS (50 pg/mL) was pre-treated for 2 h before AgNPs exposure. Results were presented as means ± SD. One-way ANOVA analysis showed significance ($p < 0.0001$) (**a**,**b**), and Student's *t*-test between certain pairs (**b**) was used for statistical analysis. Reproduced from [216] with permission of Elsevier.

From the literature, AgNPs can cross the brain blood barrier (BBB) through the blood circulation system [58,218]. An earlier study by Trickler et al. reported that AgNPs increased the BBB permeability in primary rat brain endothelial cells, and induced a size-dependent pro-inflammatory response by secreting PGE2, TNF-α and Il-1β [58]. Cramer et al. studied the effect of AgNPs' surface coatings (citrate and ethylene oxide (EO) on neurotoxicity of primary porcine brain capillary endothelial cells (PBCECs) [219]. Neutral red uptake assay revealed that cell viability decreased markedly from 100% to 58% and 71%, respectively, upon exposure to EO-AgNPs and CT-AgNPs at 50 μg/mL. Furthermore, AgNPs disturbed cell barrier integrity and tight junctions, and induced oxidative stress and DNA strand breaks. Those adverse effects were reduced to a lesser extent using citrate coating. Liu et al. examined the toxic effect of AgNPs (23 nm) on embryonic neural stem cells (NSCs) from human and rat fetuses [220]. In addition, mitochondrial metabolism (MTT assay) was substantially reduced, while LDH leakage and ROS generation were markedly increased under a dose-dependent manner. AgNPs-induced neurotoxicity was further revealed by up-regulated Bax protein expression, and an increased number of TUNEL-positively stained cells [220]. From the literature, the Trojan-horse effect in murine astrocytes and microglial cells due to AgNPs uptake also led to ROS generation [189]. As such, intracellular Ag$^+$ ions interacted with thiol-groups of cysteine (CYS) protein, producing Ag(CYS) and Ag(CYS)$_2$ species. Yin et al. studied the effects of AgNPs (34 nm) and Ag$^+$ ions in the form of silver nitrate on neurotoxicity of mouse embryonic stem cells (mESCs) [221]. They demonstrated that both AgNPs and Ag$^+$ ions perturbed mESCs global and neural progenitor cell-specific differentiation processes. AgNPs and Ag$^+$ ions induced anomalous expression of neural ectoderm marker genes at concentrations lower than 0.1 μg/mL [221]. Ma et al. studied the cytotoxic effect of AgNPs (30 nm) on murine hippocampal neuronal HT22 cells [222]. They reported that cytotoxicity is caused by mitochondrial membrane depolarization, increased ROS generation, and caspase-3 activation. Mitochondrial membrane depolarization results from a loss of mitochondrial membrane integrity, leading to a decrease of MMP. Caspase-3 is the main caspase responsible for apoptosis execution [223]. Apparently, brain tissue with high lipid content is particularly vulnerable to the oxidative stress. By treating HT22 cells with both AgNPs and sodium selenite, cell viability increases significantly due to selenium, and can suppress ROS generation and caspase-3 activation.

With the fast development of material examination techniques in recent years, atomic force microscopy (AFM) has been used increasingly in the biological field [224,225]. AFM measures the surface roughness and elastic modulus of a material by moving its tip across the specimen surface. The force between the tip and the sample is measured through the deflection of cantilever during scanning [224]. For biomaterials, changes in biophysical properties (cell height and roughness) as well as biomechanics (elastic modulus) can be analyzed accordingly. Thus, AFM is a powerful tool to analyze the interaction between the cells and AgNPs at high accuracy. In this respect, Subbiah et al. employed AFM to investigate the physicomechanical responses of A549, human bone marrow stromal cells (HS-5) and mouse fibroblasts (NIH3T3) exposed to AgNPs [226]. Bioassays (CCK-8, GSH, and lipid peroxidation) were also concurrently performed. As such, the results were compared and correlated with those of AFM. From their study, AgNPs exhibited a dose-dependent reduction in glutathione (GSH), but showed an increased manner with the MDA level. AgNPs bonded directly to GSH and inhibited the enzymes for GSH synthesis, leading to GSH depletion and ROS buildup. As recognized, GSH depletion is an early event during apoptosis, which occurs before the loss of cell viability [227]. From the AFM measurements, it was seen that treatment using AgNPs leads to a substantial change in cell morphology due to enhanced cell surface roughness. Moreover, the stiffness of AgNP-treated cells also increases markedly because of the deposition of AgNPs on the cell surfaces. Figure 25A–C show the correlation between cell viability and AFM results.

More recently, Jiang et al. combined AFM and bioassays to study cytotoxic effect of AgNPs on human embryonic kidney 293T cells (HEK293T cells) [55]. In their study, AFM was used to measure cellular viscosity from the force-displacement curve. The measurements showed that cellular viscosity decreases with increasing AgNPs concentration, demonstrating that structural changes occur in kidney cells upon exposure to AgNPs. Bioassays (comet, gene expression profiling) tests showed that severe DNA damage occurs in HEK293T cells due to downregulation of antiapoptosis Bcl2-t and Bclw genes, and upregulation of the proapoptosis Bid gene. Table 2 summarizes the cytotoxic effects of AgNPs on human cell lines.

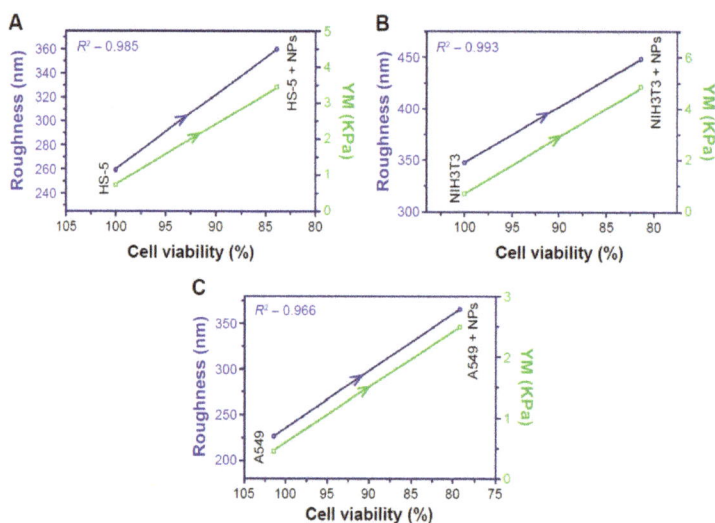

Figure 25. Correlation between cell viability and the roughness or stiffness of (**A**) HS-5, (**B**) NIH3T3 and (**C**) A549 cells before and after treatment with AgNPs. NP: nanoparticles; Y. M.: Young's modulus. Reproduced from [226] with permission of Dove Medical Press Ltd.

Table 2. Cytotoxic effects of AgNPs on human cell lines.

Synthetic Route and Size	AgNPs Dosage and Exposure Time	Cell Type	Cytotoxic Effect	Ref.
Green & chemical reduction; 15 nm	10, 20, 30, 40 and 50 μg/mL for 24 h	A549	ROS creation, MMP reduction, LDH leakage, phagocytosis	[193]
Green synthesis; 11 nm	AgNP (1 μM) + MS-275 (1 μM) for 24 h	A549	Apoptosis due to ROS creation, LDH leakage, mitochondria dysfunction, DNA fragmentation	[199]
Chemical reduction; 15.9 ± 7.6 nm	12.1 μg/mL for 24 and 48 h	A549	Exposure of AgNPs for 24 h altered the regulation of more than 1000 genes; ROS generation	[198]
Chemical reduction; 19.5 nm	1.25, 2.5, 5, 10, 20 and 40 μg/mL for 24 h	A549, HS-5; NIH3T3	AgNPs treatment increased surface roughness and stiffness of the cells.	[226]
Commercial particles; CT-AgNPs: 10, 40, 75 nm; PVP-AgNPs: 10 nm	5, 10, 20 and 50 μg/mL for 24 h	BEAS-2B	Size-dependent toxicity. AgNPs with 10 nm were more toxic, leading to DNA damage without ROS generation	[201]
Commercial particles; CT-AgNPs: 10, 30 and 60 nm	10 and 40 μg/mL for 24 h and 48 h	HaCaT	Dose-dependent ROS generation	[49]
Green synthesis; 20 nm	10, 20, 40, 60, 80 and 100 μg/mL for 24 h	CRL-2310	Dose-dependent toxicity. Cell viability was 98.76% at 10 μg/mL, but reduced to 74.5% at 100 μg/mL.	[203]
Commercial particles; Pristine AgNPs: 42 nm; PEI/PVP coated-AgNPs: 4.7 nm	AgNPs: 0.1, 0.5, 1.6 and 6.7 μg/mL. Coated AgNPs: 0.1, 0.5, 0.8, 1.6 μg/mL	HPF and NDHF	DNA strand breaks in a dose- and time-dependent manner. Smaller coated-AgNPs were more genotoxic than larger pristine AgNPs	[204]
Chemical reduction; 65 nm	0.5, 1, 1.5 and 2 μg/mL	HUVEC	Dose-dependent toxicity. ROS creation and cell dysfunction via IKK/NF-κB pathways	[54]
Commercial particles; <100 nm	5, 10, 15, 25, 35, 40 and 50 μg/mL for 24 h	HBEC5i; HUVEC; EA.hy926	Cell viability and membrane damage were dose-dependent.	[59]
Commercial particles; 15 nm	40, 80 and 160 μg/mL for 24 h and 48 h	HepG2	Dose-dependent cytotoxicity. ROS creation, MPP reduction & apoptosis	[194]
Green synthesis; 10–50 nm	1, 5, 10, 20, 40 and 80 μg/mL for 24 h	HepG2	Dose-dependent cytotoxicity; IC$_{50}$ = 20 μg/mL	[212]
Commercial particles; 60 nm	10, 20 and 40 μg/mL for 24 h	HEK293T	Decreased cell viability, increased DNA damage by exposing to AgNPs with increasing concentration	[55]
Chemical reduction; AgNPs: 30 and 100 nm AgNWs: length (1–2 μm)	100, 200, 300, 400 and 500 μg/mL for 2 h	Human erythrocyte	Size- and dose- dependent hemolysis	[50]
Commercial particles; 5, 28 and 100 nm	0.15, 3, 6, 9, 1.15, 1.25, 2.5 and 6.25 μg/mL for 6 h	PBMC	Dose-dependent cytotoxicity. AgNPs induced inflammasomes to produce IL-1β.	[216]
Green synthesis; 24.4 nm	2, 5, 6.25, 10, 12.5, 50 μg/mL for 24 h	THP-1	Cell death more than 42% at 12.5 μg/mL AgNPs. Induced cytokines IL-6 and TNF-α	[217]
Commercial particles; 10, 20, 50 and 100 nm	1, 2.5, 5, 10, 15 and 25 μg/mL for 24 h	THP-1	AgNPs (10 nm) and AgNPs (20 nm) induced DNA damage	[60]
Chemical reduction; 23 nm	1, 5, 10, 20 μg/mL for 24 h	NSC	Reduction in mitochondrial metabolism; increased LDH leakage and ROS level	[220]

5. In Vivo Animal Model

An in vivo model for AgNP-induced cytotoxicity is performed directly on the tissues of a whole living animal under a controlled environment. The tests are expensive, time consuming, and subjected to several restrictions due to ethical issues. The experiments are typically performed on rodents (rats, mice and guinea pigs) through oral administration, intravenous (i.v.) injection, intraperitoneal (i.p.) injection, intratracheal (i.t.) instillation, subcotaneous injection, etc. [228]. The in vivo cytotoxic effects of AgNPs depend on several factors such as nanoparticle size and dose, administration route, exposure time and type of animal model. From published literature reports, AgNPs accumulate mainly in the target organs of animals through several administration routes, thereby inducing toxic effects such as cell dysfunction, inflammation, DNA damage, and animal death [229–237].

Liver is one of the main target organs for administration routes involving translocation of AgNPs in the blood circulation system [229–240]. Kupffer phagocytic cells in the liver are essential for particle removal following intravenous administration. As such, AgNPs are deposited in the Kupffer cells after injection [236,237]. Accumulated AgNPs in the liver may cause several negative effects such as the generation of ROS, pathological changes in liver morphology, and enzyme activity. Dziendzikowska et al. intravenously injected AgNPs (20 and 200 nm) to male Wistar rats at a dose of 5 mg/kg [234]. AgNPs were translocated from the bloodstream to liver, spleen, kidneys, lungs and brain, with the liver being the main target organ. Silver concentrations in these organs of the rats treated with AgNPs (20 nm) were significantly higher than those treated with AgNPs (200 nm). Furthermore, silver concentrations in these organs displayed a time- and size-dependent accumulation manner. Lee et al. intraperitoneally injected AgNPs into Sprague-Dawley (SD) rats, and reported that AgNPs accumulated mainly in the liver [235]. AgNPs caused a significant increase of caspase-3 level in the liver of treated rats from day 1 until day 30. Although autophagy was induced following i.p. injection at day 1, failure to preserve autophagy in the following days led to liver dysfunction and eventual apoptosis.

Recently, Recordati et al. intravenously injected CT- and PVP-coated AgNPs as well as silver acetate into CD-1 mice [229]. Commercial CT-AgNPs and PVP-AgNPs with sizes of 10, 40 and 100 nm were used in their study. Cytotoxic effects were strongly size-dependent, while coating type (CT or PVP) had no impact on biodistribution of AgNPs in the organ tissues. Histological examination revealed that AgNPs were predominantly accumulated in the spleen and liver, and to a lesser extent in the kidney and lung (Figure 26). Very high silver concentrations were detected by inductively coupled plasma mass spectrometry (ICP-MS) in the spleen and liver, followed by lung, kidney and brain. AgNPs (10 nm) were found to be the most toxic nanoparticles (Figure 27). Silver acetate (AgAc) at the same dosage (10 mg/kg) was also detected in these organs after administration. Very recently, Yang et al. also demonstrated that AgNPs (3 nm) were mainly deposited in the liver and spleen of male mice, followed by the kidney, heart, lungs and testis, and the least accumulation was found in the stomach, intestine following *i.v.* injection [230]. RT-qPCR analysis of the liver revealed substantial changes in the gene expression profiles, i.e., upregulation of several genes such as *p*53, caspase-3, caspase-8, transferrin (Trf), and Bcl-2. As is known, caspases are enzymes that cause apoptosis by cleaving cellular proteins. Initiator caspases such as caspase 2, 8, 9 and 10 initiate the apoptotic process, leading to the activation of effector caspases, i.e., caspase 3, 6 and 7 [222]. Wen et al. intravenously injected SD rats with CT-AgNPs (6.3 nm) at a dosage of 5 mg/kg body weight (bw) respectively [231]. They reported that the lungs, spleen, and liver were enriched with Ag content on the basis of ICP-MS measurements. In addition, the silver concentration distribution in the organs from highest to lowest took the following sequence: lung > spleen > liver > kidney > thymus > heart. Furthermore, AgNPs induced chromosome aberration in bone marrow cells.

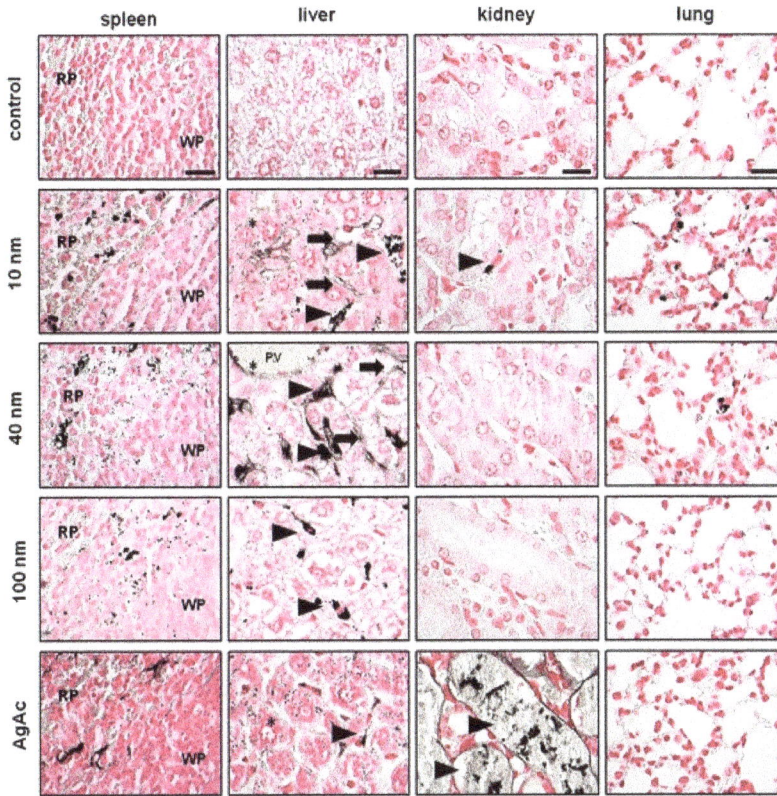

Figure 26. Histological examination of silver tissue localization by autometallography staining. Representative images of spleen, liver, kidney, and lung (scale bar = 20 μm), from AgNPs (10, 40, 100 nm) and silver acetate treated mice. In the spleen, silver was localized within the cytoplasm of macrophages especially in the spleen white pulp (WP) and red pulp (RP). Triangles indicate the accumulation of silver in organ tissues. Reproduced from [229], BioMed Central Ltd under the Creative Commons Attribution License.

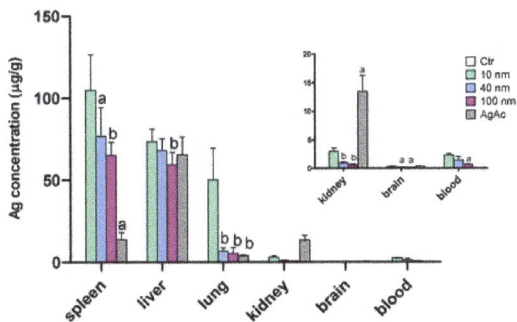

Figure 27. Silver tissue concentration after *i.v.* injection of AgNPs and AgAc in mice. Data are expressed as means ± SD. The inset illustrates a magnified view showing Ag concentration in the kidney, brain, and blood. Statistical significance: a = $p < 0.05$; b = $p < 0.01$. Reproduced from [229], BioMed Central Ltd under the Creative Commons Attribution License.

From in vitro cell cultivation, AgNPs increased the permeability of tight junctions of brain endothelial cells [219]. The ICP-MS measurements of in vivo animal model showed the presence of a small amount of AgNPs in the mice brain [229,232,234]. Thus, AgNPs can cross the brain blood barrier (BBB) through the bloodstream, thereby inducing neurotoxicity and neuronal death. Hadrup et al. reported that AgNPs (14 nm) with doses of 4.5 and 9 mg/kg bw/day and ionic silver in the form of silver acetate (9 mg/kg bw/day) increased the dopamine concentration in the brain of female rats following 28 days of oral administration, resulting in cellular apoptosis [241]. Wen et al. conducted intranasal instillation of PVP-AgNPs (26.2 ± 8.9 nm) in neonatal SD rats with doses of 0.1 and 1 mg/kg bw/day, and ionic silver in the form of silver nitrate for 4 and 12 weeks, respectively [232]. Dose-dependent silver accumulation occurred for both AgNPs and silver ions in the liver, lung and brain. The highest silver concentration was found in the liver at week 4, while it shifted to the brain after week 12. Their findings revealed the potential neuronal damage from the intranasal administration of AgNPs or silver colloid-based products [232]. Xu et al. administered intragastrically a low dose (1 mg/kg bw/day) and a high-dose (10 mg/kg bw/day) into SD rats for 14 days [242,243]. A low dosage induced neuron shrinkage and astrocyte swelling. The adverse effect of AgNPs was attributed to the presence of lymphocytes around astrocytes. More recently, Dabrowska-Bouta et al. investigated the influence of AgNPs on the toxicity of cerebral myelin [244]. In that study, Wistar rats were exposed to 0.2 mg/kg bw per day of AgNPs (10 nm) via the gastrointestinal route. They observed enhanced lipid peroxidation and decreased concentrations of protein and non-protein –SH groups in myelin membranes.

We now consider cytotoxic effects induced by AgNPs and ionic silver in mice following oral administration [238,245–248]. Liver and kidney are the main target organs for mice administered orally with AgNPs. These organs play crucial roles in the clearance of exogenous substances. Bergin et al. administered CT- and PVP-AgNPs with sizes of 20 and 110 nm, and doses of 0.1, 1 and 10 mg/kg into Black-6 mice for three days through oral gavage [245]. Nearly 70.5–98.6% of administered AgNPs was excreted in feces following oral administration. Thus, no toxicity and significant tissue accumulation of AgNPs were found in treated mice. Boudreau et al. introduced AgNPs of different sizes and dosages into SD rats via oral gavage [246]. They found low accumulation of silver in tissues of rats treated with AgNPs of larger sizes, i.e., 75 and 100 nm. In contrast, tissues from rats treated with smaller AgNPs (10 nm) at 36 mg/kg bw/day showed significant silver accumulation in the kidneys, spleen and liver. In the kidneys, silver was localized within the renal tubular epithelium. Qin et al. studied the toxicity of PVP-AgNPs and $AgNO_3$ in male and female SD rats treated with repeated oral administration at doses of 0.5 and 1 mg/kg bw daily for 28 days. They found no significant toxic effects of AgNPs and $AgNO_3$ up to 1 mg/kg in terms of the body weight, organ weight, food intake, and histopathological examination [238]. However, ICP-MS results revealed the presence of silver in the liver, kidney, spleen and, plasma (Figure 28A,B). The total Ag contents in organs were significantly lower in the AgNPs-treated rats than those in the $AgNO_3$ treated rats. In addition, silver was detected in the testis of male rats. Statistical difference in silver concentrations was found in major organs of male rats treated with AgNPs, while no difference of Ag distributions was observed in female rats. The gender-related difference in AgNPs' distribution may be related to hormonal regulation in these organs. van der Zande et al. also indicated that silver contents in the liver, spleen, testis, and kidney of rats are mainly derived from the Ag^+ ions of $AgNO_3$, and to a much lesser extent from AgNPs after oral administration for 28 days [248].

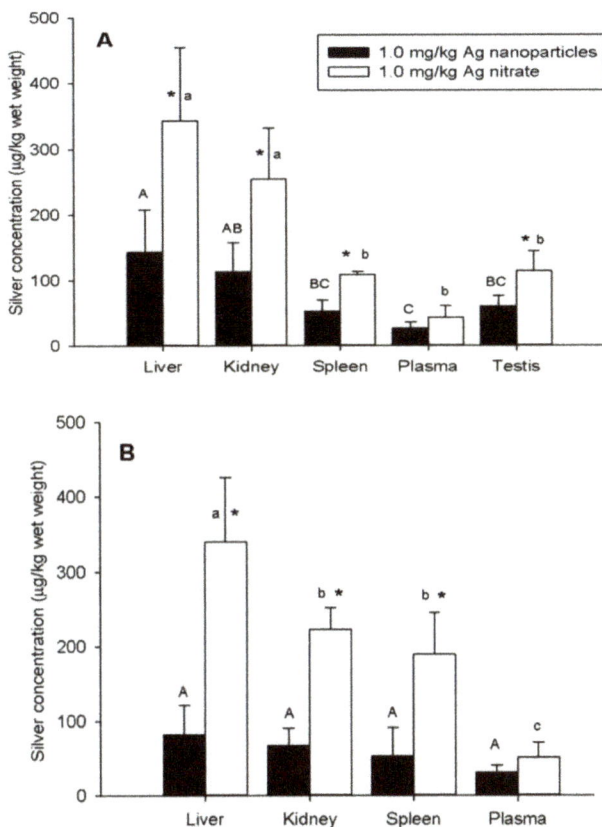

Figure 28. Silver concentrations in major organs and plasma of (**A**) male and (**B**) female rats. Values are presented as means ± SD, *n* = 5. The asterisk (*) indicates significant difference between AgNPs and AgNO$_3$ treatment groups at $p < 0.05$. Means with the same capital letters are not significantly different among AgNPs groups ($p < 0.05$) and same small letter are not statistically different among AgNO$_3$ groups by the Tukey test ($p < 0.05$). Reproduced from [238] with permission of Wiley.

Environmental airborne AgNP levels (5–289 mg/m^3) in occupational settings such as factories or laboratories are harmful to the lung tissues of humans due to the inhalation of nanoparticles [249]. Therefore, intratracheal instillation and inhalation in animal models provide relevant information for assessing toxicity arising from airborne nanoparticles [250]. Studies on intratracheally instilled AgNPs into mice have been carried out in recent years [57,251–254]. For instance, Anderson et al. studied the effects of size, surface coating, and dose on the persistence of silver in the lung of the rats through *i.t.* instillation of AgNPs for 1, 7, and 21 days. Silver retention in the lung was assessed at those mentioned timepoints. Four different AgNPs: 20 nm or 110 nm in size and coated with either citrate or PVP, at 0.5 mg/kg and 1.0 mg/kg doses were adopted in their study [251]. These dosages were chosen to simulate an environmental particle exposure (5–289 mg/m^3) in manufacturing industries [249]. CT-AgNPs was found to persist in the lung up to 21 days with retention higher than 90%, while PVP-AgNP showed lower retention in the lung, i.e., <30%. As a result, CT-AgNPs triggered lung macrophages for nanoparticle clearance. Larger nanoparticles were more rapidly cleared from the lung airways than smaller particles. Table 3 lists the recent in vivo animal studies relating cytoxicity of AgNPs through different administration routes.

Table 3. Recent in vivo studies showing biodistribution and toxicity of AgNPs in rodents through different administration routes.

Coating Type & Size of AgNPs	Model	Dosage & Exposure Time	Entry Route	Cytotoxic Effect	Ref.
CT and PVP; 10, 40 &100 nm	CD-1 Mice	10 mg/kg bw;24 h	i.v.	Biodistributed in spleen and liver followed by lung, kidney and brain. AgNPs (10 nm) are the most toxic nanoparticles	[229]
Carboxyl; 3 nm	KM Mice	11.3–13.3 mg/kg bw; 4 weeks	i.v.	Biodistributed mainly in spleen and liver, followed by kidney, lung, heart and testis	[230]
CT; 6.3 nm	SD rats	5 mg/kg bw; 24 h	i.v.	Biodistributed in the organs with decreasing Ag concentration, i.e., lung > spleen > liver > kidney > thymus > heart	[231]
PVP; 26.2 nm	SD rats	0.1 and 1 mg/kg bw per day for 4 and 12 weeks	Intranasal instillation	Dose- and time-dependent accumulation of both AgNPs and silver ion (AgNO$_3$) in liver, lung and brain	[232]
2 and 20 nm	Wistar rats	5 mg/kg bw; 1, 7and 28 days	i.v.	Time- and size-dependent accumulation of AgNPs in the liver, spleen, kidney and brain	[234]
PVP; 10–30 nm	SD rats	500 mg/kg bw; 1, 4, 7, 10 and 30 days	i.p.	AgNPs located mainly in the liver. A significant increase in caspase-3 in the liver of treated rats from day 1 to day 30	[235]
CT; 3–10 nm	SD rats	1mg/kg bw. and 10 mg/kg bw for 14 days	Intragastric	Neuron shrinkage, cytoplasmic or foot swelling of astrocytes and inflammation	[229]
CT; 10 nm	Wistar rats	0.2 mg/kg bw per day for 14 days	Gastro-intestinal	Enhanced lipid peroxidation and decreased concentrations of protein and non-protein –SH groups in myelin	[243]
CT, PVP; 20 and 110 nm	Black 6 mice	0.1, 1, 100 mg/kg bw per day for 3 days	Oral gavage	No toxicity and no significant tissue accumulation	[245]
CT; 10, 75 & 110 nm	SD rats	9, 18, 36 mg/kg bw for 13 weeks	Oral gavage	AgNPs predominantly deposited within cells of major organs	[246]
PVP; 28–43 nm	SD rats	0.5 and 1 mg/kg bw daily for 28 days	Oral admi-nistration	Biodistributed in liver, kidney, spleen and blood plasma.	[238]
CT, PVP; 20 and 110 nm	SD rats	0.5 and 1 mg/kg bw for 1, 7 and 21 days	i.t. instillation	CT-AgNPs persisted in the lung to 21 days with retention >90%, while PVP-AgNP had lower retention of less than 30%. CT-AgNPs triggered lung macrophages for clearance of AgNPs	[251]

In this review, we have discussed many cases relating the toxic effects of AgNPs in mammalian cells under in vitro and in vivo conditions. However, AgNPs would show no cytotoxicity toward mammalian cells and have high antibacterial efficacy in certain cases. AgNPs capped with appropriate polymers at certain concentrations do not exhibit cytotoxicity. Jena et al. reported that chitosan-capped AgNPs exhibit antibacterial activity against *P. aeruginosa*, *S. typhi*, and *S. aureus*, and they do not exhibit cytotoxic effects on mouse macrophage cell line (RAW264.7) at the bactericidal concentration [255]. Tam and coworkers demonstrated that CT-AgNPs promoted wound healing in mice through the modulation of fibrogenic cytokines, in addition to their antimicrobial properties [21,28]. Furthermore, AgNPs enhanced the differentiation of fibroblasts into myofibroblasts, thereby promoting wound contraction [28]. Pallavicini et al. synthesized AgNPs and coated with a biopolymer peptin acting both as a reductant and a stabilizing agent [29]. The as-synthesized AgNPs showed bactericidal activity against *E. coli* and *S. epidermidis*, and also facilitated normal human dermal fibroblasts (NHDFs) proliferation and wound healing on model cultures. Alarcon et al. functionalized AgNPs with thiol-LL37 cathelicidin peptide (LL37-SH), and then incorporated them into collagen hydrogels [26]. In in vitro experiments, the resulting hydrogel nanocomposites exhibited high antibacterial activity against *P. aeruginosa*, while showing no toxicity toward HUVEC and human corneal epithelial cell (HCEC). Finally, subcutaneous implantation of hydrogel nanocomposites into mice did not increase the secretion of pro-inflammatory cytokine IL-6 [26].

6. Conclusions

The article provides a comprehensive and state-of-the art review on the synthesis of AgNPs, their antibacterial activity, and cytotoxic effect in mammalian cells. The bactericidal activity of AgNPs has led to their widespread use in cosmetics, medical products, antimicrobial dressings, etc. However, the extensive use of AgNPs has raised significant public concerns over the safety and environment impacts of these products. In this respect, it deems necessary to study the interaction between AgNPs and biological cells in order to achieve a better understanding of the health risks arising from the use of nanoparticles. AgNPs have been shown to be toxic to numerous bacterial strains. The antibacterial activity against both gram-negative and gram-positive bacteria is found to be size-, shape-, dose-, charge- and time-dependent. Several studies have revealed that the membrane damage, mitochondrial dysfunction, ROS generation, oxidative stress and DNA damage are responsible for the cellular damage of treated bacterial cells. However, the exact bactericidal mechanisms of AgNPs remain unclear. The bactericidal effect may arise from either AgNPs themselves, released silver ions, or a combination of both.

Cell culture studies revealed that AgNPs are able to induce cytotoxicity in human cell lines including human bronchial epithelial cells, HUVECs, red blood cells, macrophages, liver cells, etc., particularly for those with sizes ≤10 nm. The cytotoxicity of AgNPs has been reported to be a dose-, size- and time-dependent manner. Similarly, there is much debate in the literature on whether AgNPs or silver ions exert toxic effects in mammalian cells. In vivo animal model tests have shown that AgNPs can pass the BBB of mice through the circulation system, thereby inducing neurotoxicity and neuronal death. Furthermore, AgNPs tend to accumulate in mice organs such as liver, spleen, kidney and brain following intravenous, intraperitoneal, and intratracheal routes of administration.

In spite of the widespred use of AgNPs in healthcare and cosmetic applications, several challenges remain to be overcome. The development of AgNPs and their nanocomposites, having both antimicrobial properties and no cytotoxic effects, is crucial for treating bacterial infections. AgNPs have been shown to be nontoxic in mouse fibroblasts, NHDFs and HCECs [21,26,28,29]. However, they are considered to be toxic to most human cell lines. More recently, biosynthesized AgNPs have been reported to be effective in killing multidrug-resistant bacteria [38,40,106,107]. Green synthesis of AgNPs generally suffers from some drawbacks, such as selection of appropriate plant extract, long reaction time, and difficulty in controlling the size and shape of AgNPs. The nature of biomolecules present in the plant extracts plays a crucial factor in the biosynthesis of AgNPs. In this respect,

Int. J. Mol. Sci. **2019**, *20*, 449

the quality of selected extract is considered to be of great importance [256]. Thus, it needs a systematic, reproducible, and scaled-up process for preparing green AgNPs with desirable antimicrobial properties and low toxicity. Finally, many in vitro studies in the literature employed several specific bioassays for evaluating cytotoxic effects of AgNPs. Standardization of bioassays is useful because it can provide a reliable and reproducible data for evaluating the mechanism responsible for cytotoxicity of AgNPs.

Author Contributions: S.C.T. conceived and designed the topic and content. C.L., Y.L. and S.C.T. wrote the article.

Funding: The authors would like to thank National Youth Science Foundation (China) for supporting this research under projects No. 21703096 and No. 51407087.

Conflicts of Interest: The authors declare no conflict of interest.

References

1. Tjong, S.C.; Chen, H. Nanocrystalline materials and coatings. *Mater. Sci. Eng. R Rep.* **2004**, *45*, 1–88. [CrossRef]
2. He, L.X.; Tjong, S.C. Aqueous graphene oxide-dispersed carbon nanotubes as inks for the scalable production of all-carbon transparent conductive films. *J. Mater. Chem. C* **2016**, *4*, 7043–7051. [CrossRef]
3. He, L.X.; Tjong, S.C. Nanostructured transparent conductive films: Fabrication, characterization and applications. *Mater. Sci. Eng. R Rep.* **2016**, *109*, 1–101. [CrossRef]
4. He, L.; Liao, C.; Tjong, S.C. Scalable fabrication of high-performance transparent conductors using graphene oxide-stabilized single-walled carbon nanotube inks. *Nanomaterials* **2018**, *8*, 224. [CrossRef] [PubMed]
5. Liu, C.; Shen, J.; Yeung, K.W.K.; Tjong, S.C. Development and antibacterial performance of novel polylactic acid-graphene oxide-silver nanoparticle hybrid nanocomposite mats prepared by electrospinning. *ACS Biomater. Sci. Eng.* **2017**, *3*, 471–486. [CrossRef]
6. Liu, C.; Chan, K.W.; Shen, J.; Wong, K.M.; Yeung, K.W.; Tjong, S.C. Melt-compounded polylactic acid composite hybrids with hydroxyapatite nanorods and silver nanoparticles: Biodegradation, antibacterial ability, bioactivity and cytotoxicity. *RSC Adv.* **2015**, *5*, 72288–72299. [CrossRef]
7. Liao, C.Z.; Wong, H.M.; Yeung, K.W.; Tjong, S.C. The development, fabrication and material characterization of polypropylene composites reinforced with carbon nanofiber and hydroxyapatite nanorod hybrid fillers. *Int. J. Nanomed.* **2014**, *9*, 1299–1310. [CrossRef]
8. Liao, C.Z.; Li, K.; Wong, H.M.; Tong, W.Y.; Yeung, K.W.K.; Tjong, S.C. Novel polypropylene biocomposites reinforced with carbon nanotubes and hydroxyapatite nanorods for bone replacements. *Mater. Sci. Eng. C* **2013**, *13*, 1380–1388. [CrossRef]
9. Tjong, S.C. *Nanocrystalline Materials: Their Synthesis-Structure-Property Relationships and Applications*, 2nd ed.; Elsevier: London, UK, 2013; ISBN 9780124077966.
10. Ng, C.T.; Baeg, G.H.; Yu, L.E.; Ong, C.N.; Bay, B.H. Biomedical applications of nanomaterials as therapeutics. *Curr. Med. Chem.* **2018**, *25*, 1409–1419. [CrossRef]
11. Kravets, V.; Almemar, Z.; Jiang, K.; Culhane, K.; Machado, R.; Hagen, G.; Kotko, A.; Dmytruk, I.; Spendier, K.; Pinchuk, A. Imaging of biological cells using luminescent silver nanoparticles. *Nanoscale Res. Lett.* **2016**, *11*, 30. [CrossRef]
12. Elahi, N.; Kamali, M.; Baghersad, M.H. Recent biomedical applications of gold nanoparticles: A review. *Talanta* **2018**, *184*, 537–556. [CrossRef] [PubMed]
13. Benyettou, F.; Rezgui, R.; Ravaux, F.; Jaber, T.; Blumer, K.; Jouiad, M.; Motte, L.; Olsen, J.C.; Platas-Iglesias, C.; Magzoub, M.; et al. Synthesis of silver nanoparticles for the dual delivery of doxorubicin and alendronate to cancer cells. *J. Mater. Chem. B* **2015**, *3*, 7237–7245. [CrossRef]
14. Kokura, S.; Handa, O.; Takagi, T.; Ishikawa, T.; Naito, Y.; Yoshikawa, T. Silver nanoparticles as a safe preservative for use in cosmetics. *Nanomedicine* **2010**, *6*, 570–574. [CrossRef] [PubMed]
15. Dakal, T.C.; Kumal, N.; Majumdal, R.; Yadav, V. Mechanistic basis of antimicrobial actions of silver nanoparticles. *Front. Microbiol.* **2016**, *7*, 1831. [CrossRef]
16. D'Agostino, A.; Taglietti, A.; Desando, R.; Bini, M.; Patrini, M.; Dacarro, G.; Cucca, L.; Pallavicini, P.; Grisoli, P. Bulk surfaces coated with triangular silver nanoplates: Antibacterial action based on silver release and photo-thermal effect. *Nanomaterials* **2017**, *7*, 7. [CrossRef]

17. Burdusel, A.C.; Gherasim, O.; Grumezescu, A.M.; Mogoanta, L.; Ficai, A.; Andronescu, E. Biomedical applications of silver nanoparticles: An up-to-date overview. *Nanomaterials* **2018**, *8*, 681. [CrossRef]

18. Orlowski, P.; Zmigrodzka, M.; Tomaszewska, E.; Ranoszek-Soliwoda, K.; Czupryn, M.; Antos-Bielska, M.; Szemraj, J.; Celichowski, G.; Grobelny, J.; Krzyzowska, M. Tannic acid-modified silver nanoparticles for wound healing: The importance of size. *Int. J. Nanomed.* **2018**, *13*, 991–1007. [CrossRef] [PubMed]

19. Konop, M.; Damps, T.; Misicka, A.; Rudnicka, L. Certain aspects of silver and silver nanoparticles in wound care: A minireview. *J. Nanomater.* **2016**, 7614753. [CrossRef]

20. Liu, C.; Shen, J.; Liao, C.Z.; Yeung, K.W.; Tjong, S.C. Novel electrospun polyvinylidene fluoride-graphene oxide-silver nanocomposite membranes with protein and bacterial antifouling characteristics. *Express Polym. Lett.* **2018**, *12*, 365–382. [CrossRef]

21. Tian, J.; Wong, K.K.; Ho, C.M.; Lok, C.N.; Yu, W.Y.; Che, C.M.; Cliu, J.F.; Tam, P.K. Topical delivery of silver nanoparticles promotes wound healing. *ChemMedChem* **2007**, *2*, 129–136. [CrossRef] [PubMed]

22. Keat, C.L.; Aziz, A.; Eid, A.M.; Elmarzugi, N.A. Biosynthesis of nanoparticles and silver nanoparticles. *Bioresour. Bioprocess.* **2015**, *2*, 47. [CrossRef]

23. Chernousova, S.; Epple, M. Silver as antibacterial agent: Ion, nanoparticle and metal. *Angew. Chem. Int. Ed.* **2013**, *52*, 1636–1653. [CrossRef] [PubMed]

24. Boca, S.C.; Potara, M.; Gabudean, A.; Juhem, A.; Baldeck, P.L.; Astilean, S. Chitosan-coated triangular silver nanoparticles as a novel class of biocompatible, highly effective photothermal transducers for in vitro cancer cell therapy. *Cancer Lett.* **2011**, *311*, 131–140. [CrossRef] [PubMed]

25. Boca, S.; Potara, M.; Simon, T.; Juhem, A.; Baldeck, P.; Astilean, S. Folic acid-conjugated, SERS-labeled silver nanotriangles for multimodal detection and targeted photothermal treatment on human ovarian cancer cells. *Mol. Pharm.* **2014**, *11*, 391–399. [CrossRef] [PubMed]

26. Alarcon, E.I.; Vulesevic, B.; Argawal, A.; Ross, A.; Bejjani, P.; Podrebarac, J.; Ravichandran, R.; Phopase, J.; Suuronen, E.J.; Griffith, M. Coloured cornea replacements with anti-infective properties: Expanding the safe use of silver nanoparticles in regenerative medicine. *Nanoscale* **2016**, *8*, 6484–6489. [CrossRef] [PubMed]

27. Rigo, C.; Ferroni, L.; Tocco, I.; Roman, M.; Munivrana, I.; Gardin, C.; Cairns, W.R.; Vindigni, V.; Azzena, B.; Barbante, C. Active silver nanoparticles for wound healing. *Int. J. Mol. Sci.* **2013**, *14*, 4817–4840. [CrossRef]

28. Liu, X.; Lee, P.Y.; Ho, C.M.; Lui, V.C.; Chen, Y.; Chi, C.M.; Tam, P.K.; Wong, K.Y. Silver nanoparticles mediate differential responses in keratinocytes and fibroblasts during skin wound healing. *ChemMedChem* **2010**, *5*, 468–475. [CrossRef]

29. Pallavicini, P.; Arciola, C.R.; Bertoglio, F.; Curtosi, S.; Dacarro, G.; D'Agostino, A.; Ferrari, F.; Merli, D.; Milanese, C.; Rossi, S.; et al. Silver nanoparticles synthesized and coated with pectin: An ideal compromise for anti-bacterial and anti-biofilm action combined with wound-healing properties. *J. Colloid Interface Sci.* **2017**, *498*, 271–281. [CrossRef]

30. Tang, B.; Li, J.; Hou, X.; Afrin, T.; Sun, L.; Wang, X. Colorful and antibacterial silk fiber from anisotropic silver nanoparticles. *Ind. Eng. Chem. Res.* **2013**, *52*, 4556–4563. [CrossRef]

31. Vukoje, I.; Lazic, V.; Vodnik, V.; Mitric, M.; Jokic, B.; Ahrenkiel, S.P.; Nedeljkovic, J.M.; Radetic, M. The influence of triangular silver nanoplates on antimicrobial activity and color of cotton fabrics pretreated with chitosan. *J. Mater. Sci.* **2014**, *49*, 4453–4460. [CrossRef]

32. Nowack, B.; Krug, H.F.; Height, M. 120 Years of nanosilver history: Implications for policy makers. *Environ. Sci. Technol.* **2011**, *45*, 1177–1183. [CrossRef] [PubMed]

33. Kedziora, A.; Speruda, M.; Krzyzewska, E.; Rybka, J.; LukowiK, A.; Bugla-Płoskonska, G. Similarities and differences between silver ions and silver in nanoforms as antibacterial agents. *Int. J. Mol. Sci.* **2018**, *19*, 444. [CrossRef] [PubMed]

34. Pallavicini, P.; Dacarro, G.; Taglietti, A. Self-assembled monolayers of silver nanoparticles: From intrinsic to switchable inorganic antibacterial surfaces. *Eur. J. Inorg. Chem.* **2018**, *2018*, 4846–4855. [CrossRef]

35. Taglietti, A.; Fernandez, Y.A.; Amato, E.; Cucca, L.; Dacarro, G.; Grisoli, P.; Necchi, V.; Pallavicini, P.; Pasotti, L.; Patrini, M. Antibacterial activity of glutathione-coated silver nanoparticles against gram positive and gram negative bacteria. *Langmuir* **2012**, *28*, 8140–8148. [CrossRef] [PubMed]

36. Frolich, E.E.; Frolich, E. Cytotoxicity of nanoparticles contained in food on intestinal cells and the gut microbiota. *Int. J. Mol. Sci.* **2016**, *17*, 509. [CrossRef]

37. Slavin, Y.N.; Asnis, J.; Hafeli, U.O.; Bach, H. Metal nanoparticles: Understanding the mechanisms behind antibacterial activity. *J. Nanobiotechnol.* **2017**, *15*, 65. [CrossRef]

38. Gurunathan, S.; Choi, Y.Z.; Kim, J.H. Antibacterial efficacy of silver nanoparticles on endometritis caused by Prevotella melaninogenica and Arcanobacterum pyogenes in dairy cattle. *Int. J. Mol. Sci.* **2018**, *19*, 1210. [CrossRef]

39. Baptista, P.V.; McCusker, M.P.; Carvalho, A.; Ferreira, D.A.; Mohan, N.; Martins, M.; Fernandes, A.R. Nano-strategies to fight multidrug resistant bacteria—"A battle of the titans". *Front. Microbiol.* **2018**, *9*, 1441. [CrossRef]

40. Katva, S.; Das, S.; Moti, H.S.; Jyoti, A.; Kaushik, S. Antibacterial synergy of silver nanoparticles with gentamicin and chloramphenicol against Enterococcus faecalis. *Pharmacogn. Mag.* **2018**, *13*, S828–S833. [CrossRef]

41. Agnihotri, S.; Mukherji, S.; Mukherji, S. Size-controlled silver nanoparticles synthesized over the range 5–100 nm using the same protocol and their antibacterial efficacy. *RSC Adv.* **2014**, *4*, 3974–3983. [CrossRef]

42. Hong, X.; Wen, J.; Xiong, X.; Hu, Y. Shape effect on the antibacterial activity of silver nanoparticles synthesized via a microwave-assisted method. *Environ. Sci. Pollut. Res.* **2016**, *23*, 4489–4497. [CrossRef] [PubMed]

43. Raza, M.A.; Kanwal, Z.; Rauf, A.; Sabri, A.N.; Riaz, S.; Naseem, S. Size- and shape-dependent antibacterial studies of silver nanoparticles synthesized by wet chemical routes. *Nanomaterials* **2016**, *6*, 74. [CrossRef] [PubMed]

44. Xia, X.; Zeng, J.; Moran, C.H.; Xia, Y. Recent developments in shape-controlled synthesis of silver nanocrystals. *J. Phys. Chem. C* **2012**, *116*, 21647–21656. [CrossRef] [PubMed]

45. Hosseinidoust, Z.; Basenet, M.; van de Ven, T.G.; Tufenkji, N. One-pot green synthesis of anisotropic silver nanoparticles. *Environ. Sci. Nano* **2016**, *3*, 1259–1264. [CrossRef]

46. Mitrano, D.M.; Rimmele, E.; Wichser, A.; Erni, R.; Height, M.; Nowack, B. Presence of nanoparticles in wash water from conventional silver and nano-silver textiles. *ACS Nano* **2014**, *8*, 7208–7219. [CrossRef] [PubMed]

47. Syafiuddin, A.; Salmiati, A.; Hadibarata, T.; Kueh, A.B.; Salim, M.R.; Zaini, M.A. Silver nanoparticles in the water environment in Malaysia: Inspection, characterization, removal, modeling, and future perspective. *Sci. Rep.* **2018**, *8*, 986. [CrossRef] [PubMed]

48. Miyayama, T.; Arai, Y.; Hirano, S. Health Effects of Silver Nanoparticles and Silver Ions. In *Biological Effects of Fibrous and Particulate Substances*; Otsuki, T., Yoshioka, Y., Holian, Y., Eds.; Springer: Tokyo, Japan, 2016; pp. 137–147, ISBN 978-4-431-55731-9.

49. Carrola, J.; Bastos, V.; Jarak, I.; Oliveira-Silva, R.; Malheiro, E.; Daniel-da-Silva, A.L.; Oliveira, H.; Santos, C.; Gil, A.M.; Duarte, I.F. Metabolomics of silver nanoparticles toxicity in HaCaT cells: Structure-activity relationships and role of ionic silver and oxidative stress. *Nanotoxicology* **2016**, *10*, 1105–1117. [CrossRef] [PubMed]

50. Kim, M.J.; Shin, S. Toxic effects of silver nanoparticles and nanowires on erythrocyte rheology. *Food Chem. Toxicol.* **2014**, *67*, 80–86. [CrossRef]

51. Sabella, S.; Carney, R.P.; Brunetti, V.; Malvindi, M.A.; Al-Juffali, N.; Vecchio, G.; Janes, S.M.; Bakr, O.M.; Cingolani, R.; Stellacci, F.; et al. A general mechanism for intracellular toxicity of metal-containing nanoparticles. *Nanoscale* **2014**, *6*, 7052–7061. [CrossRef]

52. Li, L.; Cui, J.; Liu, Z.; Zhou, X.; Li, Z.; Yu, Y.; Jia, Y.; Zuo, D.; Wu, Y. Silver nanoparticles induce SH-SY5Y cell apoptosis via endoplasmic reticulum- and mitochondrial pathways that lengthen endoplasmic reticulum–mitochondria contact sites and alter inositol-3-phosphate receptor function. *Toxicol. Lett.* **2018**, *285*, 156–167. [CrossRef]

53. Sahu, S.C.; Zheng, J.; Graham, L.; Chen, L.; Ihrie, J.; Yourick, J.J.; Sprando, R.L. Comparative cytotoxicity of nanosilver in human liver HepG2 and colon Caco2 cells in culture. *J. Appl. Toxicol.* **2014**, *34*, 1155–1166. [CrossRef] [PubMed]

54. Shi, J.; Sun, X.; Zou, X.; Li, J.; Liao, Y.; Du, M. Endothelial cell injury and dysfunction induced by silver nanoparticles through oxidative stress via IKK/NF-κB pathways. *Biomaterials* **2014**, *35*, 6657–6666. [CrossRef] [PubMed]

55. Jiang, X.; Lu, C.; Tang, M.; Yang, Z.; Jia, W.; Ma, Y.; Jia, P.; Pei, D.; Wang, H. Nanotoxicity of silver nanoparticles on HEK293T cells: A combined study using biomechanical and biological techniques. *ACS Omega* **2018**, *3*, 6770–6778. [CrossRef] [PubMed]

56. Vidanapathirana, A.K.; Thompson, L.C.; Herco, M.; Odom, J.; Sumner, S.J.; Fennell, T.R.; Brown, J.M.; Wingard, C.J. Acute intravenous exposure to silver nanoparticles during pregnancy induces particle size and vehicle dependent changes in vascular tissue contractility in Sprague Dawley rats. *Reprod. Toxicol.* **2018**, *75*, 10–22. [CrossRef] [PubMed]

57. Alessandrini, F.; Vennemann, A.; Gschwendtner, S.; Neumann, A.U.; Rothballer, M.; Seher, T.; Wimmer, M.; Kublik, S.; Traidl-Hoffmann, C.; Schloter, M.; et al. Pro-inflammatory versus immunomodulatory effects of silver nanoparticles in the lung: The critical role of dose, size and surface modification. *Nanomaterials* **2017**, *7*, 300. [CrossRef] [PubMed]

58. Trickler, W.; Lantz, S.M.; Murdock, R.C.; Schrand, A.M.; Robinson, B.L.; Newport, G.D.; Schalager, J.J.; Oldenburg, S.J.; Paule, M.G.; Slikker, W., Jr.; et al. Silver nanoparticle induced blood-brain barrier inflammation and increased permeability in primary brain microvessel endothelial cells. *Toxicol. Sci.* **2010**, *118*, 160–170. [CrossRef] [PubMed]

59. Sokolowska, P.; Bialkowska, K.; Siatkowskaa, M.; Rosowski, M.; Kucinska, M.; Komorowski, P.; Makowski, K.; Walkowiak, B. Human brain endothelial barrier cells are distinctly less vulnerable to silver nanoparticles toxicity than human blood vessel cells—A cell-specific mechanism of the brain barrier? *Nanomed. Nanotechnol.* **2017**, *13*, 2127–2130. [CrossRef]

60. Butler, K.S.; Peeler, D.J.; Casey, B.J.; Dair, B.J.; Elespuru, R.K. Silver nanoparticles: Correlating nanoparticle size and cellular uptake with genotoxicity. *Mutagenesis* **2015**, *30*, 577–591. [CrossRef]

61. Jeong, Y.; Lim, D.W.; Choi, J. Assessment of size-dependent antimicrobial and cytotoxic properties of silver nanoparticles. *Adv. Mater. Sci. Eng.* **2014**, 763807. [CrossRef]

62. Dayem, A.A.; Hossain, M.K.; Lee, S.B.; Kim, K.; Saha, S.K.; Yang, G.M.; Choi, H.Y.; Cho, S.G. The role of reactive oxygen species (ROS) in the biological activities of metallic nanoparticles. *Int. J. Mol. Sci.* **2017**, *18*, 120. [CrossRef]

63. Iravani, S.; Korbekandi, H.; Mirmohammadi, S.V.; Zolfaghari, B. Synthesis of silver nanoparticles: Chemical, physical and biological methods. *Res. Pharm. Sci.* **2014**, *9*, 385–406. [PubMed]

64. Jung, J.; Oh, H.; Noh, H.; Ji, J.; Kim, S. Metal nanoparticle generation using a small ceramic heater with a local heating area. *J. Aerosol. Sci.* **2006**, *37*, 1662–1670. [CrossRef]

65. Tsuji, T.; Iryo, K.; Watanabe, N.; Tsuji, M. Preparation of silver nanoparticles by laser ablation in solution: Influence of laser wavelength on particle size. *Appl. Surf. Sci.* **2002**, *202*, 80–85. [CrossRef]

66. Perito, B.; Giorgetti, E.; Marsili, P.; Muniz-Miranda, M. Antibacterial activity of silver nanoparticles obtained by pulsed laser ablation in pure water and in chloride solution. *Beilstein J. Nanotechnol.* **2016**, *7*, 465–473. [CrossRef] [PubMed]

67. Sportelli, M.C.; Izzi, M.; Volpe, A.; Clemente, M.; Picca, R.A.; Ancona, A.; Lugara, P.M.; Palazzo, G.; Cioffi, N. The pros and cons of the use of laser ablation synthesis for the production of silver nano-antimicrobials. *Antibiotics* **2018**, *7*, 67. [CrossRef] [PubMed]

68. Sato-Berru, R.; Redon, R.; Vazquez-Olmos, A.; Saniger, J.M. Silver nanoparticles synthesized by direct photoreduction of metal salts. Application in surface-enhanced Raman spectroscopy. *J. Raman Spectrosc.* **2009**, *40*, 376–380. [CrossRef]

69. Krol-Gracz, A.; Michalak, E.; Nowak, P.; Dyonizy, A. Photo-induced chemical reduction of silver bromide to silver nanoparticles. *Cent. Eur. J. Chem.* **2011**, *9*, 982–989. [CrossRef]

70. Beyene, H.D.; Werkneh, A.A.; Bezabha, H.K.; Ambaye, T.G. Synthesis paradigm and applications of silver nanoparticles (AgNPs), a review. *Sustain. Mater. Technol.* **2017**, *13*, 18–23. [CrossRef]

71. Pacioni, N.L.; Borsarelli, C.D.; Rey, V.; Veglia, A.V. Synthetic Routes for the Preparation of Silver Nanoparticles: A Mechanistic Perspective. In *Silver nanoparticle Applications*; Alarcon, E.I., Griffith, M., Udekwu, K.I., Eds.; Springer: Cham, Switzerland, 2015; pp. 13–46, ISBN 978-3-319-11261-9.

72. Helmlinger, J.; Sengstock, C.; Groß-Heitfeld, C.; Mayer, C.; Schildhaue, T.A.; Koller, M.; Epple, M. Silver nanoparticles with different size and shape: Equal cytotoxicity, but different antibacterial effects. *RSC Adv.* **2016**, *6*, 18490–18501. [CrossRef]

73. Ranoszek-Soliwoda, K.; Tomaszewska, E.; Socha, E.; Krzyczmonik, P.; Ignaczak, A.; Orlowski, P.; Krzyzowska, M.; Celichowski, G.; Grobeln, J. The role of tannic acid and sodium citrate in the synthesis of silver nanoparticles. *J. Nanopart. Res.* **2017**, *19*, 273. [CrossRef]

74. Kytsya, A.; Bazylyak, L.; Hrynda, Y.; Horechyy, A.; Medvedevdkikh, Y. The kinetic rate law for the autocatalytic growth of citrate-stabilized silver nanoparticles. *Int. J. Chem. Kinet.* **2015**, *47*, 351–360. [CrossRef]

75. Malassis, L.; Dreyfus, R.; Murphy, R.J.; Hough, L.A.; Donnio, B.; Murray, C.B. One-step green synthesis of gold and silver nanoparticles with ascorbic acid and their versatile surface post-functionalization. *RSC Adv.* **2016**, *6*, 33092–33100. [CrossRef]

76. Gharibshahi, L.; Saion, E.; Gharibshahi, E.; Shaari, A.H.; Matori, K.A. Influence of poly(vinylpyrrolidone) concentration on properties of silver nanoparticles manufactured by modified thermal treatment method. *PLoS ONE* **2017**, *12*, e0186094. [CrossRef] [PubMed]

77. Kvitek, L.; Panacek, A.; Soukupova, J.; Kolar, M.; Vecerova, R.; Prucek, R.; Holecova, M.; Zboril, R. Effect of surfactants and polymers on stability and antibacterial activity of silver nanoparticles (NPs). *J. Phys. Chem. C* **2008**, *112*, 5825–5834. [CrossRef]

78. Sun, Y. Controlled synthesis of colloidal silver nanoparticles in organic solutions: Empirical rules for nucleation engineering. *Chem. Soc. Rev.* **2013**, *42*, 2497–2511. [CrossRef] [PubMed]

79. Fievet, F.; Ammar-Merah, S.; Brayner, R.; Chao, F.; Giraud, M.; Mammeri, F.; Peron, J.; Piquemal, J.Y.; Sicard, L.; Viau, G. The polyol process: A unique method for easy access to metal nanoparticles with tailored sizes, shapes and compositions. *Chem. Soc. Rev.* **2018**, *47*, 5187–5233. [CrossRef] [PubMed]

80. Dang, T.M.; Le, T.T.; Fribourg-Blanc, E.; Dang, M.C. Influence of surfactant on the preparation of silver nanoparticles by polyol method. *Adv. Nat. Sci. Nanosci. Nanotechnol.* **2012**, *3*, 035004. [CrossRef]

81. Nguyen, N.T.; Nguyen, B.H.; Ba, D.T.; Pham, D.G.; Kahi, T.V.; Nguyen, L.T.; Tran, L.D. Microwave-assisted synthesis of silver nanoparticles using chitosan: A novel approach. *Mater. Manuf. Process.* **2014**, *29*, 418–421. [CrossRef]

82. Ajitha, B.; Ashok Kumar Reddy, Y.; Sreedhara Reddy, P. Enhanced antimicrobial activity of silver nanoparticles with controlled particle size by pH variation. *Powder Technol.* **2015**, *269*, 110–117. [CrossRef]

83. Zhang, Y.; Newton, B.; Lewis, E.; Fu, P.P.; Kafoury, R.; Ray, P.C.; Yu, H. Cytotoxicity of organic surface coating agents used for nanoparticles synthesis and stability. *Toxicol. In Vitro* **2015**, *29*, 762–768. [CrossRef]

84. Alam, M.S.; Siddiq, A.M.; Balamurugan, S.; Mandal, A.B. Role of cloud point of the capping agent (nonionic surfactant, Triton X-100) on the synthesis of silver nanoparticles. *J. Disper. Sci. Technol.* **2016**, *37*, 853–859. [CrossRef]

85. Sosa, Y.D.; Laberelo, M.; Trevini, M.E.; Saade, H.; Lopez, R.G. High-yield synthesis of silver nanoparticles by precipitation in a high-aqueous phase content reverse microemulsion. *J. Nanomater.* **2010**, *2010*, 392572. [CrossRef]

86. Rivera-Rangela, R.D.; Gonzalez-Muñoza, M.P.; Avila-Rodriguez, M.; Razo-Lazcano, T.A.; Solans, C. Green synthesis of silver nanoparticles in oil-in-water microemulsion and nano-emulsion using geranium leaf aqueous extract as a reducing agent. *Colloids Surf. A* **2018**, *536*, 60–67. [CrossRef]

87. Das, M.; Patowary, K.; Vidya, R.; Malipeddi, H. Microemulsion synthesis of silver nanoparticles using biosurfactant extracted from Pseudomonas aeruginosa MKVIT3 strain and comparison of their antimicrobial and cytotoxic activities. *IET Nanobiotechnol.* **2016**, *10*, 411–418. [CrossRef] [PubMed]

88. Baig, N.; Nadagouda, M.N.; Polshettiwar, V. Sustainable Synthesis of Metal Oxide Nanostructures. In *Sustainable Inorganic Chemistry*; Atwood, D.A., Ed.; Wiley: Hoboken, NJ, USA, 2016; pp. 483–494, ISBN 978-1-118-70342-7.

89. Lu, Y.; Zhang, C.; Hao, R.; Zhang, D.; Fu, Y.; Moeendarbari, S. Morphological transformations of silver nanoparticles in seedless photochemical synthesis. *Mater. Res. Exp.* **2016**, *3*, 055014. [CrossRef]

90. Xu, G.N.; Qiao, X.L.; Qiu, X.L.; Chen, J.G. Preparation and characterization of stable monodisperse silver nanoparticles via photoreduction. *Colloids Surf. A* **2008**, *320*, 222–226. [CrossRef]

91. Omrani, A.A.; Taghavinia, N. Photo-induced growth of silver nanoparticles using UV sensitivity of cellulose fibers. *Appl. Surf. Sci.* **2012**, *258*, 2373–2377. [CrossRef]

92. Ashraf, J.M.; Ansari, M.A.; Khan, H.M.; Alzohairy, M.A.; Choi, I. Green synthesis of silver nanoparticles and characterization of their inhibitory effects on AGEs formation using biophysical techniques. *Sci. Rep.* **2016**, *6*, 20414. [CrossRef]

93. Siddiqi, K.S.; Husen, A.; Rao, R.A. A review on biosynthesis of silver nanoparticles and their biocidal properties. *J. Nanobiotechnol.* **2018**, *16*, 14. [CrossRef]

94. Vasquez, R.D.; Apostol, J.G.; Leon, J.D.; Mariano, J.D.; Mirhan, C.N.; Pangan, S.S.; Reyes, A.G.; Zamora, E.T. Polysaccharide-mediated green synthesis of silver nanoparticles from Sargassum siliquosum J.G. Agardh: Assessment of toxicity and hepatoprotective activity. *OpenNano* **2016**, *1*, 16–24. [CrossRef]

95. Logaranjan, K.; Raiza, A.J.; Gopinath, S.C.; Chen, Y.; Pandia, K. Shape- and size-controlled synthesis of silver nanoparticles using aloe vera plant extract and their antimicrobial activity. *Nanoscale Res. Lett.* **2016**, *11*, 520. [CrossRef] [PubMed]

96. Verma, A.; Mehata, M.S. Controllable synthesis of silver nanoparticles using Neem leaves and their antimicrobial activity. *J. Radiat. Res. Appl. Sci.* **2016**, *9*, 109–115. [CrossRef]

97. Singh, R.; Rawat, D.; Isha. Microwave-assisted synthesis of silver nanoparticles from Origanum majorana and citrus sinensis leaf and their antibacterial activity: A green chemistry approach. *Bioresour. Bioprocess.* **2016**, *3*, 14. [CrossRef]

98. Venkatesan, J.; Kim, S.K.; Shim, M.S. Antimicrobial, antioxidant, and anticancer activities of biosynthesized silver nanoparticles using marine algae Ecklonia cava. *Nanomaterials* **2016**, *6*, 235. [CrossRef] [PubMed]

99. Sanyasi, S.; Majhi, R.K.; Kumar, S.; Mishra, M.; Ghosh, A.; Suar, M.; Satyam, P.V.; Mohapatra, H.; Goswami, C.; Goswami, L. Polysaccharide-capped silver nanoparticles inhibit biofilm formation and eliminate multidrug-resistant bacteria by disrupting bacterial cytoskeleton with reduced cytotoxicity towards mammalian cells. *Sci. Rep.* **2016**, *6*, 24929. [CrossRef]

100. Xia, Q.H.; Ma, Y.J.; Wang, J.W. Biosynthesis of silver nanoparticles using Taxus yunnanensis Callus and their antibacterial activity and cytotoxicity in human cancer cells. *Nanomaterials* **2016**, *6*, 160. [CrossRef]

101. Skandalis, N.; Dimopoulou, A.; Georgopoulou, A.; Gallios, N.; Papadopoulos, D.; Tsipas, D.; Theologidis, I.; Michailidis, N.; Chatzinikolaidou, M. The effect of silver nanoparticles size, produced using plant extract from Arbutus unedo, on their antibacterial efficacy. *Nanomaterials* **2017**, *7*, 178. [CrossRef]

102. Mohammed, A.E.; Al-Qahtani, A.; al-Mutairi, A.; Al-Shamri, B.; Aabed, K. Antibacterial and cytotoxic potential of biosynthesized silver nanoparticles by some plant extracts. *Nanomaterials* **2018**, *8*, 382. [CrossRef]

103. Chain, R.B.; Monzo-Cabrera, J.; Solyom, K. Microwave-Assisted Plant Extraction Processes. In *Alternative Energy Sources for Green Chemistry*; Stefanidis, G., Stankiewicz, A., Eds.; RSC Publishing: London, UK, 2016; Chapter 2; pp. 34–63, ISBN 978-1-78262-140-9.

104. Peng, H.; Yang, A.; Xiong, J. Green, microwave assisted synthesis of silver nanoparticles using bamboo hemicelluloses and glucose in an aqueous medium. *Carbohydr. Polym.* **2013**, *91*, 348–355. [CrossRef]

105. Kahrilas, G.A.; Wally, L.M.; Fredrick, S.J.; Hiskey, M.; Prieto, A.L.; Owens, J.E. Microwave-assisted green synthesis of silver nanoparticles using orange peel extract. *ACS Sustain. Chem. Eng.* **2014**, *2*, 367–376. [CrossRef]

106. Ali, K.; Ahmed, B.; Dwivedi, S.; Saquib, Q.; Al-Kedhairy, A.A.; Musarrat, J. Microwave accelerated green synthesis of stable silver nanoparticles with Eucalyptus globulus leaf extract and their antibacterial and antibiofilm activity on clinical isolates. *PLoS ONE* **2015**, *10*, e0131178. [CrossRef] [PubMed]

107. Yuan, Y.G.; Peng, Q.L.; Gurunathan, S. Effects of silver nanoparticles on multiple drug-resistant strains of staphylococcus aureus and pseudomonas aeruginosa from mastitis-infected goats: An alternative approach for antimicrobial therapy. *Int. J. Mol. Sci.* **2017**, *18*, 569. [CrossRef] [PubMed]

108. Fatimah, I. Green synthesis of silver nanoparticles using extract of Parkia speciosa Hassk pods assisted by microwave irradiation. *J. Adv. Res.* **2016**, *7*, 961–969. [CrossRef] [PubMed]

109. Velusamy, P.; Su, C.H.; Kumar, G.V.; Adhikary, S.; Pandian, K.; Gopinath, S.C.; Chen, Y.; Anbu, P. Biopolymers regulate silver nanoparticle under microwave irradiation for effective antibacterial and antibiofilm activities. *PLoS ONE* **2016**, *11*, e0157612. [CrossRef] [PubMed]

110. Kumar, S.V.; Bafana, A.P.; Pawar, P.; Rahman, A.; Dahoumane, S.; Jeffreys, C.S. High conversion synthesis of <10 nm starch-stabilized silver nanoparticles using microwave technology. *Sci. Rep.* **2018**, *8*, 5106. [CrossRef] [PubMed]

111. Meng, Y.Z.; Tjong, S.C.; Hay, A.S.; Wang, S.J. Synthesis and proton conductivities of phosphonic acid containing poly-(arylene ether)s. *J. Polym. Sci. A Polym. Chem.* **2001**, *39*, 3218–3226. [CrossRef]

112. Meng, Y.Z.; Hay, A.S.; Jian, X.G.; Tjong, S.C. Synthesis and properties of poly (aryl ether sulfone)s containing the phthalazinone moiety. *J. Appl. Polym. Sci.* **1998**, *68*, 137–143. [CrossRef]

113. Kelly, F.M.; Johnston, J.H. Colored and functional silver nanoparticle–wool fiber composites. *ACS Appl. Mater. Interfaces* **2011**, *3*, 1083–1092. [CrossRef]

114. Tjong, S.C.; Meng, Y.Z. Morphology and mechanical characteristics of compatibilized polyamide 6-liquid crystalline polymer composites. *Polymer* **1997**, *38*, 4609–4615. [CrossRef]

115. Meng, Y.Z.; Tjong, S.C. Rheology and morphology of compatibilized polyamide 6 blends containing liquid crystalline copolyesters. *Polymer* **1998**, *39*, 99–107. [CrossRef]

116. Liang, J.Z.; Li, R.K.Y.; Tjong, S.C. Tensile properties and morphology of PP/EPDM/glass bead ternary composites. *Polym. Compos.* **1999**, *20*, 413–422. [CrossRef]

117. Li, R.K.Y.; Liang, J.Z.; Tjong, S.C. Morphology and dynamic mechanical properties of glass beads filled 1139 low density polyethylene composites. *J. Mater. Process. Technol.* **1998**, *79*, 59–65. [CrossRef]

118. Tjong, S.C.; Liu, S.L.; Li, R.K.Y. Mechanical properties of injection moulded blends of polypropylene with thermotropic liquid crystalline polymer. *J. Mater. Sci.* **1996**, *31*, 479–484. [CrossRef]

119. He, L.; Tjong, S.C. Facile synthesis of silver-decorated reduced graphene oxide as a hybrid filler material for electrically conductive polymer composites. *RSC Adv.* **2015**, *5*, 15070–15076. [CrossRef]

120. Tjong, S.C.; Meng, Y.Z. Preparation and characterization of melt-compounded polyethylene/vermiculite nanocomposites. *J. Polym. Sci. B Polym. Phys.* **2003**, *41*, 1476–1484. [CrossRef]

121. Liu, C.; Chan, K.W.; Shen, J.; Liao, C.Z.; Yeung, K.W.K.; Tjong, S.C. Polyetheretherketone hybrid composites with bioactive nanohydroxyapatite and multiwalled carbon nanotube fillers. *Polymers* **2016**, *8*, 425. [CrossRef]

122. Li, K.; Cui, S.; Hua, J.; Zhou, Y.; Liu, Y. Crosslinked pectin nanofibers with well-dispersed Ag nanoparticles: Preparation and characterization. *Carbohydr. Polym.* **2018**, *199*, 68–74. [CrossRef] [PubMed]

123. Stauffer, S.R.; Peppast, N.A. Poly(vinyl alcohol) hydrogels prepared by freezing-thawing cyclic processing. *Polymer* **1992**, *33*, 3932–3936. [CrossRef]

124. Hassan, C.M.; Peppas, N.A. Structure and morphology of freeze/thawed PVA hydrogels. *Macromolecules* **2000**, *33*, 2472–2479. [CrossRef]

125. Figueroa-Pizano, M.D.; Velaz, I.; Penas, F.J.; Zavala-Rivera, P.; Rosas-Durazo, A.J.; Maldonado-Arce, A.D.; Martinez-Barbos, M.E. Effect of freeze-thawing conditions for preparation of chitosan-poly (vinyl alcohol) hydrogels and drug release studies. *Carbohydr. Polym.* **2018**, *195*, 476–485. [CrossRef]

126. Loo, C.Y.; Young, P.M.; Lee, W.H.; Cavaliere, R.; Witchurch, C.B.; Rohanizadeh, R. Non-cytotoxic silver nanoparticle-polyvinyl alcohol hydrogels with anti-biofilm activity: Designed as coatings for endotracheal tube materials. *Biofouling* **2014**, *30*, 773–788. [CrossRef] [PubMed]

127. Agnihotri, S.; Mukherji, S. Antimicrobial chitosan–PVA hydrogel as a nanoreactor and immobilizing matrix for silver nanoparticles. *Appl. Nanosci.* **2012**, *2*, 179–188. [CrossRef]

128. El-Shishtawy, R.M.; Asiri, A.M.; Abdelwahed, N.A.; Al-Otaibi, M.M. In situ production of silver nanoparticle on cotton fabric and its antimicrobial evaluation. *Cellulose* **2011**, *18*, 75–82. [CrossRef]

129. Babaahmadi, V.; Montazer, A. A new route to synthesis silver nanoparticles on polyamide fabric using stannous chloride. *J. Text. Inst.* **2015**, *106*, 970–977. [CrossRef]

130. Shinde, V.V.; Jadhav, P.R.; Kim, J.H.; Patil, P.S. One-step synthesis and characterization of anisotropic silver nanoparticles: Application for enhanced antibacterial activity of natural fabric. *J. Mater. Sci.* **2013**, *48*, 8393–8401. [CrossRef]

131. Babu, K.F.; Dhandapani, P.; Maruthamuthu, S.; Kulandainathan, M.A. One pot synthesis of polypyrrole silver nanocomposite on cotton fabrics for multifunctional property. *Carbohydr. Polym.* **2012**, *90*, 1557–1563. [CrossRef] [PubMed]

132. Kim, T.S.; Cha, J.R.; Gong, M.S. Investigation of the antimicrobial and wound healing properties of silver nanoparticle-loaded cotton prepared using silver carbamate. *Text. Res. J.* **2018**, *88*, 766–776. [CrossRef]

133. Montazer, M.; Shamei, A.; Alimohammadi, F. Synthesis of nanosilver on polyamide fabric using silver/ammonia complex. *Mater. Sci. Eng. C* **2014**, *38*, 170–176. [CrossRef] [PubMed]

134. Montazer, M.; Shamei, A.; Alimohammadi, F. Synthesizing and stabilizing silver nanoparticles on polyamide fabric using silver-ammonia/PVP/UVC. *Prog. Org. Coat.* **2012**, *75*, 379–385. [CrossRef]

135. Breitwiesera, D.; Moghaddama, M.M.; Spirkc, S.; Baghbanzadeha, M.; Pivecd, T.; Fasl, H.; Ribitscha, V.; Kappe, C.O. In situ preparation of silver nanocomposites on cellulosic fibers–Microwave vs. conventional heating. *Carbohydr. Polym.* **2013**, *94*, 677–686. [CrossRef]

136. Choudhary, U.; Dey, E.; Bhattacharyya, R.; Ghosh, S.K. A brief review on plasma treatment of textile materials. *Adv. Res. Text. Eng.* **2018**, *3*, 1019.

137. Zille, A.; Fernandes, M.M.; Francesko, A.; Tzanov, T.; Fernandes, M.; Oliveira, F.R.; Almeida, L.; Amorim, T.; Carneiro, N.; Esteves, M.F.; et al. Size and aging effects on antimicrobial efficiency of silver nanoparticles coated on polyamide fabrics activated by atmospheric DBD plasma. *ACS Appl. Mater. Interfaces* **2015**, *7*, 13731–13744. [CrossRef] [PubMed]

138. Ilic, V.; Saponjic, Z.; Vodnik, V.; Lazovic, S.; Dimitrijevic, S.; Jovancic, P.; Nedeljkovic, J.M.; Radetic, M. Bactericidal efficiency of silver nanoparticles deposited onto radio frequency plasma pretreated polyester fabrics. *Ind. Eng. Chem. Res.* **2010**, *49*, 7287–7293. [CrossRef]

139. Sondi, I.; Salopek-Sondi, B. Silver nanoparticles as antimicrobial agent: A case study on *E. coli* as a model for Gram negative bacteria. *J. Colloid Interf. Sci.* **2004**, *275*, 177–182. [CrossRef] [PubMed]

140. Gahlawat, G.; Shikha, S.; Chaddha, B.S.; Chaudhuri, S.R.; Mayilraj, S.; Choudhury, A.R. Microbial glycolipoprotein-capped silver nanoparticles as emerging antibacterial agents against cholera. *Microb. Cell Fact.* **2016**, *15*, 25. [CrossRef] [PubMed]

141. Lok, C.N.; Ho, C.M.; Chen, R.; He, Q.; Yu, W.Y.; Sun, H.; Tam, P.K.; Chiu, J.F.; Che, C.M. Proteomic analysis of the mode of antibacterial action of silver nanoparticles. *J. Proteome Res.* **2006**, *5*, 916–924. [CrossRef] [PubMed]

142. Barros, C.H.; Fulaz, S.; Stanisic, D.; Tasic, L. Biogenic nanosilver against multidrug-resistant bacteria (MRDB). *Antibiotics* **2018**, *7*, 69. [CrossRef]

143. Morones, J.R.; Elechiguerra1, J.L.; Camacho, A.; Holt, K.; Kouri, J.B.; Ramirez, J.T.; Yacaman, M.J. The bactericidal effect of silver nanoparticles. *Nanotechnology* **2005**, *16*, 2346–2353. [CrossRef]

144. Tjong, S.C.; Hoffman, R.W.; Yeager, E.B. Electron and ion spectroscopic iron-chromium alloys. *J. Electrochem. Soc.* **1982**, *129*, 1662–1668. [CrossRef]

145. Tjong, S.C.; Yeager, E. ESCA and SIMS studies of the passive film on iron. *J. Electrochem. Soc.* **1981**, *128*, 2251–2254. [CrossRef]

146. Behra, R.; Sigg, L.; Clift, M.J.; Herzog, F.; Minghetti, M.; Johnston, B.; Petri-Fink, A.; Rothen-Rutishauser, B. Bioavailability of silver nanoparticles and ions: From a chemical and biochemical perspective. *J. R. Soc. Interface* **2013**, *10*, 20130396. [CrossRef]

147. Xiu, Z.M.; Zhang, Q.B.; Puppala, H.L.; Colvin, V.L.; Alvarez, P.J. Negligible particle-specific antibacterial activity of silver nanoparticles. *Nano Lett.* **2012**, *12*, 4271–4275. [CrossRef]

148. Jung, W.K.; Koo, H.C.; Kim, K.W.; Shin, S.; Kim, S.Y.; Park, Y.H. Antibacterial activity and mechanism of action of the silver ion in Staphylococcus aureus and Escherichia coli. *Appl. Environ. Microbiol.* **2008**, *74*, 2171–2178. [CrossRef]

149. Hsueh, Y.H.; Lin, K.S.; Ke, W.J.; Hsieh, C.T.; Chiang, C.L.; Tzou, D.Y.; Liu, S.T. The antimicrobial properties of silver nanoparticles in bacillus subtilis are mediated by released Ag^+ ions. *PLoS ONE* **2015**, *10*, e0144306. [CrossRef]

150. Khalandi, B.; Asadi, N.; Milani, M.; Davaran, S.; Abadi, A.J.; Abasi, E.; Akbarzadeh, A. A review on potential role of silver nanoparticles and possible mechanisms of their action on bacteria. *Drug Res.* **2016**, *67*, 70–76. [CrossRef]

151. Bondarenko, O.; Ivask, A.; Kakinen, A.; Kurvet, I.; Kahru, A. Particle-cell contact enhances Antibacterial sctivity of silver nanoparticles. *PLoS ONE* **2013**, *8*, e64060. [CrossRef]

152. Ivask, A.; ElBadawy, A.; Kaweeteerawat, C.; Boren, D.; Fischer, H.; Ji, Z.; Chang, C.H.; Liu, R.; Tolaymat, T.; Telesca, D.; et al. Toxicity mechanisms in Escherichia coli vary for silver nanoparticles and differ from ionic silver. *ACS Nano* **2014**, *8*, 374–386. [CrossRef]

153. Quinteros, M.A.; Aristizabal, V.C.; Dalmasso, P.R.; Paraje, M.G.; Paez, P.L. Oxidative stress generation of silver nanoparticles in three bacterial genera and its relationship with the antimicrobial activity. *Toxicol. In Vitro* **2016**, *36*, 216–223. [CrossRef]

154. Pareek, V.; Gupta, R.; Panwar, J. Do physico-chemical properties of silver nanoparticles decide their interaction with biological media and bactericidal action? A review. *Mater. Sci. Eng. C* **2018**, *90*, 739–749. [CrossRef]

155. Gao, M.; Sun, L.; Wang, Z.; Zhao, Y. Controlled synthesis of Ag nanoparticles with different morphologies and their antibacterial properties. *Mater. Sci. Eng. C* **2013**, *33*, 397–404. [CrossRef]

156. Abbaszadegan, A.; Ghahramani, Y.; Gholami, A.; Hemmateenejad, B.; Dorostkar, S.; Nabavizadeh, M.; Shargh, H. The effect of charge at the surface of silver nanoparticles on antimicrobial activity against Gram-positive and Gram-negative bacteria: A preliminary study. *J. Nanomater.* **2015**, *2015*, 720654. [CrossRef]

157. Lu, Z.; Rong, K.; Li, J.; Yang, H.; Chen, R. Size-dependent antibacterial activities of silver nanoparticles against oral anaerobic pathogenic bacteria. *J. Mater. Sci. Mater. Med.* **2013**, *24*, 1465–1471. [CrossRef] [PubMed]

158. Lee, H.J.; Lee, S.G.; Oh, E.J.; Chung, H.Y.; Han, S.I.; Kim, E.J.; Seo, S.Y.; Ghim, H.D.; Yeum, J.H.; Choi, J.H. Antimicrobial polyethyleneimine-silver nanoparticles in a stable colloidal dispersion. *Colloids Surf. B* **2011**, *88*, 505–511. [CrossRef] [PubMed]

159. Kim, D.H.; Park, J.C.; Jeon, G.E.; Kim, C.S.; Seo, J.H. Effect of the size and shape of silver nanoparticles on bacterial growth and metabolism by monitoring optical density and fluorescence intensity. *Biotechnol. Bioprocess Eng.* **2017**, *22*, 210–217. [CrossRef]

160. Pal, S.; Tak, Y.K.; Song, J.M. Does the antibacterial activity of silver nanoparticles depend on the shape of the nanoparticle? A study of the Gram-negative bacterium Escherichia coli. *Appl. Environ. Microbiol.* **2007**, *73*, 1712–1720. [CrossRef] [PubMed]

161. Acharya, D.; Singha, K.M.; Pandey, P.; Mohanta, B.; Rajkumari, J.; Singha, L.P. Shape dependent physical mutilation and lethal effects of silver nanoparticles on bacteria. *Sci. Rep.* **2018**, *8*, 201. [CrossRef] [PubMed]

162. Kidd, T.J.; Mills, G.; Sa-Pessoa, J.; Dumigan, A.; Frank, C.G.; Insua, J.; Ingram, R.; Hobley, L.; Bengoeche, J.A. A Klebsiella pneumoniae antibiotic resistance mechanism that subdues host defenses and promotes virulence. *EMBO Mol. Med.* **2017**, *9*, 430–447. [CrossRef] [PubMed]

163. Van der Wal, A.; Norde, W.; Zehnder, A.; Lyklema, J. Determination of the surface charge in the cell walls of gram-positive bacteria. *Colloid Surf. B* **1997**, *9*, 81–100. [CrossRef]

164. Badawy, A.M.; Silva, R.; Morris, B.; Scheckel, K.G.; Suidan, M.T.; Tolaymat, T.M. Surface charge-dependent toxicity of silver nanoparticles. *Environ. Sci. Technol.* **2011**, *45*, 283–287. [CrossRef] [PubMed]

165. Rivero, P.J.; Urrutia, A.; Goicoechea, J.; Arregui, F.J. Nanomaterials for functional textiles and Fibers. *Nanoscale Res. Lett.* **2015**, *10*, 501. [CrossRef] [PubMed]

166. Lin, J.; Chen, X.Y.; Chen, C.Y.; Hu, J.T.; Zhou, C.; Cai, X.F.; Wang, W.; Zheng, C.; Zhang, P.; Cheng, J.; et al. Durably antibacterial and bacterially antiadhesive cotton fabrics coated by cationic fluorinated polymers. *ACS Appl. Mater. Interfaces* **2018**, *10*, 6124–6136. [CrossRef] [PubMed]

167. El-Rafie, M.H.; Shaheen, T.I.; Mohamed, A.A.; Hebeish, A. Bio-synthesis and applications of silver nanoparticles onto cotton fabrics. *Carbohydr. Polym.* **2012**, *90*, 915–920. [CrossRef]

168. Liu, H.; Lv, M.; Deng, B.; Li, J.; Yu, M.; Huang, Q.; Fan, C. Laundering durable antibacterial cotton fabrics grafted with pomegranate-shaped polymer wrapped in silver nanoparticle aggregations. *Sci. Rep.* **2014**, *4*, 5920. [CrossRef] [PubMed]

169. Deng, X.; Nikifolov, A.Y.; Coenye, T.; Cools, P.; Aziz, G.; Morent, R.; De Geyter, N.; Leys, C. Antimicrobial nano-silver non-woven polyethylene terephthalate fabric via an atmospheric pressure plasma deposition process. *Sci. Rep.* **2015**, *5*, 10138. [CrossRef] [PubMed]

170. Han, J.W.; Ruiz-Garcia, L.; Qian, J.P.; Yang, X.T. Food packaging: A comprehensive review and future trends. *Compr. Rev. Food Sci. Food Saf.* **2018**, *17*, 860–877. [CrossRef]

171. Bumbudsanpharoke, N.; Choi, J.; Ko, S. Applications of nanomaterials in food packaging. *J. Nanosci. Nanotechnol.* **2015**, *15*, 6357–6372. [CrossRef] [PubMed]

172. Mousavi, F.P.; Pour, H.H.; Nasab, A.H.; Rajabalipour, A.A.; Barouni, M. Investigation into shelf life of fresh dates and pistachios in a package modified with nano-silver. *Glob. J. Health Sci.* **2015**, *8*, 134–144. [CrossRef] [PubMed]

173. Tavakoli, H.; Rastegar, H.; Taherian, M.; Somadi, M.; Rostami, H. The effect of nano-silver packaging in increasing the shelf life of nuts: An in vitro model. *Ital. J. Food Saf.* **2017**, *6*, 6874. [CrossRef] [PubMed]

174. Huang, Y.; Mei, L.; Chen, X.; Wang, Q. Recent developments in food packaging based on nanomaterials. *Nanomaterials* **2018**, *8*, 830. [CrossRef]

175. Martínez-Abad, A.; Lagaron, J.M.; Ocio, M.J. Development and characterization of silver-based antimicrobial ethylene—vinyl alcohol copolymer (EVOH) films for food-packaging applications. *J. Agric. Food Chem.* **2012**, *60*, 5350–5359. [CrossRef]

176. Russel, A.D. Challenge testing: Principles and practice. *Int. J. Cosmet. Sci.* **2003**, *25*, 147–153. [CrossRef] [PubMed]

177. Chaloupka, K.; Malam, Y.; Seifalian, A.M. Nanosilver as a new generation of nanoproduct in biomedical applications. *Trends Biotechnol.* **2010**, *28*, 580–588. [CrossRef] [PubMed]

178. Lopez-Carballo, G.; Higueras, L.; Gavara, R.; Hernandez-Muñoz, P. Silver ions release from antibacterial chitosan films containing in situ generated silver nanoparticles. *J. Agric. Food Chem.* **2013**, *61*, 260–267. [CrossRef]

179. Huang, Y.; Chen, S.; Bing, X.; Gao, C.; Wang, T.; Yuan, B. Nanosilver migrated into food-simulating solutions from commercially available food fresh containers. *Packag. Technol. Sci.* **2011**, *24*, 291–297. [CrossRef]

180. Echegoyen, Y.; Nerin, C. Nanoparticle release from nano-silver antimicrobial food containers. *Food Chem. Technol.* **2013**, *62*, 16–22. [CrossRef] [PubMed]

181. You, C.; Li, Q.; Wang, X.; Wu, P.; Ho, J.K.; Jin, R.; Zhang, L.; Shao, H.; Han, C. Silver nanoparticle loaded collagen/chitosan scaffolds promote wound healing via regulating fibroblast migration and macrophage activation. *Sci. Rep.* **2017**, *7*, 10489. [CrossRef] [PubMed]

182. Bhowmick, S.; Koul, V. Assessment of PVA/silver nanocomposite hydrogel patch as antimicrobial dressing scaffold: Synthesis, characterization and biological evaluation. *Mater. Sci. Eng. C* **2016**, *59*, 109–119. [CrossRef] [PubMed]

183. Gonzalez-Sanchez, M.I.; Perni, S.; Tommasi, G.; Morris, N.J.; Hawkins, K.; Lopez-Cabarcos, E.; Prokopovich, P. Silver nanoparticle based antibacterial methacrylate hydrogels potential for bone graft applications. *Mater. Sci. Eng. C* **2015**, *50*, 332–340. [CrossRef]

184. Oliveira, R.N.; Rouze, R.; Quilty, B.; Alves, G.G.; Soares, G.D.A.; Thire, R.M.S.; McGuinness, G.B. Mechanical properties and in vitro characterization of polyvinyl alcohol-nano-silver hydrogel wound dressings. *Interf. Focus* **2014**, *4*, 20130049. [CrossRef]

185. De Matteis, V. Exposure to inorganic nanoparticles: Routes of entry, immune response, biodistribution and in vitro/in vivo toxicity evaluation. *Toxics* **2017**, *5*, 29. [CrossRef]

186. Zhang, S.; Gao, H.; Bao, G. Physical process of nanoparticle cellular endocytosis. *ACS Nano* **2015**, *9*, 8655–8671. [CrossRef] [PubMed]

187. Akter, M.; Sikder, M.T.; Rahman, M.; Ullah, A.K.; Hossain, K.F.; Banik, S.; Hosokawa, T.; Saito, T.; Kurasaki, M. A systematic review on silver nanoparticles-induced cytotoxicity: Physicochemical properties and perspectives. *J. Adv. Res.* **2018**, *9*, 1–16. [CrossRef] [PubMed]

188. Fu, P.P.; Xia, Q.; Huang, H.M.; Ray, P.C.; Yu, H. Mechanisms of nanotoxicity: Generation of reactive oxygen species. *J. Food Drug. Anal.* **2014**, *22*, 64–75. [CrossRef] [PubMed]

189. Grzelak, A.; Wojewodzka, M.; Meczynska-Wielgosz, S.; Zuberek, M.; Wojciechowska, D.; Kruszewski, M. Crucial role of chelatable iron in silver nanoparticles induced DNA damage and cytotoxicity. *Redox Biol.* **2018**, *15*, 435–440. [CrossRef] [PubMed]

190. AshaRani, P.V.; Hande, M.P.; Valiyaveettil, S. Anti-proliferative activity of silver nanoparticles. *BMC Cell Biol.* **2009**, *10*, 65. [CrossRef] [PubMed]

191. Zhang, X.F.; Shen, W.; Gurunathan, S. Silver nanoparticle-mediated cellular responses in various cell lines: An in vitro model. *Int. J. Mol. Sci.* **2016**, *17*, 1603. [CrossRef] [PubMed]

192. Riaz Ahmed, K.B.; Nagy, A.M.; Brown, R.P.; Zhang, Q.; Malghan, S.G.; Goering, P.L. Silver nanoparticles: Significance of physicochemical properties and assay interference on the interpretation of in vitro cytotoxicity studies. *Toxicol. In Vitro* **2017**, *38*, 179–192. [CrossRef]

193. Han, J.W.; Gurunathan, S.; Jeong, J.K.; Choi, Y.J.; Kwon, D.N.; Park, J.K.; Kim, J.H. Oxidative stress mediated cytotoxicity of biologically synthesized silver nanoparticles in human lung epithelial adenocarcinoma cell line. *Nanoscle Res. Lett.* **2014**, *9*, 459. [CrossRef]

194. Xue, Y.; Zhang, T.; Zhang, B.; Gong, F.; Huang, Y.; Tang, M. Cytotoxicity and apoptosis induced by silver nanoparticles in human liver HepG2 cells in different dispersion media. *J. Appl. Toxicol.* **2015**, *36*, 352–360. [CrossRef]

195. Hsiao, I.L.; Hsieh, Y.K.; Wang, C.F.; Chen, I.C.; Huang, Y.J. Trojan-horse mechanism in the cellular uptake of silver nanoparticles verified by direct intra- and extracellular silver speciation analysis. *Environ. Sci. Technol.* **2015**, *49*, 3813–3821. [CrossRef]

196. Pourzahedi, L.; Vance, M.; Eckelman, M.J. Life cycle assessment and release studies for nanosilver-enabled consumer products: Investigating hotspots and patterns of contribution. *Environ. Sci. Technol.* **2017**, *51*, 7148–7158. [CrossRef]

197. Theodorou, I.G.; Ryan, M.P.; Tetley, T.D.; Porter, A.E. Inhalation of silver nanomaterials—Seeing the risk. *Int. J. Mol. Sci.* **2014**, *15*, 23936–23974. [CrossRef]

198. Fordbjerg, R.; Irwing, E.S.; Hayashi, Y.; Sutherland, D.S.; Thorsen, K.; Autrup, S.; Beer, C. Global gene expression profiling of human lung epithelial cells after exposure to nanosilver. *Toxicol. Sci.* **2012**, *130*, 145–157. [CrossRef] [PubMed]

199. Gurunathan, S.; Kang, M.H.; Kim, J.H. Combination effect of silver nanoparticles and histone deacetylases inhibitor in human alveolar basal epithelial cells. *Molecules* **2018**, *23*, 2046. [CrossRef]

200. Nymark, P.; Catalan, J.; Suhonen, S.; Jarventaus, H.; Bilkeda, R.; Clausen, P.A.; Jensen, K.A.; Vippola, M.; Savolainen, K.; Norppa, H. Genotoxicity of polyvinylpyrrolidone-coated silver nanoparticles in BEAS 2B cells. *Toxicology* **2013**, *313*, 38–48. [CrossRef] [PubMed]

201. Gliga, A.R.; Skoglund, S.; Wallinder, I.O.; Fadee, B.; Karlsson, H.L. Size-dependent cytotoxicity of silver nanoparticles in human lung cells: The role of cellular uptake, agglomeration and Ag release. *Part. Fibre Toxicol.* **2014**, *11*, 11. [CrossRef] [PubMed]

202. Kim, H.R.; Kim, M.J.; Lee, S.Y.; Oh, S.M.; Chung, K.H. Genotoxic effects of silver nanoparticles stimulated by oxidative stress in human normal bronchial epithelial (BEAS-2B) cells. *Mutat Res.* **2011**, *726*, 129–135. [CrossRef]

203. Sapkota, K.; Narayanan, K.B.; Han, S.S. Environmentally sustainable synthesis of catalytically active silver nanoparticles and their cytotoxic effect on human keratinocytes. *J. Clust. Sci.* **2017**, *28*, 1605–1616. [CrossRef]

204. Avalos, A.S.; Haza, A.I.; Morales, P. Manufactured silver nanoparticles of different sizes induced DNA strand breaks and oxidative DNA damage in hepatoma and leukaemia cells and in dermal and pulmonary fibroblasts. *Folia Biol.* **2015**, *61*, 33–42.

205. Avalos, A.S.; Haza, A.I.; Mateo, D.; Morales, P. Interactions of manufactured silver nanoparticles of different sizes with normal human dermal fibroblasts. *Int. Wound J.* **2016**, *13*, 101–109. [CrossRef]

206. Huo, L.; Chen, R.; Zhao, L.; Shi, X.; Bai, R.; Long, D.; Chen, F.; Zhao, Y.; Chang, Y.Z.; Chen, C. Silver nanoparticles activate endoplasmic reticulum stress signaling pathway in cell and mouse models: The role in toxicity evaluation. *Biomaterials* **2015**, *61*, 307–315. [CrossRef] [PubMed]

207. Matuszak, J.; Baumgartner, J.; Zaloga, J.; Juenet, M.; da Silva, A.E.; Franke, D.; Almer, G.; Texier, I.; Faivre, D.; Metselaar, J.M.; et al. Nanoparticles for intravascular applications: Physicochemical characterization and cytotoxicity testing. *Nanomedicine* **2016**, *11*, 597–616. [CrossRef] [PubMed]

208. Gromnicova, R.; Kaya, M.; Romero, I.A.; Williams, P.; Satchell, S.; Sharrack, B.; Male, D. Transport of gold nanoparticles by vascular endothelium from different human tissues. *PLoS ONE* **2016**, *11*, e0161610. [CrossRef] [PubMed]

209. Cao, Y.; Gong, Y.; Liu, L.; Zhou, Y.; Fang, X.; Zhang, C.; Li, Y.; Li, J. The use of human umbilical vein endothelial cells (HUVECs) as an in vitro model to assess the toxicity of nanoparticles to endothelium: A review. *J. Appl. Toxicol.* **2017**, *37*, 1359–1369. [CrossRef] [PubMed]

210. Guo, H.; Zhang, Z.; Boudreau, M.; Meng, J.; Ying, J.J.; Liu, J.; Xu, H. Intravenous administration of silver nanoparticles causes organ toxicity through intracellular ROS-related loss of inter-endothelial junction. *Part Fibre Toxicol.* **2016**, *13*, 21. [CrossRef] [PubMed]

211. Zuberek, M.; Wojciechowska, D.; Krzyzanowski, D.; Meczynska-Wielgosz, S.; Kruszewski, M.; Grzelak, A. Glucose availability determines silver nanoparticles toxicity in HepG2. *J. Nanobiotechnol.* **2015**, *13*, 72. [CrossRef] [PubMed]

212. Singh, A.; Dar, M.Y.; Joeshi, B.; Sharma, B.; Shrisvatava, S.; Sukla, S. Phytofabrication of silver nanoparticles: Novel drug to overcome hepatocellular ailments. *Toxicol. Rep.* **2018**, *5*, 333–342. [CrossRef] [PubMed]

213. Chen, L.Q.; Fang, L.; Ling, J.; Ding, C.Z.; Kang, B.; Huang, C.Z. Nanotoxicity of silver nanoparticles to red blood cells: Size-dependent adsorption, uptake and hemolytic activity. *Chem. Res. Toxicol.* **2015**, *28*, 501–509. [CrossRef] [PubMed]

214. Ferdous, Z.; Beegam, S.; Tariq, S.; Ali, B.H.; Nemmar, A. The in vitro effect of polyvinylpyrrolidone and citrate coated silver nanoparticles on erythrocytic oxidative damage and eryptosis. *Cell Physiol. Biochem.* **2018**, *49*, 1577–1588. [CrossRef] [PubMed]

215. Carlson, C.; Hussain, S.M.; Schrand, A.M.; Braydich-Stolle, L.K.; Hess, K.L.; Jones, R.L.; Schlager, J.J. Unique cellular interaction of solver nanoparticles: Size-dependent generation of reactive oxygen species. *J. Phys. Chem. B* **2008**, *112*, 13608–13619. [CrossRef] [PubMed]

216. Yang, E.J.; Kim, S.; Kim, J.S.; Choi, I.H. Inflammasome formation and IL-1β release by human blood monocytes in response to silver nanoparticles. *Biomaterials* **2012**, *33*, 6858–6867. [CrossRef] [PubMed]

217. Martinez-Gutierrez, F.; Thi, E.P.; Silverman, J.M.; De Oliveira, C.C.; Svensson, S.L.; Vanden Hoek, A.; Sanchez, E.M.; Reiner, N.E.; Gaynor, E.C.; Pryzdial, E.L.; et al. Antibacterial activity, inflammatory response, coagulation and cytotoxicity effects of silver nanoparticles. *Nanomedicine* **2012**, *8*, 328–336. [CrossRef] [PubMed]

218. Feng, X.; Chen, A.; Zhang, Y.; Wang, J.; Shao, L.; Wei, L. Central nervous system toxicity of metallic nanoparticles. *Int. J. Nanomed.* **2015**, *10*, 4321–4340. [CrossRef]

219. Cramer, S.; Tacke, S.; Bornhors, J.; Klingauf, J.; Schwerdtl, T.; Galla, H.J. The influence of silver nanoparticles on the blood-brain and the blood-cerebrospinal fluid barrier in vitro. *J. Nanomed. Nanotechnol.* **2014**, *5*, 225. [CrossRef]

220. Liu, F.; Mahmood, M.; Xu, Y.; Watanabe, F.; Biris, A.S.; Hansen, D.K.; Inselman, A.; Casciano, D.; Patterson, T.A.; Paule, M.G.; et al. Effects of silver nanoparticles on human and rat embryonic neural stem cells. *Front. Neurosci.* **2015**, *9*, 115. [CrossRef] [PubMed]

221. Yin, N.; Hu, B.; Yang, R.; Liang, S.; Liang, S.X.; Faiola, F. Assessment of the developmental neurotoxicity of silver nanoparticles and silver ions with mouse embryonic stem cells in vitro. *JOIN* **2018**, *3*, 133–145. [CrossRef]

222. Ma, W.; Jing, L.; Valladares, A.; Mehta, S.L.; Wang, Z.; Li, P.A.; Bang, J.J. Stress, caspase-3 activation and cell death: Amelioration by sodium selenite. *Int. J. Biol. Sci.* **2015**, *11*, 86–867. [CrossRef]

223. McIlwain, D.R.; Berger, T.; Mak, T.W. Caspase functions in cell death and disease. *Cold Spring Harb. Perspect. Biol.* **2013**, *5*, a008656. [CrossRef]

224. Iturri, J.; Toca-Herrera, J.L. Characterization of cell scaffolds by atomic force microscopy. *Polymers* **2017**, *9*, 383. [CrossRef]

225. Variola, F. Atomic force microscopy in biomaterials surface science. *Phys. Chem. Chem. Phys.* **2015**, *17*, 2950–2959. [CrossRef]

226. Subbiah, R.; Jeon, S.B.; Park, K.; Ahn, S.J.; Yun, K. Investigation of cellular responses upon interaction with silver nanoparticles. *Int. J. Nanomed.* **2015**, *10*, 191–201. [CrossRef]

227. Franco, R.; Cidlowski, J.A. Apoptosis and glutathione: Beyond an antioxidant. *Cell Death Differ.* **2009**, *16*, 1303–1314. [CrossRef] [PubMed]

228. Turner, P.V.; Brabb, T.; Pekow, C.; Vasbinde, M.A. Administration of substances to laboratory animals: Routes of administration and factors to consider. *J. Am. Assoc. Lab. Anim. Sci.* **2011**, *50*, 600–613. [PubMed]

229. Recordati, C.; De Maglie, M.; Bianchessi, S.; Argentiere, S.; Cella, C.; Mattiello, S.; Cubadda, F.; Aureli, F.; D'Amato, M.; Raggi, A.; et al. Tissue distribution and acute toxicity of silver after single intravenous administration in mice: Nano-specific and size-dependent effects. *Part. Fibre Toxicol.* **2016**, *13*, 12. [CrossRef] [PubMed]

230. Yang, L.; Kuang, H.; Zhang, W.; Aguilar, Z.P.; Wei, H.; Xu, H. Comparisons of the biodistribution and toxicological examinations after repeated intravenous administration of silver and gold nanoparticles in mice. *Sci. Rep.* **2017**, *7*, 3303. [CrossRef] [PubMed]

231. Wen, H.; Dan, M.; Yang, Y.; Lyu, J.; Shao, A.; Cheng, X.; Chen, L.; Xu, L. Acute toxicity and genotoxicity of silver nanoparticle in rats. *PLoS ONE* **2017**, *12*, e0185554. [CrossRef] [PubMed]

232. Wen, H.; Yang, X.; Hu, L.; Sun, C.; Zhou, Q.; Jiang, G. Brain-targeted distribution and high retention of silver by chronic intranasal instillation of silver nanoparticles and ions in Sprague–Dawley rats. *J. Appl. Toxicol.* **2015**, *36*, 445–453. [CrossRef] [PubMed]

233. Gaiser, B.K.; Hirn, S.; Kermanizadeh, A.; Kanase, N.; Fytianos, K.; Wenk, A.; Harbel, N.; Brunelli, A.; Kreyling, W.G.; Stove, V. Effects of silver nanoparticles on the liver and hepatocytes in vitro. *Toxicol. Sci.* **2013**, *131*, 537–547. [CrossRef]

234. Dziendzikowska, K.; Gromadzka-Ostrowska, J.; Lankoff, A.; Oczkowski, M.; Krawczynska, A.; Chwastowska, J.; Sadowska-Bratek, M.; Chajduk, E.; Wojewodzka, M.; Kruszewski, M. Time-dependent biodistribution and excretion of silver nanoparticles in male Wistar rats. *J. Appl. Toxicol.* **2012**, *32*, 920–928. [CrossRef]

235. Lee, T.Y.; Liu, M.S.; Huang, L.J.; Lue, S.I.; Lin, L.C.; Kwan, A.L.; Yang, R.C. Bioenergetic failure correlates with autophagy and apoptosis in rat liver following silver nanoparticle intraperitoneally administration. *Part. Fibre Toxicol.* **2013**, *10*, 40. [CrossRef]

236. Sadauskas, E.; Wallin, H.; Stoltenberg, M.; Vogel, U.; Doering, P.; Larsen, A.; Danscher, G. Kupffer cells are central in the removal of nanoparticles from the organism. *Part. Fibre Toxicol.* **2007**, *4*, 10. [CrossRef] [PubMed]

237. Sarhan, O.M.; Hussein, R.M. Effects of intraperitoneally injected silver nanoparticles on histological structures and blood parameters in the albino rat. *Int. J. Nanomed.* **2014**, *9*, 1505–1517. [CrossRef]

238. Qin, G.; Tang, S.; Li, S.; Lu, H.; Wang, Y.; Zhao, P.; Li, B.; Zhang, J.; Peng, L. Toxicological evaluation of silver nanoparticles and silver nitrate in rats following 28 days of repeated oral exposure. *Environ. Toxicol.* **2017**, *32*, 609–618. [CrossRef] [PubMed]

239. Zhang, X.F.; Park, J.H.; Choi, Y.J.; Kang, M.H.; Gurunathan, S.; Kim, J.H. Silver nanoparticles cause complications in pregnant mice. *Int. J. Nanomed.* **2015**, *10*, 7057–7071. [CrossRef]

240. Fennell, T.R.; Mortensen, N.P.; Black, S.R.; Snyder, R.W.; Levine, K.E.; Poitras, E.; Harrington, J.M.; Wingard, C.J.; Holland, N.A.; Patmasiri, W. Disposition of intravenously or orally administered silver nanoparticles in pregnant rats and the effect on the biochemical profile in urine. *J. Appl. Toxicol.* **2017**, *37*, 530–544. [CrossRef] [PubMed]

241. Hadrup, N.; Loeschner, K.; Mortensen, A.; Sharma, A.K.; Qvortrup, K.; Larsen, E.H.; Lam, H.R. The similar neurotoxic effects of nanoparticulate and ionic silver in vivo and in vitro. *Neurotoxicol.* **2012**, *33*, 416–423. [CrossRef] [PubMed]

242. Xu, L.; Dan, M.; Shao, A.; Cheng, X.; Zhang, C.; Yokel, R.A.; Takemura, T.; Hanagata, N.; Niwa, M.; Wanatabe, D. Silver nanoparticles induce tight junction disruption and astrocyte neurotoxicity in a rat blood–brain barrier primary triple coculture model. *Int. J. Nanomed.* **2015**, *10*, 6105–6119. [CrossRef]

243. Xu, L.; Shao, A.; Zhao, Y.; Wang, Z.; Zhang, C.; Sun, Y.; Deng, J.; Cho, L.L. Neurotoxicity of silver nanoparticles in rat brain after intragastric exposure. *J. Nanosci. Nanotechnol.* **2014**, *9*, 1–14. [CrossRef]

244. Dabrowska-Bouta, B.; Sulkowski, G.; Struzynski, W.; Struzynska, L. Prolonged exposure to silver nanoparticles results in oxidative stress in cerebral myelin. *Neurotox. Res.* **2018**, *34*, 1–10. [CrossRef]

245. Bergin, I.L.; Wilding, L.A.; Morishita, M.; Walacavage, K.; Ault, A.P.; Axson, J.L.; Stark, D.I.; Hashway, S.A.; Capracotta, S.S.; Leroueil, P.R.; et al. Effects of particle size and coating on toxicologic parameters, fecal elimination kinetics and tissue distribution of acutely ingested silver nanoparticles in a mouse model. *Nanotoxicology* **2016**, *10*, 352–360. [CrossRef]

246. Boudreau, M.D.; Imam, M.S.; Paredes, A.M.; Bryant, M.S.; Cunningham, C.K.; Felton, R.P.; Jones, M.Y.; Davis, K.J.; Olson, G.R. Differential effects of silver nanoparticles and silver ions on tissue accumulation, distribution, and toxicity in the Sprague Dawley rat following daily oral gavage administration for 13 weeks. *Toxicol. Sci.* **2016**, *150*, 131–160. [CrossRef] [PubMed]

247. Park, K. Toxicokinetic differences and toxicities of silver nanoparticles and silver ions in rats after single oral administration. *J. Toxicol. Environ. Health Part A* **2013**, *76*, 1246–1260. [CrossRef] [PubMed]

248. Van der Zande, M.; Vandebriel, R.J.; Doren, E.V.; Kramer, E.; Rivera, Z.H.; Serrano-Rojero, C.S.; Gremmer, E.R.; Mast, J.; Peters, R.J.; Hollman, P.C.; et al. Distribution, elimination, and toxicity of silver nanoparticles and silver ions in rats after 28-day oral exposure. *ACS Nano* **2012**, *6*, 7427–7442. [CrossRef] [PubMed]

249. Li, N.; Georas, S.; Alexis, N.; Fritz, P.; Xia, T.; Williams, M.A.; Horner, E.; Nel, A. A work group report on ultrafine particles (American Academy of Allergy, Asthma & Immunology): Why ambient ultrafine and engineered nanoparticles should receive special attention for possible adverse health outcomes in human subjects. *J. Allergy Clin. Immunol.* **2016**, *138*, 386–396. [CrossRef] [PubMed]

250. Morimoto, Y.; Izumi, H.; Yoshiura, Y.; Fujishima, K.; Yatera, K.; Yamamoto, K. Usefulness of intratracheal instillation studies for estimating nanoparticle-induced pulmonary toxicity. *Int. J. Mol. Sci.* **2016**, *17*, 165. [CrossRef] [PubMed]

251. Anderson, D.S.; Silva, R.M.; Lee, D.; Edwards, P.C.; Sharmah, A.; Guo, T.; Pinkerton, K.E.; Van Winkle, L.S. Persistence of silver nanoparticles in the rat lung: Influence of dose, size, and chemical composition. *Nanotoxicology* **2015**, *9*, 591–602. [CrossRef] [PubMed]

252. Seiffert, J.; Hussain, F.; Wiegman, C.; Li, F.; Bey, L.; Baker, W.; Porter, A.; Ryan, M.P.; Chang, Y.; Gow, A.; et al. Pulmonary toxicity of instilled silver nanoparticles: Influence of size, coating and rat strain. *PLoS ONE* **2015**, *10*, e0119726. [CrossRef] [PubMed]

253. Roda, E.; Barni, S.; Milzani, A.; Dalle-Donne, I.; Colombo, G.; Coccini, T. Single silver nanoparticle instillation induced early and persisting moderate cortical damage in rat kidneys. *Int. J. Mol. Sci.* **2017**, *18*, 2115. [CrossRef] [PubMed]

254. Wiemann, M.; Vennemann, A.; Blaske, F.; Sperling, M.; Karst, U. Silver nanoparticles in the lung: Toxic effects and focal accumulation of silver in remote organs. *Nanomaterials* **2017**, *7*, 441. [CrossRef] [PubMed]

255. Jena, P.; Mohanty, S.; Mallick, R.; Jacob, B.; Sonawane, A. Toxicity and antibacterial assessment of chitosan coated silver nanoparticles on human pathogens and macrophage cells. *Int. J. Nanomed.* **2012**, *7*, 1805–1818. [CrossRef]

256. Sanchez, G.R.; Catilla, C.L.; Gomez, N.B.; Garcia, A.; Marcos, R.; Carmona, E.R. Leaf extract from the endemic plant Peumus boldus as an effective bioproduct for the green synthesis of silver nanoparticles. *Mater. Lett.* **2016**, *183*, 255–260. [CrossRef]

International Journal of
Molecular Sciences

MDPI

Article

Size-Dependent Effect of Silver Nanoparticles on the Tumor Necrosis Factor α-Induced DNA Damage Response

Alaa Fehaid [1,2,3] and Akiyoshi Taniguchi [1,2,*]

[1] Cellular Functional Nanobiomaterials Group, Research Center for Functional Materials, National Institute for Materials Science, 1-1 Namiki, Tsukuba, Ibaraki 305-0044, Japan; ahmed.alaa@nims.go.jp
[2] Graduate School of Advanced Science and Engineering, Waseda University, 3-4-1 Okubo, Shinjuku-ku, Tokyo 169-8555, Japan
[3] Forensic Medicine and Toxicology Department, Faculty of Veterinary Medicine, Mansoura University, Dakahlia 35516, Egypt
* Correspondence: taniguchi.akiyoshi@nims.go.jp; Tel.: +81-29-860-4505

Received: 16 January 2019; Accepted: 23 February 2019; Published: 27 February 2019

Abstract: Silver nanoparticles (AgNPs) are widely used in many consumer products due to their anti-inflammatory properties. Therefore, the effect of exposure to AgNPs should be investigated in diseased states in addition to healthy ones. Tumor necrosis factor-α (TNFα) is a major cytokine that is highly expressed in many diseased conditions, such as inflammatory diseases, sepsis, and cancer. We investigated the effects of two different sizes of AgNPs on the TNFα-induced DNA damage response. Cells were exposed to 10 and 200 nm AgNPs separately and the results showed that the 200 nm AgNPs had a lower cytotoxic effect with a higher percent of cellular uptake compared to the 10 nm AgNPs. Moreover, analysis of reactive oxygen species (ROS) generation and DNA damage indicated that TNFα-induced ROS-mediated DNA damage was reduced by 200 nm AgNPs, but not by 10 nm AgNPs. Tumor necrosis factor receptor 1 (TNFR1) was localized on the cell surface after TNFα exposure with or without 10 nm AgNPs. In contrast, the expression of TNFR1 on the cell surface was reduced by the 200 nm AgNPs. These results suggested that exposure of cells to 200 nm AgNPs reduces the TNFα-induced DNA damage response via reducing the surface expression of TNFR1, thus reducing the signal transduction of TNFα.

Keywords: silver nanoparticles; tumor necrosis factor; DNA damage; TNFR1

---

1. Introduction

Nanotechnology is an advanced field that studies very small materials ranging from 0.1 to 100 nm [1]. Silver nanoparticles (AgNPs) are a high-demand nanomaterial for consumer products [2]. Because of their potent antimicrobial activity, AgNPs are incorporated into many products such as textiles, paints, biosensors, electronics, and medical products including deodorant sprays, catheter coatings, wound dressings, and surgical instruments [3–6]. Most of the medical applications create concerns over human exposure, due to the properties of AgNPs which allow them to cross the blood brain barrier easily [7].

The characteristics of AgNPs, including morphology, size, size distribution, surface area, surface charge, stability, and agglomeration, have a significant impact on their interaction with biological systems [8–10]. All of these physicochemical characteristics affect nanoparticle–cellular interactions, including cellular uptake, cellular distribution, and various cellular responses such as inflammation, proliferation, DNA damage, and cell death [11–13]. Therefore, to address safety and improve quality, each characteristic of AgNPs should be clearly determined and separately assessed for its effects on different cellular responses. In this study, we focused on the effect of AgNP size on the cellular response.

Int. J. Mol. Sci. **2019**, *20*, 1038

Several research groups have investigated the effects of AgNPs with sizes ranging from 5 to 100 nm on different cell lines; the cytotoxic effect of AgNPs on human cell lines (A549, SGC-7901, HepG2, and MCF-7) is size-dependent, with 5 nm being more toxic than 20 or 50 nm and inducing elevated reactive oxygen species (ROS) levels and S phase cell cycle arrest [14]. In RAW 264.7 macrophages and L929 fibroblasts, 20 nm AgNPs are more potent in decreasing metabolic activity compared to the larger 80 and 113 nm nanoparticles, acting by inhibiting stem cell differentiation and promoting DNA damage [15]. Because of the importance of nanoparticle size and its impact on cellular uptake and response, in this study we hypothesized that larger AgNPs with sizes above 100 nm might induce different cellular responses than those of less than 100 nm because of different cellular uptake ratios and mechanisms. Therefore, we investigated the size-dependent effect of AgNPs on a lung epithelial cell line in vitro to elucidate the molecular mechanisms underlying the pulmonary cellular response.

To increase applications for AgNPs, we should consider their effects on diseased subjects as well as healthy ones. Inflammatory diseases, asthma, infections, and cancer are common diseases for which the effects of exposure to AgNPs should be investigated. Tumor necrosis factor-α (TNFα), a pro-inflammatory cytokine and a regulator of immunological reactions in many physiological and pathological conditions [16], is a common molecule that is enhanced in most diseased conditions. TNFα is involved in many signal transduction pathways, such as NF-KB activation, MAPK activation, and cell death signaling, resulting in different cellular responses such as inflammation, DNA damage, proliferation, differentiation, and cell death [17–19]. TNFα cellular responses are mainly mediated by one of the two tumor necrosis factor (TNF) receptors (TNFR1 and TNFR2), which elicit different intracellular signals and are without any significant domain homology [20,21]. DNA damage is a very important response because it regulates the cell fate toward death, proliferation, or carcinogenesis; TNFα-induced DNA damage is mostly oxidative and mediated by ROS generation in many cell types [22]. In this study, we hypothesized that AgNPs affect DNA damage along with their known anti-apoptotic and anti-inflammatory effects, so we focused on the TNFα-induced DNA damage response.

We investigated the size-dependent effect of AgNPs, and our results revealed that the expression of TNFR1 on the cell surface was reduced by 200 nm AgNPs but not by 10 nm AgNPs, suggesting a reduction in TNFα-induced DNA damage by 200 nm AgNPs.

2. Results

2.1. Effect of AgNPs on Cell Viability

The size of AgNPs is one of their most important characteristics and influences their uptake by cells and the cellular response. Our aim was to clarify the size-dependent cytotoxic effect of AgNPs. Many studies have investigated the effect of AgNPs in particle sizes ranging from 10 to 100 nm; however, nanoparticles larger than 100 nm might have different effects because they can induce different mechanisms of cellular uptake or have a different uptake ratio. We therefore conducted a cell viability assay to determine the differences between 10 nm and 200 nm AgNPs on the viability of NCI-H292 cells. As shown in Figure 1, the percentage of viable cells decreased in a dose-dependent manner in cells exposed to 10 nm and 200 nm AgNPs (increasing the concentration of AgNPs reduced the percentage of viable cells). Cells exposed to 200 nm AgNPs showed lower cytotoxic effects compared to the 10 nm AgNP-exposed cells; the percentages of viable cells after 24 h exposure to 1, 2.5, 5, 10, 25, 50, 75, and 100 μg/mL of 200 nm and 10 nm AgNPs were 110.1%, 109.8%, 109.3%, 107.2%, 98.2%, 87.4%, 74.5%, and 73.1%; and 98.2%, 99.7%, 94.2%, 86,1%, 59.9%, 38.8%, 29.4%, and 26.2%, respectively. These results demonstrated that the 200 nm AgNPs had a lower cytotoxic effect than the 10 nm AgNPs, showing the impact of nanoparticle size on cytotoxicity.

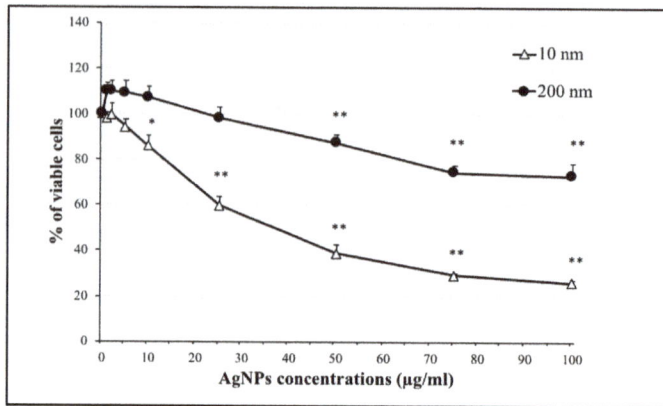

Figure 1. Effect of silver nanoparticles (AgNPs) (10 nm and 200 nm) on the viability of NCI-H292 cells. Viability of cells exposed to 10 nm and 200 nm AgNPs separately at concentrations of 0, 5, 10, 25, 50, 75, and 100 μg/mL. Cells were exposed to AgNPs for 24 h, then cell viability was determined using a CellTiter-Glo® luminescent cell viability assay. The results are shown as means ± SD, n ≥ 3, for each group; *$0.01 < P < 0.05$, ** $P < 0.01$. * Represents significance compared to the control group.

2.2. Cellular Uptake of AgNPs

Cellular uptake of nanoparticles plays an important role in cellular responses including proliferation, inflammation, DNA damage, and cell death. We therefore estimated the cellular uptake of 10 nm and 200 nm AgNPs, and the results are shown in Figure 2. The percentage of cells incorporated with 200 nm AgNPs was higher than the percentage of cells incorporated with 10 nm AgNPs, resulting in an increase in cell density as expressed by side scatter (SSC) as shown in the right panel of Figure 2A. After 24 h of exposure, uptake of 200 nm AgNPs occurred in 30.5% of cells, while uptake of 10 nm AgNPs occurred in only 11.5% of cells, as shown in Figure 2B. These results revealed that larger AgNP size (200 nm) induced higher cellular uptake than a smaller size (10 nm).

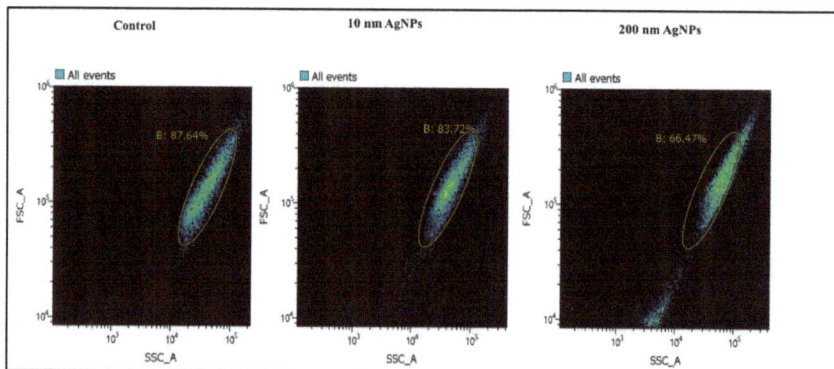

(a)

Figure 2. *Cont.*

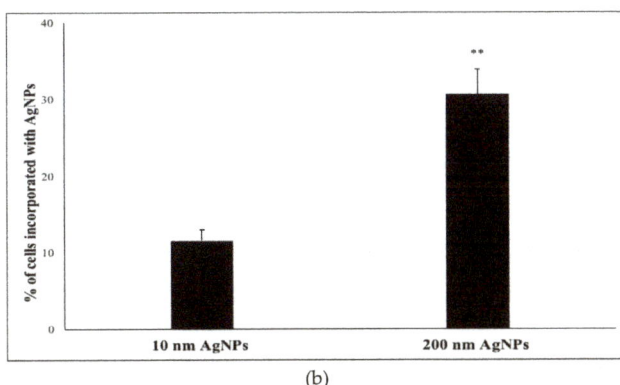

(b)

Figure 2. Uptake ratios of 10 nm and 200 nm AgNPs by NCI-H292 cells. Cells were incubated with AgNPs at a concentration of 100 μg/mL for 24 h. Cellular uptake of AgNPs was calculated using FACS based on the side scatter (SSC). (**a**) Gated forward and side scatter plot of most cells depending on the control population; the cells exposed to 200 nm AgNPs showed higher SSC in the right panel. (**b**) Percentage of cells incorporated with 10 nm and 200 nm AgNPs. The results are shown as means ± SD, $n \geq 3$, for each group; ** $P < 0.01$.

2.3. Interference of AgNPs with TNFα-Induced ROS Generation

In many disease states such as inflammatory disease, infections, and cancer, TNFα acts as a major cytokine. TNFα has been reported to be involved in ROS generation resulting in DNA damage and cell death [23]. Therefore, we conducted a DCF assay to understand how different sizes of AgNPs affect TNFα-induced ROS generation. As shown in Figure 3, cells exposed to TNFα (20 ng/mL) only, 10 nm AgNPs (100 μg/mL) only, or both showed highly significant increases in ROS generation compared to the negative control group. Moreover, cells exposed to TNFα (20 ng/mL) + 200 nm AgNPs (100 μg/mL) showed a significant decrease in ROS generation compared to the TNFα-exposed group. These data suggested that the 200 nm AgNPs, but not the 10 nm AgNPs, reduced TNFα-induced ROS generation. Also, only 10 nm AgNPs induced ROS generation on their own.

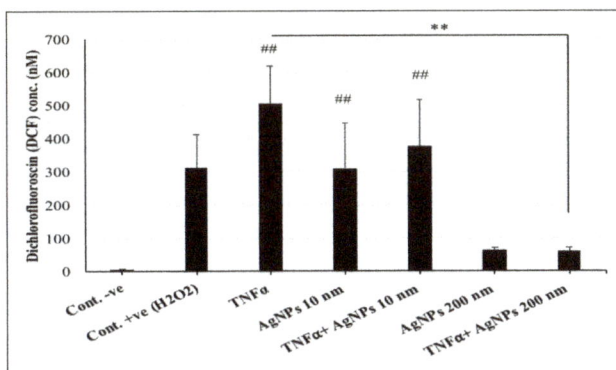

Figure 3. Reactive oxygen species (ROS) production in NCI-H292 cells. Cells were exposed to tumor necrosis factor-α (TNFα) (20 ng/mL) and AgNPs 10 nm (100 μg/mL) or AgNPs 200 nm (100 μg/mL) separately and together for 24 h. ROS production is expressed by the produced DCF amount. The results are shown as means ± SD, $n \geq 3$, for each group; ## and ** indicate $P < 0.01$. ## represents a significant increase compared to the Control -ve group, ** represents a significant decrease compared to the TNFα-exposed group. Cont. -ve are the non-treated cells and Cont. +ve are the cells exposed to H2O2 to induce ROS generation.

2.4. Effect of AgNPs on TNFα-Induced DNA Damage

ROS-mediated DNA damage is well known to be induced by TNFα. Since TNFα-induced ROS generation was affected by AgNPs, DNA damage was also evaluated as an important cellular response that is influenced by ROS generation and regulates cell fate.

For this purpose, we transfected a B-cell Translocation Gene 2 (*BTG2*) promoter-reporter plasmid to detect the DNA damage response. As shown in Figure 4, the fold inductions of the *BTG2* response in cells exposed to TNFα (20 ng/mL) only, 10 nm AgNPs (100 μg/mL) only, or both, were significantly increased compared to the control group. However, cells exposed to TNFα (20 ng/mL) + 200 nm AgNPs (100 μg/mL) showed a significant decrease in the fold induction of the response compared to the TNFα-exposed group. These results suggest that 200 nm AgNPs might regulate TNFα-induced DNA damage.

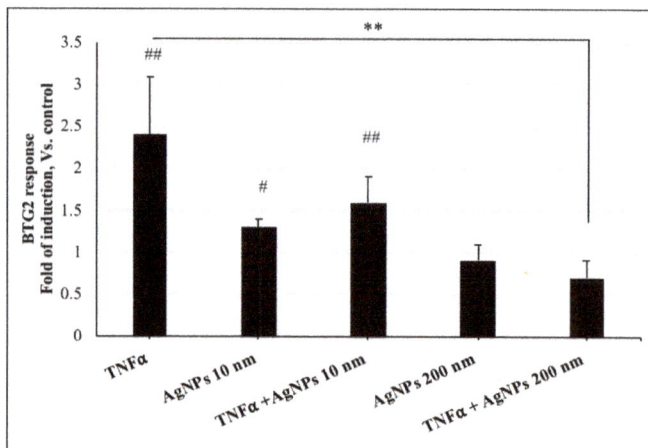

Figure 4. *BTG2* response in NCI-H292 cells. Cells were transfected with *BTG2* promoter-reporter plasmid, and then the transfected cells were exposed to TNFα (20 ng/mL) and AgNPs 10 nm (100 μg/mL) or AgNPs 200 nm (100 μg/mL) separately and together for 24 h. The results are shown as means ± SD, $n \geq 3$, for each group; # indicates $0.01 < P < 0.05$, ##, ** indicates $P < 0.01$. # represents a significant increase compared to the control group. * represents a significant decrease compared to the TNFα -exposed group.

DNA damage is regulated by the DNA damage response (DDR) signaling pathway, which uses signal sensors, transducers, and effectors; the ataxia-telangiectasia mutated (ATM) and ATM- and Rad3-Related (ATR) proteins are the most upstream DDR kinases that are activated by the sensors of the DDR pathway [24]. Therefore, we used PCR array to detect the regulation of genes associated with ATM/ATR signaling as a featured pathway for DDR. Genes that were upregulated by more than 1-fold are listed in Table 1. In particular, the expressions of *ATM*, the CHK1 checkpoint homolog of *Schizosaccharomyces pombe* (*CHEK1*), the RAD21 homolog of *S. pombe* (*RAD21*), structural maintenance of chromosomes 1A (*SMC1A*), and tumor protein p53 (*TP53*) genes were increased by ≥ 1.9-fold in cells exposed to TNFα (20 ng/mL). Also, these same genes exhibited ≥ 1.2-fold induction in cells exposed to TNFα (20 ng/mL) + 10 nm AgNPs (100 μg/mL). However, in cells exposed to TNFα (20 ng/mL) + 200 nm AgNPs (100 μg/mL), the expression of these genes showed downregulation of ≤ 0.8-fold, indicating a reduction in TNFα-induced DNA damage by 200 nm AgNPs. To confirm the induction of the above five genes, we conducted real-time PCR analysis. As shown in Figure 5, none of the genes in cells exposed to TNFα + 10 nm AgNPs showed any significant difference in expression compared to the TNFα-exposed group, except for *SMC1A*, which showed a significant decrease from 2.5- to 1.8-fold induction. In contrast, for cells exposed to TNFα (20 ng/mL) + 200 nm AgNPs (100 μg/mL), all five genes showed significant downregulated expression compared to the TNFα-exposed group, especially *TP53*, *RAD21*, and *CHEK1*, which were downregulated from 2.5 to 0.9, 1.6 to 0.4, and 2.3 to 0.5, respectively. The mRNA expressions of these genes involved in the DDR signaling demonstrated that most of the TNFα-induced upregulated genes were downregulated by 200 nm but not 10 nm AgNPs, suggesting that 200 nm AgNPs reduced the TNFα-induced DNA damage.

Table 1. Induction of mRNA expression of DNA-damage genes in NCI-H292 cells.

Symbol of Genes	Description of the Genes	Fold Regulation Vs. Control		
		TNFα	TNFα + 10 nm AgNPs	TNFα + 200 nm AgNPs
ATM	Ataxia telangiectasia mutated	2.1	1.9	0.8
CDK7	Cyclin-dependent kinase 7	3.8	6.9	1
CHEK1	CHK1 checkpoint homolog (*S. pombe*)	2.9	4.4	0.4
DDIT3	DNA-damage-inducible transcript 3	2.1	2.7	1.8
RAD21	RAD21 homolog (*S. pombe*)	1.9	2.9	0.3
RAD51	RAD51 homolog (*S. cerevisiae*)	1.5	1.6	0.6
SIRT1	Sirtuin 1	1.3	3.5	0.8
SMC1A	Structural maintenance of chromosomes 1A	1.9	1.2	0.7
SUMO1	SMT3 suppressor of mif two 3 homolog 1 (*S. cerevisiae*)	2.5	4.1	1.6
TP53	Tumor protein p53	2.6	2.8	0.6

Cells were exposed to TNFα (20 ng/mL) only, or TNFα + 10 nm AgNPs (100 μg/mL), or TNFα + 200 nm AgNPs (100 μg/mL) for 8 h. Expressions of mRNAs were measured using a DNA damage RT2 Profiler PCR Array. Fold regulation values more than 1 were considered as positive regulation (upregulation).

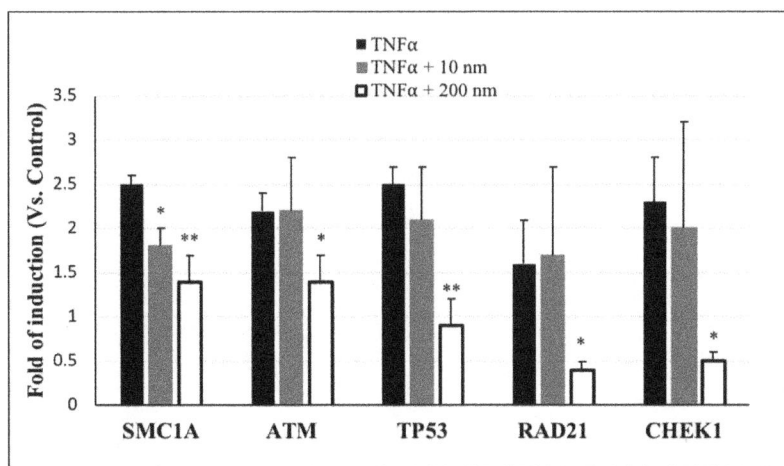

Figure 5. Expression of DNA damage marker mRNAs in NCI-H292 cells. Cells were exposed to TNFα (20 ng/mL) only, TNFα + AgNPs 10 nm (100 μg/mL), or TNFα + AgNPs 200 nm (100 μg/mL) for 8 h. Expressions of mRNAs were measured using RT-PCR. The results are shown as means ± SD, $n \geq 3$, for each group; *$0.01 < P < 0.05$, **$P < 0.01$, and each represents significant differences compared to the corresponding TNFα-exposed group.

2.5. Localization of Tumor Necrosis Factor-α Receptor 1 (TNFR1)

TNFα has two receptors, TNFR1 and TNFR2. TNFR1 is the major receptor, exists in most cell types, and mediates the NF-KB activation signaling pathway, which is involved in ROS generation and DNA damage [25]. Our above-mentioned results indicated that the 200 nm AgNPs, but not the 10 nm AgNPs, affected TNFα-induced ROS production and DNA damage. Also, the cellular uptake of the 200 nm was higher than that of the 10 nm AgNPs. We hypothesized that TNFR1 might have a role in this effect. We examined the localization of TNFR1 by immunofluorescence staining using confocal microscopy as shown in Figure 6. The results revealed that TNFR1 is slightly aggregated and distributed on the membranes of TNFα-exposed cells as shown in Figure 6a. Also, it was distributed over the entire membrane of cells exposed to TNFα (20 ng/mL) + 10 nm AgNPs (100 µg/mL) as shown in Figure 6b. In contrast, Figure 6c shows that TNFR1 was localized inside the cells with very few localized on the membranes of cells exposed to TNFα (20 ng/mL) + 200 nm AgNPs (100 µg/mL). These data suggest that 200 nm AgNPs reduced the expression level of TNFR1 on the cell membrane, and this reduction in surface expression of TNFR1 reduced the signal transduction of TNFα, resulting in a reduction in TNFα-induced DNA damage.

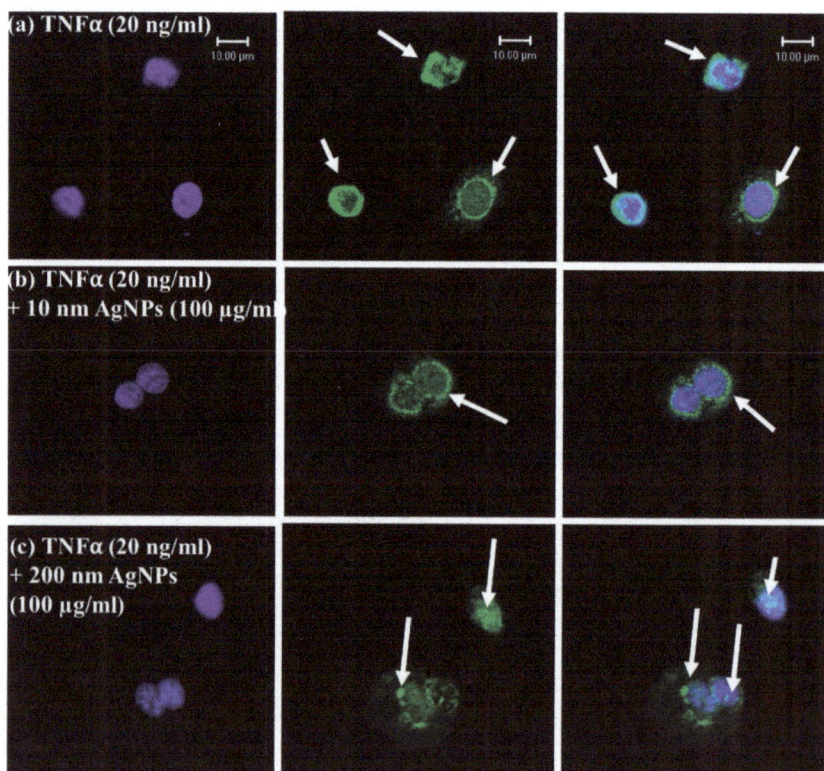

Figure 6. Localization of TNFR1 in NCI-H292 cells using confocal microscopy. Blue shows the nucleus, green shows the receptors (TNFR1), and blue and green together are the merged form. White arrows show TNFR1. (**a**) NCI-H292 cells were exposed to TNFα (20 ng/mL), and TNFR1 was distributed on the cell membrane with some aggregations. (**b**) NCI-H292 cells were exposed to both TNFα (20 ng/mL) + 10 nm AgNPs (100 µg/mL), and TNFR1 localization was scattered over the entire cell membrane. (**c**) NCI-H292 cells were exposed to both TNFα (20 ng/mL) and 200 nm AgNPs (100 µg/mL), and TNFR1 was localized inside cells with very few receptors on the cell membrane. Exposure was 24 h for all experiments. Scale bar is 10 µm for all panels.

3. Discussion

AgNPs are considered to be a double-edged sword that can induce opposing effects. AgNPs have a well-known potential anti-inflammatory effect [26,27], but they can also induce inflammatory responses [28–30]. Moreover, our previous research found an anti-apoptotic effect of AgNPs [31], while some other reports have found that AgNPs can induce apoptosis [32,33]. The size of AgNPs is one of the most important characteristics that modulates their opposing effects. Therefore, size should be clearly determined, and each effect specified for each size. Generally, after the internalization of AgNPs into cells, many different cellular responses are seen such as proliferation, inflammation, DNA damage, and cell death. The determination of specific cellular responses to specific sizes would provide better details about the molecular mechanisms of the induced responses.

Here, we investigated the size-dependent effects of polyvinylpyrrolidone (PVP)-coated AgNPs. We used 10 and 200 nm particles, hypothesizing that they would have different behaviors when interacting with lung epithelial cells. Interestingly, our results showed that the 200 nm particles were less cytotoxic (Figure 1), despite the significant increase in their cellular uptake (Figure 2) compared to the 10 nm AgNPs. These results suggest that thorough uptake of the 200 nm particles by cells might occur via endocytosis of their spheres, and while being held in endosomes they are not easily ionized, which results in their low cytotoxic effect. In contrast, uptake of the 10 nm AgNPs occurred easily through the cell membrane to the cytoplasm. However, the cytoplasmic environment would enhance the ionization of AgNPs, allowing the Ag ions to induce a strong cytotoxic effect. By the same mechanism, the results shown in Figure 3 indicated that ROS generation in cells exposed to 10 nm AgNPs was significantly increased compared to control cells because of this ionization. Dissolution of AgNPs and ion release are always related to their cytotoxicity; it has been found that the smaller nanoparticles are more toxic because of their larger surface area which induces faster dissolution and ion release [34,35]. On the other hand, the PVP coating of AgNPs could increase the stability of the nanoparticles (NPs) and reduce the amount of released Ag ions in the culture medium [36]. Therefore, the difference in the produced cytotoxic effect of 10 nm and 200 nm AgNPs could be due to a combination of both ion release from the nanoparticles and different ways of cellular uptake and uptake ratios.

TNFα is highly expressed and is involved in many acute and chronic inflammatory diseases and cancer; it also induces many different signal transduction pathways that regulate cellular responses [37,38]. Since our goal was to investigate the effects of exposure to different sizes of AgNPs under diseased states, we used TNFα as a DNA damage-inducing agent. The relationship between AgNPs of different sizes and the TNFα-induced DNA damage response was analyzed. The results of DNA damage analysis by *BTG2* response (Figure 4), gene expression by PCR array (Table 1), and RT-PCR (Figure 5) were all consistent with the ROS generation after exposure of the cells to 10 and 200 nm AgNPs. All results confirmed that the 200 nm AgNPs reduced TNFα-induced DNA damage. In contrast, 10 nm AgNPs could induce DNA damage by their own action without affecting that induced by TNFα. These results suggest that the 200 nm AgNPs can reduce DNA damage in diseased conditions that occurs via TNFα.

In order to understand the molecular mechanism of the change in TNFα-induced DNA damage response by the differently sized AgNPs, TNFR1 localization was determined by confocal microscopy. TNFR1 is a receptor of TNFα, and when they bind together TNFα signal transduction is induced. Therefore, TNFR1 might play a role in the different effects of the 10 and 200 nm AgNPs. As shown in Figure 6, in cells exposed to TNFα only, TNFR1 was distributed on the cell membrane surface with few aggregations. Also, in cells exposed to TNFα and 10 nm AgNPs together, TNFR1 was distributed homogenously on the cell membrane. In contrast, TNFR1 was localized mainly inside cells with very few receptors scattered on the membrane surface during exposure to both TNFα and 200 nm AgNPs. These results prompted us to propose the molecular mechanism shown in Figure 7. In cells exposed to TNFα only, TNFα specifically binds to TNFR1 by receptor/ligand binding, and they move together into cells to release TNFα and free the receptors to return to the cell membrane to bind more TNFα. This normal binding cycle induces TNFα signal transduction, leading to the ROS-mediated DNA damage response. However, in cells exposed to both TNFα and 200 nm AgNPs, the nanoparticles might attach to TNFR1/TNFα to form a TNFR1/TNFα/NP complex, which is then endocytosed inside of cells by receptor-mediated endocytosis. TNFα would then be easily released from the receptor and induce signal transduction, while TNFR1 might remain in complex with the 200 nm AgNPs. This complex might change the receptor properties such as shape, molecular weight, and surface characteristics, resulting in a disturbance of the receptor's normal pathway of returning to the cell membrane, causing less TNFR1 localization on the cell membrane and more inside cells, as shown in Figure 6C. In cells exposed to both TNFα and 10 nm AgNPs, TNFα enters cells by normal TNFR1/TNFα binding. Then, TNFα is released from the receptors and they return to the surface of the cell membrane. Meanwhile, the 10 nm AgNPs uptake by cells occurs through the membrane without any receptor-dependent uptake, so they do not affect the localization of TNFR1 on the membrane surface, as shown in Figure 6B. This mechanism would explain the role of TNFR1 in the size-dependent effect of AgNPs on TNFα-induced DNA damage. It indicates that the 200 nm AgNPs hinder recycling of the TNFR1 to the cell membrane, resulting in a decrease in TNFα signal transduction followed by a decrease in the DNA damage response. In contrast, the 10 nm AgNPs have no effect on localization of TNFR1 to the cell membrane. Therefore, TNFα signal transduction and DNA damage are not affected by 10 nm AgNPs. This indicates an independent mode of action for 10 nm AgNPs.

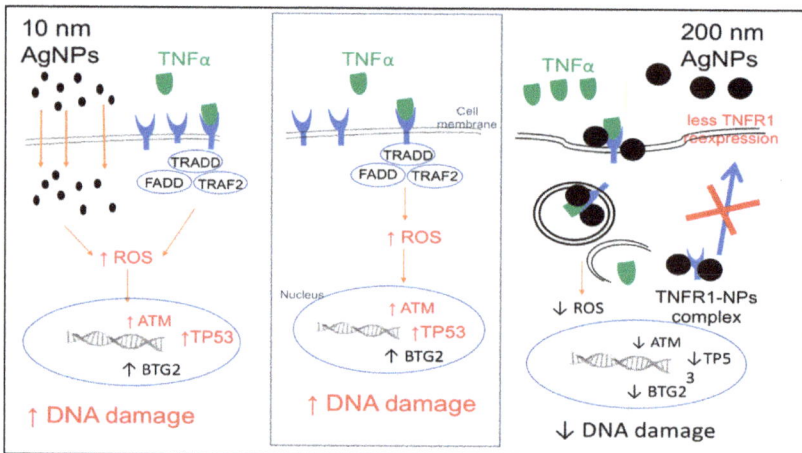

Figure 7. Proposed molecular mechanism explaining why TNFα-induced DNA damage was reduced by 200 nm AgNPs but not by 10 nm AgNPs, and how 200 nm AgNPs decreased membrane localization of TNFR1.

In this study, we investigated the cellular response of the lung epithelial cell line after exposure to TNFα and AgNPs. However, AgNPs not only affect the epithelial cells but also induce changes in the cellular responses of the immune cells specially the macrophages [39,40], therefore, we suggest further testing of the effect of AgNPs on the cellular responses of TNFα in macrophage cell lines.

In addition, the properties of TNFα imply that TNFα blockers are useful as a therapy for many different diseases like Alzheimer's disease [41] or as an adjuvant for cancer treatment [42]. There are currently successful applications in the treatment of chronic inflammatory diseases such as rheumatoid arthritis [43,44] using TNFα blockers. Our findings suggest that 200 nm AgNPs could serve as a promising TNFα blocker. Further in vivo testing is needed to discover their therapeutic potential as a new strategy to block TNFα using a laboratory animal model of inflammatory diseases to support our in vitro findings.

4. Materials and Methods

4.1. Cell Culture

Human pulmonary epithelial cell line NCI-H292 (ATCC CRL-1848TM) cells were cultured in an incubator with a humidified atmosphere containing 5% CO2 at 37°C. RPMI-1640 medium (L-glutamine with phenol red, Nacalai Tesque, Japan) supplemented with 10% (v/v) heat-inactivated fetal bovine serum (HFBS, Biowest, USA), 100 µg/mL penicillin, and 10 µg/mL streptomycin (Nacalai Tesque) was used to culture the cells.

4.2. Silver Nanoparticles (AgNPs)

Polyvinylpyrrolidone (PVP)-coated AgNPs with two different sizes of 10 nm and 200 nm (Cat. Nos. 795925 and 796026, respectively; Sigma-Aldrich, USA) were used in this study. Electronic light scattering (zeta potential and particle size analyzer ELSZ-2000, Otsuka Electronics, Japan) was used to analyze the particle sizes and zeta potentials. The average hydrodynamic diameter of 10 nm AgNPs in deionized water was 12.0 ± 1.8 nm, the polydispersity index was 0.191, and the zeta potential was -21.45 mV. For the 200 nm AgNPs, the average hydrodynamic diameter in deionized water was 221.9 ± 50.9 nm, the polydispersity index was 0.026, and the zeta potential was -27.59 mV. AgNP suspensions were concentrated and sterilized by autoclaving (121 °C for 20 min), then the working solutions were prepared by resuspension in RPMI 1640 medium for cell exposure.

4.3. Tumor Necrosis Factor-α (TNFα)

Recombinant human TNFα (Peprotech, USA) was reconstituted in water to 100 µg/mL. The working dilutions were prepared using sterilized culture medium (RPMI-1640) containing a carrier protein.

4.4. Cell Viability Assay

The viability of NCI-H292 cells was measured using a CellTiter-Glo® luminescent cell viability assay (Promega, Madison, WI, USA) according to the manufacturer's protocol. Cells were seeded at a density of 1×104 cells/well in an opaque 96-well plate and incubated at 37 °C and 5% CO$_2$ overnight, then the cells were checked for 80–90% confluency. Cells were exposed to 10 nm and 200 nm AgNPs (final concentrations of 0, 5, 10, 25, 50, 75, and 100 µg/mL) separately for 24 h. CellTiter-Glo® reagent was added to each well, and the percent of viable cells was calculated based on quantification of adenosine triphosphate (ATP) using a luminometer (TECAN, Japan).

4.5. Cellular Uptake Assay

To check the percentage of cells that incorporated AgNPs, NCI-H292 cells were seeded at a concentration of 8×104 cells/well in a 24-well plate (Costar, Washington, DC, USA). After overnight incubation, the cells were exposed to a final concentration of 100 µg/mL of both 10 nm and 200 nm AgNPs. After 24 h of exposure, the cells were washed twice with phosphate buffered saline (PBS) and collected by trypsinization using trypsin EDTA (Wako, Japan) and centrifugation. Then, the cells were resuspended in 1 mL PBS supplemented with 6% HFBS and kept on ice until analysis.

The percentage of cells taking up AgNPs was analyzed using a flow cytometer (FACS, SP6800 spectral analyzer, Sony Biotechnology, Japan). Forward scatter (FSC) is the laser light scattered at narrow angles to the axis of the laser beam and is proportional to the cell size. Side scatter (SSC) is the laser light scattered at a 90° angle to the axis of the laser and is proportional to the intracellular density, which is increased by the uptake of nanoparticles. The mean SSC for each group of cells was calculated depending on the peak intensities of treated cells compared to the control cells using the software supplied with the instrument.

4.6. DCF Assay for Oxidative Stress Determination

To quantify intracellular ROS, a ROS assay kit (OxiSelectTM, Cell Biolabs, Inc., USA) was used. This assay is based on the cell permeable fluorogenic probe 2′,7′-dichlorodihydrofluorescein diacetate (DCFH-DA), which diffuses into cells and is deacetylated by intracellular esterases to the nonfluorescent 2′,7′-dichlorodihydrofluorescein (DCFH). In the presence of ROS, DCFH is rapidly oxidized to the highly fluorescent 2′,7′-dichlorodihydrofluorescein (DCF). The fluorescence intensity is proportional to the intracellular ROS levels.

According to the manufacturer's protocol, cells were cultured at a density of 1×104 cells/well in a black 96-well plate and incubated overnight. Subsequently, media were removed, and cells were washed twice gently with DPBS (14190, GIBCO, Invitrogen, Carlsbad, CA, USA) and incubated with DCFH-DA/media solution for 30 min in the dark at 37 °C. Then, after removing the solution and washing the cells with DPBS, the DCFH-DA-loaded cells were exposed to TNFα (20 ng/mL) and 10 nm AgNPs (100 μg/mL) or 200 nm AgNPs (100 μg/mL) separately and together for 24 h. Parallel sets of wells containing DCFH-DA-loaded cells without any further exposure were used as a negative control. Another set of DCFH-DA-loaded cells were exposed to hydrogen peroxide (H_2O_2) and used as a positive control. The fluorescence of DCF was measured at regular intervals at an excitation/emission wavelength of 480 nm/530 nm using a fluorometric plate reader (Microplate Fluorometer, Twinkle LB 970, BERTHOLD TECHNOLOGIES, BadWildbad, Germany). The amounts of produced DCF were calculated based on a DCF standard curve.

4.7. Dual-Luciferase Reporter Assay for BTG2 Response Assessment

4.7.1. Plasmids Employed

pGL3-Control vector (E1741, Promega) was used as an empty control reporter plasmid. *BTG2* promoter-reporter plasmid (the region from nt −100 to −20 bp of the *BTG2* gene containing a p53 binding site mutation) was used to detect DNA damage. Both reporter plasmids contain SV40 promoters and enhancer sequences that result in strong expression of the luciferase encoding gene (luc+) in different types of mammalian cells. Also, pRL-CMV vector (E2261, Promega), which is a Renilla luciferase-encoding control plasmid, was used as an internal control for variations in the transfection efficiency.

4.7.2. Transfection

pGL3 blank control reporter plasmid or *BTG2* promoter-reporter plasmid and pRL-CMV internal control plasmid were co-transfected into NCI-H292 cells. LipofectamineTM LTX reagent with a PlusTM reagent kit (Invitrogen) was used to perform the transfection according to the manufacturer's protocol. Cells were cultured at a density of 2×105/mL in a 24-well plate and incubated overnight at 37 °C. Opti-MEM medium (Life Technologies, Carlsbad, CA, USA) was used to dilute the Lipofectamine LTX reagent and plasmids, then the Plus reagent was added to the diluted plasmids. Diluted plasmids with Plus reagent were added to the diluted Lipofectamine LTX reagent at 1:1 ratio and incubated for 5 min at room temperature. Finally, the plasmid–lipid complexes were added to the cells and incubated at 37 °C for at least 24 h before exposure to TNFα and AgNPs.

4.7.3. Assessment of Luciferase Activity

After exposure of transfected cells to TNFα (20 ng/mL) and 10 nm AgNPs (100 μg/mL) or 200 nm AgNPs (100 μg/mL) separately and together for 24 h, the luciferase activities were assessed using a Dual-Luciferase Reporter Assay System (E1910, Promega) according to the manufacturer's protocol. Cells were lysed using $1\times$ passive lysis buffer and gentle shaking for 10 min, then cell lysates were transferred to tubes containing luciferase assay reagent II (LAR II). Firefly luciferase (F) signals were measured, then Stop & Glo reagent was added to the tubes and the Renilla luciferase (R) signals were also measured. The firefly and Renilla luciferase signals were recorded using a luminometer (Lumat

LB9507, BERTHOLD TECHNOLOGIES) according to the instrument manual. The changes in luciferase activities were calculated using the following equation:

$$\Delta \text{ Fold activity} = (\text{F}/\text{R}) \text{ sample} \div (\text{F}/\text{R}) \text{ control.}$$

4.8. Gene Expression Analysis

4.8.1. Polymerase Chain Reaction (PCR) Array

To analyze the expression of genes involved in DNA damage, PCR array analysis was conducted as follows. NCI-H292 cells were seeded at a concentration of 4×10^5 cells/60 mm cell culture dish. After overnight incubation, the cells were exposed to TNFα (20 ng/mL) only, or together with 10 nm AgNPs (100 µg/mL) or 200 nm AgNPs (100 µg/mL). After 8 h of exposure, the cells were detached by trypsinization and collected by centrifugation, and then the total cellular RNA was extracted using an RNeasy kit (Qiagen, Germantown, MD, USA) according to the manufacturer's protocol. An aliquot (1 µg) of the extracted total RNA was reverse transcribed into cDNA using a RT2 First Strand kit (SABiosciences/Qiagen), and the expression of 89 Human DNA Damage Signaling Pathway genes was measured using a RT2 profiler PCR array kit (SABiosciences/Qiagen) according to the manufacturer's protocol. PCR array analysis was performed using an ABI PRISM 7000 sequence detection system (Applied Biosystems, Singapore, Singapore).

4.8.2. Real-Time (RT) PCR

For mRNA expression analysis, cells were seeded and exposed to TNFα and AgNPs, then total RNA and cDNA were synthetized as mentioned for the PCR array. The PCR primers for human *SMC1A*, *ATM*, *TP53*, *RAD21*, and *CHEK1* were purchased from SABiosciences/Qiagen. The reaction mixture was composed of 12.5 µL RT2 SYBR Green qPCR Master Mix (SABiosciences/Qiagen), 1 µL 10 µM gene-specific RT2 qPCR forward and reverse primers, 2 µL cDNA, and nuclease-free water to a final volume of 25 µL. Glyceraldehyde-3-phosphate dehydrogenase (*GAPDH*) was used as a house-keeping gene to normalize the data. RT-PCR analysis was performed using the same machine used for PCR array, and the thermocycling conditions were 95°C for 10 min, followed by 40 cycles of 95 °C for 15 s and 60 °C for 1 min.

4.9. Immunostaining and Confocal Laser Scanning Microscopy

To localize tumor necrosis factor receptor 1 (TNFR1), NCI-H292 cells were seeded in a CELLview cell culture dish (Greiner Bio-one North America Inc., Monroe, NC, USA) at a density of 1.5×10^4 cells/compartment and incubated for 24 h. The cells were exposed to TNFα (20 ng/mL) only, or together with 10 nm AgNPs (100 µg/mL) or 200 nm AgNPs (100 µg/mL). After 24 h of exposure, the cells were washed with $1\times$ PBS fixed with 4% formaldehyde solution in PBS (Wako) at room temperature, permeabilized with 0.1% Triton X-100, and then blocked with 10% normal goat serum in PBS for 1 h. The cells were then incubated overnight at 4°C with rabbit polyclonal anti-TNF receptor 1 antibody (Abcam, Cambridge, UK) followed by incubation with labeled goat anti-rabbit IgG H&L (Alexa Fluor 488) (Abcam) for 1 h at room temperature. Nuclear DNA was stained with DAPI (4′, 6-diamidino-2-phenylindole) (Dojindo, Kumamoto, Japan) for 5 min at room temperature. Microscopic observations and images were acquired using a confocal laser-scanning microscope (LSM510 META, Carl Zeiss Inc., Jena, Germany) with a 63×1.4 Plan-Apochromat oil immersion lens.

4.10. Statistical Analysis

Statistical analysis was performed using Student's t-test. Differences and significances between means of different groups were determined using one-way ANOVA with Duncan's multiple comparison tests. P values less than 0.05 were considered statistically different. Data are presented as means $\pm$ standard deviation (SD) with at least three independent replicates ($n \geq 3$).

5. Conclusions

In this study, we found that 200 nm AgNPs, but not 10 nm AgNPs, reduced DNA damage in NCI-H292 cells and proposed a mechanism for this effect. This mechanism works by reducing membrane localization of TNFR1 and thus decreasing TNFα signal transduction, leading to a reduction in TNFα-induced DNA damage. Also, the mechanism explains why 10 nm AgNPs induced ROS-mediated DNA damage by their own action without affecting TNFR1 and TNFα signal transduction.

Author Contributions: A.F. did most of experiments and wrote the initial draft of the manuscript. A.T. contributed to design the study and prepare the manuscript. Both authors have contributed to data interpretation and manuscript revision. Both authors approved the final version of the manuscript and agree to be responsible for the accuracy and integrity of the work.

Acknowledgments: This work was partially supported by the NIMS Molecule & Material Synthesis Platform for use of flow cytometry. Also, the Namiki foundry research-support system enabled the zeta potential and particle size analyses.

Conflicts of Interest: The authors declare no conflicts of interest.

References

1. Nikalje, A.P. Nanotechnology and its Applications in Medicine. *Med. Chem.* **2015**, *5*, 81–89. [CrossRef]
2. Edwards-Jones, V. The benefits of silver in hygiene, personal care and healthcare. *Lett. Appl. Microbiol.* **2009**, *49*, 147–152. [CrossRef] [PubMed]
3. Benn, T.M.; Westerhoff, P. Nanoparticle silver released into water from commercially available sock fabrics. *Environ. Sci. Technol.* **2008**, *42*, 4133–4139. [CrossRef] [PubMed]
4. Ahamed, M.; Karns, M.; Goodson, M.; Rowe, J.; Hussain, S.M.; Schlager, J.J.; Hong, Y. DNA damage response to different surface chemistry of silver nanoparticles in mammalian cells. *Toxicol. Appl. Pharmacol.* **2008**, *233*, 404–410. [CrossRef] [PubMed]
5. Wiegand, C.; Heinze, T.; Hipler, U.C. Comparative in vitro study on cytotoxicity, antimicrobial activity, and binding capacity for pathophysiological factors in chronic wounds of alginate and silver-containing alginate. *Wound Repair Regen.* **2009**, *17*, 511–521. [CrossRef] [PubMed]
6. Fichtner, J.; Güresir, E.; Seifert, V.; Raabe, A. Efficacy of silver- bearing external ventricular drainage catheters: A retro- spective analysis. *J. Neurosurg.* **2010**, *112*, 840–846. [CrossRef] [PubMed]
7. Tang, J.; Xiong, L.; Zhou, G.; Wang, S.; Wang, J.; Liu, L.; Li, J.; Yuan, F.; Lu, S.; Wan, Z.; et al. Silver nanoparticles crossing through and distribution in the blood-brain barrier in vitro. *J. Nanosci. Nanotechnol.* **2010**, *10*, 6313–6317. [CrossRef] [PubMed]
8. Lin, P.C.; Lin, S.; Wang, P.C.; Sridhar, R. Techniques for physicochemical characterization of nanomaterials. *Biotechnol. Adv.* **2014**, *32*, 711–726. [CrossRef] [PubMed]
9. Pleus, R. *Nanotechnologies-Guidance on Physicochemical Characterization of Engineered Nanoscale Materials for Toxicologic Assessment*; ISO: Geneva, Switzerland, 2012.
10. Murdock, R.C.; Braydich-Stolle, L.; Schrand, A.M.; Schlager, J.J.; Hussain, S.M. Characterization of nanomaterial dispersion in solution prior to in vitro exposure using dynamic light scattering technique. *Toxicol. Sci.* **2008**, *101*, 239–253. [CrossRef] [PubMed]
11. Jo, D.H.; Kim, J.H.; Lee, T.G.; Kim, J.H. Size, surface charge, and shape determine therapeutic effects of nanoparticles on brain and retinal diseases. *Nanomedicine* **2015**, *11*, 1603–1611. [CrossRef] [PubMed]
12. Duan, X.P.; Li, Y.P. Physicochemical characteristics of nanoparticles affect circulation, biodistribution, cellular internalization, and trafficking. *Small* **2013**, *9*, 1521–1532. [CrossRef] [PubMed]
13. Albanese, A.; Tang, P.S.; Chan, W.C. The effect of nanoparticle size, shape, and surface chemistry on biological systems. *Annu. Rev. Biomed. Eng.* **2012**, *14*, 1–16. [CrossRef] [PubMed]
14. Liu, W.; Wu, Y.; Wang, C.; Li, H.C.; Wang, T.; Liao, C.Y.; Cui, L.; Zhou, Q.F.; Yan, B.; Jiang, G.B. Impact of silver nano- particles on human cells: Effect of particle size. *Nanotoxicology* **2010**, *4*, 319–330. [CrossRef] [PubMed]
15. Park, M.V.; Neigh, A.M.; Vermeulen, J.P.; de la Fonteyne, L.J.; Verharen, H.W.; Briedé, J.J.; van Loveren, H.; de Jong, W.H. The effect of particle size on the cytotoxicity, inflammation, developmental toxicity and genotoxicity of silver nanoparticles. *Biomaterials* **2011**, *32*, 9810–9817. [CrossRef] [PubMed]

16. Liu, Z.G.; Han, J. Cellular responses to tumor necrosis factor. *Curr. Issues Mol. Biol.* **2001**, *3*, 79–90. [PubMed]

17. Osbor, L.; Kunkel, S.; Nabel, G.J. Tumor necrosis factor alpha and interleukin 1 stimulate the human immunodeficiency virus enhancer by activation of the nuclear factor kappa B. *Proc. Natl. Acad. Sci. USA* **1989**, *86*, 2336–2340. [CrossRef]

18. McLeish, K.R.; Knall, C.; Ward, R.A.; Gerwins, P.; Coxon, P.Y.; Klein, J.B.; Johnson, G.L. Activation of mitogen-activated protein kinase cascades during priming of human neutrophils by TNF-alpha and GM-CSF. *J. Leukoc. Biol.* **1998**, *64*, 537–545. [CrossRef] [PubMed]

19. Lin, Y.; Choksi, S.; Shen, H.; Yang, Q.; Hur, G.M.; Kim, Y.S.; Tran, J.H.; Sergei, A.; Liu, Z. Tumor Necrosis Factor-induced Nonapoptotic Cell Death Requires Receptor-interacting Protein-mediated Cellular Reactive Oxygen Species Accumulation. *J. Biol. Chem.* **2004**, *279*, 10822–10828. [CrossRef] [PubMed]

20. Smith, C.A.; Farrah, T.; Goodwin, R.G. The TNF receptor superfamily of cellular and viral proteins: Activation, costimulation, and death. *Cell* **1994**, *76*, 959–962. [CrossRef]

21. Nagata, S.; Golstein, P. The Fas death factor. *Science* **1995**, *267*, 1449–1456. [CrossRef] [PubMed]

22. Wheelhouse, N.; Chan, Y.S.; Gillies, S.; Caldwell, H.; Ross, J.; Harrison, D.; Prost, S. TNF-α induced DNA damage in primary murine hepatocytes. *Int. J. Mol. Med.* **2003**, *12*, 889–894. [CrossRef] [PubMed]

23. Corda, S.; Laplace, C.; Vicaut, E.; Duranteau, J. Rapid reactive oxygen species production by mitochondria in endothelial cells exposed to tumor necrosis factor-alpha is mediated by ceramide. *Am. J. Respir. Cell Mol. Biol.* **2001**, *24*, 762–768. [CrossRef] [PubMed]

24. Zhou, B.B.; Elledge, S.J. The DNA damage response: Putting checkpoints in perspective. *Nature* **2000**, *408*, 433–439. [CrossRef] [PubMed]

25. Lai, C.F.; Shao, J.S.; Behrmann, A.; Krchma, K.; Cheng, S.L.; Towler, D.A. TNFR1-activated reactive oxidative species signals up-regulate osteogenic Msx2 programs in aortic myofibroblasts. *Endocrinology* **2012**, *153*, 3897–3910. [CrossRef] [PubMed]

26. Tian, J.; Wong, K.K.; Ho, C.M.; Lok, C.N.; Yu, W.Y.; Che, C.M.; Chiu, J.F.; Tam, P.K. Topical delivery of silver nanoparticles promotes wound healing. *Chem. Med. Chem.* **2007**, *2*, 129–136. [CrossRef] [PubMed]

27. Wong, K.K.; Cheung, S.O.; Huang, L.; Niu, J.; Tao, C.; Ho, C.M.; Che, C.M.; Tam, P.K. Further evidence of the anti-inflammatory effects of silver nanoparticles. *Chem. Med. Chem.* **2009**, *4*, 1129–1135. [CrossRef] [PubMed]

28. Fehaid, A.; Hamed, M.F.; Abouelmagd, M.M.; Taniguchi, A. Time- dependent toxic effect and distribution of silver nano- particles compared to silver nitrate after intratracheal instillation in rats. *Am. J. Nanomater.* **2016**, *4*, 12–19.

29. Wu, T.; Tang, M. The inflammatory response to silver and titanium dioxide nanoparticles in the central nervous system. *Nanomedicine* **2018**, *13*, 233–249. [CrossRef] [PubMed]

30. Trickler, W.J.; Lantz, S.M.; Murdock, R.C.; Schrand, A.M.; Robinson, B.L.; Newport, G.D.; Schlager, J.J.; Oldenburg, S.J.; Paule, M.G.; Slikker, W.; et al. Silver Nanoparticle Induced Blood-Brain Barrier Inflammation and Increased Permeability in Primary Rat Brain Microvessel Endothelial Cells. *Toxicol. Sci.* **2010**, *118*, 160–170. [CrossRef] [PubMed]

31. Fehaid, A.; Taniguchi, A. Silver nanoparticles reduce the apoptosis induced by tumor necrosis factor-α. *Sci. Technol. Adv. Mater.* **2018**, *19*, 526–534. [CrossRef] [PubMed]

32. Zhu, B.; Li, Y.; Lin, Z.; Zhao, M.; Xu, T.; Wang, C.; Deng, N. Silver nanoparticles induce HePG-2 cells apoptosis through ROS-mediated signaling pathways. *Nanoscale Res. Lett.* **2016**, *11*, 198–206. [CrossRef] [PubMed]

33. Xue, Y.; Zhang, T.; Zhang, B.; Gong, F.; Huang, Y.; Tang, M. Cytotoxicity and apoptosis induced by silver nanoparticles in human liver HepG2 cells in different dispersion media. *J. Appl. Toxicol.* **2016**, *36*, 352–360. [CrossRef] [PubMed]

34. Choi, O.; Hu, Z. Size dependent and reactive oxygen species related nanosilver toxicity to nitrifying bacteria. *Environ. Sci. Technol.* **2008**, *42*, 4583–4588. [CrossRef] [PubMed]

35. Johnston, H.J.; Hutchison, G.; Christensen, F.M.; Peters, S.; Hankin, S.; Stone, V. A review of the in vivo and in vitro toxicity of silver and gold particulates: Particle attributes and biological mechanisms responsible for the observed toxicity. *Crit. Rev. Toxicol.* **2010**, *4*, 328–346. [CrossRef] [PubMed]

36. Nymark, P.; Catalán, J.; Suhonen, S.; Järventaus, H.; Birkedal, R.; Clausen, P.A.; Jensen, K.A.; Vippola, M.; Savolainen, K.; Norppa, H. Genotoxicity of polyvinylpyrrolidone-coated silver nanoparticles in BEAS 2B cells. *Toxicology* **2013**, *313*, 38–48. [CrossRef] [PubMed]

37. Beutler, B.; Cerami, A. Tumor necrosis, cachexia, shock, and inflammation: A common mediator. *Ann. Rev. Biochem.* **1988**, *57*, 505–518. [CrossRef] [PubMed]

38. Tracey, K.J.; Cerami, A. Tumor necrosis factor, other cytokines and disease. *Annu. Rev. Cell Biol.* **1993**, *9*, 317–343. [CrossRef] [PubMed]

39. Santoro, C.M.; Duchsherer, N.L.; Grainger, D.W. Minimal in vitro antimicrobial efficacy and ocular cell toxicity from silver nanoparticles. *Nanobiotechnology* **2007**, *3*, 55–65. [CrossRef] [PubMed]

40. Ahlberg, S.; Antonopulos, A.; Diendorf, J.; Dringen, R.; Epple, M.; Flöck, R.; Goedecke, W.; Graf, C.; Haberl, N.; Helmlinger, J.; et al. PVP-coated, negatively charged silver nanoparticles: A multi-center study of their physicochemical characteristics, cell culture and in vivo experiments. *Beilstein J. Nanotechnol.* **2014**, *5*, 1944–1965. [CrossRef] [PubMed]

41. Van Horssen, R.; Ten Hagen, T.L.; Eggermont, A.M. TNF-alpha in Cancer Treatment: Molecular Insights, Antitumor Effects, and Clinical Utility. *Oncologist* **2006**, *11*, 397–408. [CrossRef] [PubMed]

42. Anderson, G.; Nakada, M.T.; DeWitte, M. Tumor necrosis factoralpha in the pathogenesis and treatment of cancer. *Curr. Opin. Pharmacol.* **2004**, *4*, 314–320. [CrossRef] [PubMed]

43. van Dullemen, H.M.; van Deventer, S.J.; Hommes, D.W.; Bijl, H.A.; Jansen, J.; Tytgat, G.N.; Woody, J. Treatment of Crohn's disease with anti-tumor necrosis factor chimeric monoclonal antibody (cA2). *Gastroenterology* **1995**, *109*, 129–135. [CrossRef]

44. Bathon, J.M.; Martin, R.W.; Fleischmann, R.M.; Tesser, J.R.; Schiff, M.H.; Keystone, E.C.; Genovese, M.C.; Wasko, M.C.; Moreland, L.W.; Weaver, A.L.; et al. A comparison of etanercept and methotrexate in patients with early rheumatoid arthritis. *N. Engl. J. Med.* **2000**, *343*, 1586–1593. [CrossRef] [PubMed]

International Journal of
Molecular Sciences

MDPI

Review

Impacts of Silver Nanoparticles on Plants: A Focus on the Phytotoxicity and Underlying Mechanism

An Yan and Zhong Chen *

Natural Sciences and Sciences Education, National Institute of Education, Nanyang Technological University, Singapore 637616, Singapore; an.yan@nie.edu.sg
* Correspondence: zhong.chen@nie.edu.sg; Tel.: +65-67903822

Received: 31 January 2019; Accepted: 21 February 2019; Published: 26 February 2019

Abstract: Nanotechnology was well developed during past decades and implemented in a broad range of industrial applications, which led to an inevitable release of nanomaterials into the environment and ecosystem. Silver nanoparticles (AgNPs) are one of the most commonly used nanomaterials in various fields, especially in the agricultural sector. Plants are the basic component of the ecosystem and the most important source of food for mankind; therefore, understanding the impacts of AgNPs on plant growth and development is crucial for the evaluation of potential environmental risks on food safety and human health imposed by AgNPs. The present review summarizes uptake, translocation, and accumulation of AgNPs in plants, and exemplifies the phytotoxicity of AgNPs on plants at morphological, physiological, cellular, and molecular levels. It also focuses on the current understanding of phytotoxicity mechanisms via which AgNPs exert their toxicity on plants. In addition, the tolerance mechanisms underlying survival strategy that plants adopt to cope with adverse effects of AgNPs are discussed.

Keywords: plants; AgNPs; phytotoxicity; uptake; reactive oxygen species (ROS)

1. Introduction

Due to their small size (between 1 and 100 nm) and unique chemical and physical characteristics, engineered nanomaterials (ENMs) were developed and expanded for application in many industrial sectors and daily life. Among various types of ENMs, silver nanoparticles (AgNPs) are the most commonly applied nanomaterial. It is reported that nearly 25% of all nanotechnology consumer products involve AgNPs [1]. Because of their well-known antibacterial and antifungal properties, they can be used in household products, food packaging, textiles, medical devices, antiseptics in healthcare delivery, and personal healthcare [2–6]. AgNPs can also be used in electronic devices and wastewater treatment because of their good electrical conductivity and photochemical properties [6,7].

In the agriculture sector, AgNPs were developed as plant-growth stimulators [8,9], fungicides to prevent fungal diseases [10], or agents to enhance fruit ripening [11,12]. The growing consumption of AgNPs inevitably increases the chance of release into the environment during AgNP synthesis and incorporation into products, as well as handling and recycling or disposal of these products [13–15]. AgNPs are expected to flow into environment as surface waters (e.g., lakes, streams, and rivers) [16], and the main pathway is through biosolids from wastewater treatment [17,18]. Indeed, AgNPs are detected widely in water and soil; they accumulate in the soil or water reservoirs in large quantities [19–21]. An analysis of the wastewater from a sewage treatment plant indicated existence of AgNPs with a size of 9.3 nm and a concentration of 1900 ng/L [22]. Moreover, the concentrations of AgNPs in surface water and sewage treatment are increasing significantly [21,23–25]. In agriculture, AgNP-contaminated water may permeate into fields through fertilization and irrigation [26]. The released AgNPs have the ability to permeate different media and eventually enter the plant rhizosphere [27,28]. Therefore, the AgNPs are inevitably taken up by crops and easily enter into the food chain [29], not only posing impacts on food production and food quality, but also posing a risk to human health [30–33].

Silver is the second most toxic metal to aquatic organisms after mercury [34]. Actually, AgNPs can leach silver ions (Ag^+), which are persistent, bioaccumulative, and highly toxic to organisms [35]. Therefore, the release of AgNPs into ecosystems raises great concerns about their safety and environmental toxicity. As plants are a vital part of ecosystem and the primary trophic level in ecosystems, representing the base of the food chain [36,37], a good understanding of the impacts of AgNPs on plants is of paramount importance for assessing their toxicity [38]. Hence, the present review describes the uptake and translocation of AgNPs, and gives a detailed summarization of the impacts of AgNPs on plants. The phytotoxicity mechanisms via which AgNPs cause impacts on plants and the tolerance mechanisms through which plants alleviate the detrimental effects of AgNPs are discussed for a better understanding of interactions between plants and AgNPs.

2. Uptake and Translocation of AgNPs in Plants

In plants, AgNPs are transported via the intercellular spaces (short-distance transport) and via vascular tissue (long-distance transport) [29,39–41]. After exposure to plants, NPs penetrate cell walls and plasma membranes of epidermal layers in roots, followed by a series of events to enter plant vascular tissues (xylem), and move to the stele. Xylem is the most important vehicle in the distribution and translocation of NPs [42]. Through xylem, AgNPs can be taken up and translocated to leaves. In *Arabidopsis thaliana*, AgNPs can be taken up by the roots and transported to the shoots [29]. Geisler-Lee et al. found that AgNPs was taken up and progressively accumulated in the root tips, from border cells to root cap, epidermis, columella, and initials of the root meristem [39]. A further study indicated that AgNPs attached to the surface of primary roots in *Arabidopsis* and then entered root tips at an early stage after exposure. After 14 days, AgNPs gradually moved into roots and entered lateral root primordia and root hairs. After multiple lateral roots were developed, AgNPs were present in vascular tissue and throughout the whole plant from root to shoot [40].

The cell wall of the root cells is the main site through which AgNPs enter in plant cells [43]. In order to enter into the plant, AgNPs need to penetrate the cell wall and plasma membranes of epidermal layer of roots. The cell wall is a porous network of polysaccharide fiber matrices and, thus, acts as natural sieve [44,45]. The small-sized AgNPs can pass through the pores, whereas larger AgNPs are unable to enter into plant cells and are thereby sieved out [43].

Interestingly, AgNPs can induce the formation of new and large-sized pores, which permits the internalization of large AgNPs through the cell wall [44]. AgNPs can also be transported within the plant cell through the plasmodesmata process [29,46,47]. Plasmodesmata are pores of 50–60 nm in diameter and connect adjacent neighboring plant cells. In *Arabidopsis*, AgNPs are found to aggregate in plasmodesmata and in the cell wall [39], suggesting that there may be blockage of intercellular communication, which may be caused by the mechanical presence of AgNPs at these sites and may affect nutrient intercellular transport [40].

In addition to the root pathway, AgNPs can also be taken up through plant leaves. Geisler-Lee et al. found that if cotyledons of the *Arabidopsis* seedlings were immersed in AgNP-containing medium, AgNPs could be taken up and accumulated in stomatal guard cells [40]. Larue et al. found that AgNPs were effectively trapped on lettuce leaves by the cuticle after foliar exposure, and AgNPs could penetrate the leaf tissue through stomata [48]. In addition, Li et al. compared the uptake of AgNPs in soybean and rice following root versus foliar exposure, and found that foliar exposure resulted in 17–200 times more Ag bioaccumulation than root exposure [49].

Once the AgNPs enter into vascular tissues of crops, they can be taken up and transported to the leaves or other organs through long-distance transport [27,29,40]. Therefore, it is possible that the fruits, seeds, and other edible parts of plants may also be subjected to contamination by AgNPs through translocation.

3. Phytotoxicity of AgNPs

3.1. Phytotoxicity at the Morphological Level

After exposure to AgNPs, significant changes in the morphology of plants were observed. Growth potential, seed germination, biomass, and leaf surface area are the commonly used parameters for assessing the phytotoxicity of AgNPs in plants [27,42,43]. It was demonstrated that AgNP exposure could inhibit seed germination and root growth, and reduce biomass and leaf area. Jiang et al. found that AgNPs significantly decreased plant biomass, inhibited shoot growth, and resulted in root abscission in *Spirodela polyrrhiza* [50]. Kaveh et al. showed that exposure to higher concentrations (from 5 to 20 mg/L) of AgNPs resulted in reduction of the biomass in *Arabidopsis* [51]. Dimkpa et al. found that AgNPs reduced the length of shoots and roots of wheat in a dose-dependent manner in wheat [52]. Similarly, Nair and Chung showed that AgNPs significantly reduced root elongation, and shoot and root fresh weights in rice [53]. Stampoulis et al. demonstrated that AgNPs (>100 mg/L) inhibited seed germination and reduced biomass in zucchini (*Cucurbita pepo*) [54]. Similar results regarding the toxicity on seed germination, biomass accumulation, and root and shoot growth by AgNPs were reported in other studies involving various plant species, including *Arabidopsis* [55], *Brassica nigra* [56], *Lemna* [57], *Phaseolus radiatus* and *Sorghum bicolor* [58], *Lolium multiflorum* [59], rice [60], wheat [61], *Lupinus termis* L. [62], and so on. A summary of compiled descriptions of the effects of AgNPs in plants is shown in Table 1.

Table 1. Summary of studies on phytotoxicity of silver nanoparticles (AgNPs) in plants.

Size (Diameter in nm)	Concentration	Species	Impacts	References
25–70; 7.5–25.0	10, 20, 40, 50 ppm	Wheat (*Triticumaestivum* L.)	Caused various types of chromosomal aberrations	[63]
5–10	0, 0.1, 0.3, 0.5 mg/L	*Lupinus termis* L.	Reduction in shoot and root elongation, shoot and root fresh weights, total chlorophyll, and total protein contents; Decreased sugar contents and caused significant foliar proline accumulation; Caused metabolic disorders	[62]
37.4 ± 13.4 (AgNP-B [a]); 29.0 ± 6.0 (AgNP-PVP [b]); 21.5 ± 4.2 (AgNP-Citrate)	5, 10 µg/mL	Bryophyte (*Physcomitrella patens*)	Inhibited the growth of the protonema; Changed the thylakoid and chlorophyll contents	[64]
79.0 ± 8.0	0.05–2 mg/L	*Lemna minor*	Caused decays on growth rate and fronds per colony; Induced oxidative stress	[65]
3.1–8.7	20, 200, 2000 mg/kg	Wheat (*Triticum aestivum* L.)	Caused lower biomass, shorter plant height, and lower grain weight; Decreases in the contents of micronutrients (Fe, Cu, and Zn); Decreased the contents of arginine and histidine	[61]
17.2 ± 0.3	0, 1, 10 and 30 mg/L (soybean); 0, 0.1, 0.5, 1 mg/L (rice)	Soybean; Rice	Significantly reduced plant biomass; Increased the malondialdehyde and H_2O_2 contents of leaves	[49]
12.9 ± 9.1	0.01, 0.05, 0.1, 0.5, 1 mg/L	*Capsicum annuum*	Decreased plant height and biomass; Causued a significant increase in total cytokinins in the leaves	[12]
20	1000, 3000 µM	*Pisum sativum*	Declined growth, photosynthetic pigments, and chlorophyll fluorescence; Inhibited activities of glutathione reductase (GR) and dehydroascorbate reductase (DHAR).	[66]
61.2 ± 33.9 (AgNP-Citrate); 9.4 ± 1.3 (AgNP-PVP); 5.6 ± 2.1 (AgNP-CTAB[c])	25, 50, 75, 100 µM	*Allium cepa*	Caused oxidative stress; Led to strong reduction of the root growth	[67]
20	5, 10, 20 mg/L	*Allium cepa*	Induced various chromosomal aberrations in both mitotic and meiotic cells	[68]
200–800	1 mg/L	*Trigonella foenum-graecum* L.	Enhancement in plant growth and diosgenin synthesis	[69]
20	10–150 mg/L	*Arabidopsis thaliana*	Inhibited root gravitropism; Reduced auxin accumulation in root tips; Downregulated expression of auxin receptor-related genes	[70]
47	1, 3 mM	Mustard (*Brassica* sp.)	Declined growth of *Brassica* seedlings; Induced oxidative stress	[71]
35, 73	10 mg/L	Cucumber (*Cucumis sativus*); Wheat (*Triticum aestivum* L.)	Teduced growth; Upregulation of genes involved in the ethylene signalling pathway	[72]

Table 1. *Cont.*

Size (Diameter in nm)	Concentration	Species	Impacts	References
5–50	800 µg/kg	*Vicia faba* L.	Declined germination; Decreased shoot and root length; Retarded the process of nodulation; Caused early senescence of root nodules	[73]
20	0, 2, 10, 20 mg/L	Potato (*Solanum tuberosum* L.)	Total reactive oxygen species (ROS) and superoxide anions were increased; Significant increases in the activities of superoxide dismutase (SOD), catalase, ascorbate peroxidase, and glutathione reductase (GR); Higher ion leakage and cell death	[74]
35–40	50, 75 mg/L	*Triticum aestivum*; *Vigna sinensis*; *Brassica juncea*	50 ppm treatment promoted growth and increased root nodulation in cowpea; Improved shoot parameters at 75 ppm in *Brassica*	[75]
2	0, 125, 250, 500 mg/L	*Raphanus sativus*	Water content was reduced; Root and shoot lengths were reduced at 500 mg/L treatment; Significantly less Ca, Mg, B, Cu, Mn, and Zn	[76]
41	100–5000 mg/L	*Arabidopsis thaliana*	Reduced root length, leaf expansion and photosynthetic efficiency; Induced ROS accumulation; Induced Ca^{2+} in cytoplasm, inhibited plasma membrane K^+ efflux and Ca^{2+} influx currents	[77]
100	50–100 µM	*Arabidopsis thaliana*	Accumulated more amino acids	[78]
10	1, 2, 5, 8, 10 mg/L	*Wolffia globosa*	Caused oxidative damage, higher malondialdehyde (MDA) content and an upregulation of SOD activity; Decreased contents of chlorophyll a, carotenoids and soluble protein	[79]
20	5 mg/L	*Arabidopsis thaliana*	111 genes were unique in AgNPs and enriched in three biological functions: response to fungal infection, anion transport, and cell wall/plasma membrane related.	[80]
10, 20, 40, 80	0.2 µg/L	*Arabidopsis thaliana*	Inhibition of root hair development; Repressed transcriptional responses to microbial pathogens, resulting in increased bacterial colonization	[81]
60–100 (Ag_2S-NPs); 15–20 (AgNPs)	0–20mg/L (Ag_2S-NPs); 0–1.6 mg/L (Ag-NPs)	Cowpea (*Vigna unguiculata* L. Walp.); Wheat (*Triticum aestivum* L.)	Ag_2S-NPs reduced growth by up to 52%; Ag accumulated as Ag_2S in the root and shoot tissues after exposed to Ag_2S-NPs	[82]
20	75–300 µg/L	*Arabidopsis thaliana*	Prolonged vegetative and shortened reproductive growth; Decreased germination rates of offspring	[40]
6, 20	0.5, 5, 10 mg/L	*Spirodela polyrhiza*	Dose dependent increase in levels of ROS, SOD, peroxidase, and the antioxidant glutathione content; Chloroplasts accumulated starch grains and had reduced intergranal thylakoids.	[83]
20	0, 0.2, 0.5, 1 mg/L	*Arabidopsis thaliana*	Significantly reduced total chlorophyll and increased anthocyanin content; Increased lipid peroxidation; a dose-dependent increase in ROS production; Significant upregulated the expression of sulfur assimilation, glutathione biosynthesis, glutathione *S*-transferase, and glutathione reductase genes	[84]
20	0, 0.2, 0.5, 1 mg/L	*Oryza sativa* L.	Significant reduction in root elongation, shoot and root fresh weights, total chlorophyll, and carotenoids contents; Caused significant increase in H_2O_2 formation and lipid peroxidation in shoots and roots, increased foliar proline accumulation, and decreased sugar contents; Caused a dose dependent increase in ROS generation; Changes in mitochondrial membrane potential in the roots of seedlings	[53]
8, 45, 47	2–100 µM	*Arabidopsis thaliana*	Induced root growth promotion (RGP) and Cu/Zn superoxide dismutase (CSD2) accumulation; Inhibited ethylene (ET) perception and could interfere with ET biosynthesis	[85]
20,30–60, 70–120, 150	0.1, 1, 10, 100, 1000 mg/L	*Oryza sativa* L.	Seed germination and seedling growth were decreased	[86]
20, 40, 80	67–535 µg/L	*Arabidopsis thaliana*	Inhibited seedling root elongation; AgNPs were apoplastically transported in the cell wall and aggregated at plasmodesmata	[39]
20	5–25 mg/L	*Arabidopsis thaliana*	Upregulation of stress related genes, downregulation of pathogen and hormonal stimuli genes; Oxidative stress	[51]
10	0.2, 0.5, 3 mg/L	*Arabidopsis thaliana*	Root growth inhibition; Disrupted the thylakoid membrane structure and decreased chlorophyll content; Caused alteration of transcription of antioxidant and aquaporin related genes	[55]
11 ± 0.7 (Citrate)	0.05, 0.1, 1, 18.3, 36.7, 73.4 mg/L	*Zea mays Brassica oleracea*	Cell erosion in maize root apical meristem	[87]

Table 1. *Cont.*

Size (Diameter in nm)	Concentration	Species	Impacts	References
18.34	0.30–60 mg/L	*Oryza sativa* L.	Damage the cell morphology and its structural features; Total soluble carbohydrates significantly declined; Caused production of the ROS and local root tissue death	[88]
10	0.5, 1.5, 2.5, 3.5, 5 mg/kg	*Triticum aestivum*	Teduced the length of shoots and roots; Caused oxidative stress in roots; Induced expression of a metallothionein gene involved in detoxification	[52]
10–15	0, 100, 1000 mg/L	Tomatoes (*Lycopersicon esculentum*)	Significant decreases in root growth; Decreased chlorophyll contents and Higher SOD activity; Less fruit productivity,	[89]
5, 10, 25	0.01–100 mg/L	*Arabidopsis thaliana*; poplars	Stimulatory effect on root elongation, fresh weight, and evapotranspiration at sublethal concentrations; Toxicity increased with decreasing AgNPs size	[90]
20 (AgNP-PVP) 6 (AgNP-GA^d)	1, 10, 40 mg/L	Eleven species of wetland plants	40 mg/L AgNPs-GA exposure significantly reduced the germination rate of three species and enhanced the germination rate of one species.	[91]
<100	250, 750 mg/L	*Cucurbita pepo*	Reduction in plant biomass and transpiration	[92]
5–25	0, 5, 10, 20, 40 mg/L	*Phaseolus radiates*; *Sorghum bicolor*	Inhibition of plant growth	[58]
<100	0, 100, 500 mg/L	*Cucurbita pepo*	Decreased rate of transpiration	[93]
60	12.5, 25, 50, 100 mg/L	*Vicia faba*	Increased the number of chromosomal aberrations and micronuclei, and decreased the mitotic index	[94]
190–1100	0, 25, 50, 100, 200 or 400 mg/L	*Brassica juncea*	Increase in root length and increase in vigor index; Improved photosynthetic quantum efficiency and higher chlorophyll contents; Induced the activities of antioxidant enzymes, resulting in reduced reactive oxygen species levels	[95]
20, 100	5 µg/L	*Lemna minor* L.	Inhibition of plant growth	[57]
25	50, 500, 1000 mg/L	*Oryza sativa*	Broken the cell wall and damaged the vacuoles of root cells	[96]
24–55	0–80 mg/L	*Allium cepa*	Induced cell death and DNA damage through generation of ROS	[97]
<100	100 ppm	*Allium cepa*	Disturbed mitosis, reduction in mitotic index, declined metaphase, sticky chromosome, disintegration and breakdown of cell wall	[98]
20	100 mg/L	Green asparagus	Higher ascorbic acid and total chlorophyll contents	[99]

[a] AgNP-B: AgNPs without surface coating; [b] AgNP-PVP: polyvinylpyrrolidone-coated silver nanoparticles; [c] AgNP-CTAB: cetyltrimethylammonium bromide-coated silver nanoparticles; [d] AgNP-GA: gum arabic-coated silver nanoparticles.

3.2. Phytotoxicity at Physiological Level

Phytotoxicity of AgNPs to plants at the physiological level is predicted by reduction of chlorophyll and nutrient uptake, decline of transpiration rate, and alteration of hormone. AgNPs can disrupt the synthesis of chlorophyll in leaves and, thus, affect the photosynthetic system of the plants [43]. Qian et al. showed that AgNPs could accumulate in *Arabidopsis* leaves, further disrupt the thylakoid membrane structure, and decrease chlorophyll content, leading to the inhibition of plant growth [55]. Nair and Chung reported that, after exposure to AgNPs for one week, total chlorophyll and carotenoids contents were decreased significantly in rice (*Oryza sativa* L.) seedlings [53]. Vishwakarma et al. found that AgNPs could accumulate in mustard (*Brassica* sp.) seedlings and caused severe inhibition in photosynthesis [71]. A recent study showed that AgNP exposure changed the thylakoid in *Physcomitrella patens*, and AgNPs decreased the chlorophyll b content and disturbed the balance of some essential elements in the leafy gametophytes [64]. In *Lupinus termis* L. seedlings, after exposure to AgNPs for ten days, the shoot and root elongation and fresh weights, total chlorophyll, and total protein contents were significantly reduced [62]. In *Cucurbita pepo*, the rate of transpiration was remarkably reduced after AgNP exposure [54,92,93].

In addition, AgNPs can affect the fluidity and permeability of the membrane and, consequently, influence water and nutrient uptake. Zuverza-Mena et al. demonstrated that AgNP exposure on radish (*Raphanus sativus*) sprout caused a decrease in water content in a dose-dependent manner; the nutrient content (Ca, Mg, B, Cu, Mn, and Zn) was also significantly reduced, suggesting that AgNPs may affect plant growth by changing water and nutrient content [76].

It was reported that AgNPs also affect plant hormones. Sun et al. found that the root gravitropism of *Arabidopsis* seedling was inhibited by exposure to AgNPs in a dose-dependent manner. Further analysis indicated that AgNPs reduced auxin accumulation, while gene expression analysis suggested that auxin receptor-related genes were downregulated upon AgNP exposure [70]. Vinković et al. conducted hormonal analysis using ultra-high-performance liquid chromatography electrospray, and found that AgNP accumulation in pepper tissue resulted in a significant increase in total cytokinin levels, suggesting the importance of cytokinin in the plant's response to AgNPs stress [12]. Wang et al. found that Ag_2S-NPs could reduce the growth of cucumber and wheat; expressions of six genes involved in ethylene signalling pathway were significantly upregulated in cucumber after exposure to Ag_2S-NPs, suggesting that Ag_2S-NPs could affect plant growth through an interface with the ethylene signaling pathway [72].

3.3. Cytotoxicity and Genotoxicity

AgNPs can also cause toxicity at the cellular and molecular level in plants. Many studies showed that the inhibition of plant growth after AgNP exposure is accompanied with alteration of cell structure and cell division. Yin et al. found that *Lolium multiflorum* seedlings failed to develop root hair, and the cortical cells were highly vacuolated and collapsed, while the epidermis and root cap were also damaged after exposure to 40 mg/L AgNPs [59]. Pokhrel and Dubey observed that AgNPs could reduce the size of the vacuole and lead to the reduction of cell turgidity and cell size in maize (*Zea mays* L.) and cabbage (*Brassica oleracea* var. *capitata* L.) [87,100]. Similarly, Mazumdar found that after AgNPs enter the cell of *Brassica campestris*; vacuoles and cell wall integrity were damaged, and other organelles might also be affected [63,101]. Likewise, Mirzajani et al. found that AgNPs with a concentration of to 60 μg/mL could penetrate the cell wall, and damage the cell morphology and its structure in rice [88]. In addition, Kumari et al. reported that AgNP exposure in *Allium cepa* significantly decreased the mitotic index and impaired cell division, resulting in chromatin bridge, stickiness, disturbed metaphase, multiple chromosomal breaks, and cell disintegration [98]. Similarly, Patlolla et al. demonstrated that AgNP treatment significantly increased the chromosomal aberrations and micronuclei, and decreased the mitotic index (MI) in root tip cells of broad bean (*Vicia faba* L.), suggesting that cell cycle and mitosis in root tip cells was disrupted by AgNPs [94]. A recent study confirmed that the root tip cells of wheat could readily internalize the AgNPs. After AgNP internalization, the root tip cells exhibited various types of chromosomal aberrations, such as incorrect orientation at metaphase, chromosomal breakage, spindle dysfunction, fragmentation, unequal separation, and distributed and lagging chromosomes, which seriously interfered with cell function [63]. The uptake, translocation, and major phytotoxicity of AgNPs in plants are illustrated in Figure 1.

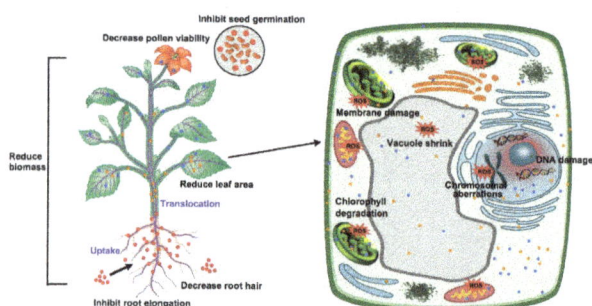

Figure 1. Schematic diagram representing uptake, translocation, and major phytotoxicity of silver nanoparticles (AgNPs) in plant (modified from Reference [102]). Generally, AgNPs are taken up by underground tissues (primary roots and lateral roots), then translocated to aboveground parts (stem, leaf, flower, etc.), where they can reduce biomass, decrease leaf area, affect pollen viability, and inhibit seed germination. At the cellular level, AgNPs enter into various organelles, leading to the production of excess reactive oxygen species (ROS), thereby causing cytotoxicity and genotoxicity, such as membrane damage, chlorophyll degradation, vacuole shrinkage, DNA damage, and chromosomal aberrations.

4. Toxicity Mechanisms

4.1. AgNP-Induced Oxidative Stress

The main mechanism underlying the phytotoxicity of AgNPs is the production of excess reactive oxygen species (ROS) induced by AgNPs, resulting in oxidative stress in plant cells [100,103]. A number of studies demonstrated that ROS production is significantly elevated in plants after exposure to AgNPs. There are four types of ROS produced in plant cells, including singlet oxygen ($^1O_2$), superoxide ($O_2^{\bullet-}$), hydrogen peroxide (H_2O_2), and hydroxyl radical ($HO^\bullet$) [36,104]. Under normal environmental conditions, ROS are generated as byproducts of normal metabolic pathways in organelles such as chloroplasts, mitochondrion, and peroxisomes [36,105]. Under stressed conditions, however, excessive amounts of ROS are generated and cause severe oxidative damage to plant biomolecules through electron transfer [106]. The production of excess ROS induced by AgNP exposure can subsequently lead to oxidative stress, cause peroxidation of polyunsaturated fatty acids (known as lipid peroxidation), and damage the cell membrane permeability and alter cell structure, directly damaging protein and DNA, resulting in potential cell death and growth inhibition in plants (Figure 1) [36,100,107–109]. For example, Panda et al. reported that AgNP-P (phyto-synthesized from silver nitrate AgNO$_3$) or AgNP-S (commercial AgNPs from Sigma–Aldrich) application in *Allium cepa* significantly increased the generation of superoxide ($O_2^{\bullet-}$) and H_2O_2; they also induced cell death to different extents in a dose-dependent fashion, following an order of AgNP-S > AgNP-P at doses $\geq$20 mg/L. Moreover, AgNP-P significantly decreased the mitotic index. Comet assay suggested that DNA damage was significantly enhanced after AgNP-P and AgNP-S treatments in a dose-dependent manner, whereby AgNP-S (threshold dose $\geq$ 10 mg/L) is more genotoxic than AgNP-P (threshold dose $\geq$ 20 mg/L) [97]. Qian et al. found that AgNPs could accumulate in *Arabidopsis* leaves and change the transcription of antioxidant and aquaporin genes, suggesting that AgNPs can change the balance between oxidant and antioxidant systems [55]. Similarly, Speranza et al. checked the in vitro toxicity of AgNPs to kiwifruit pollen, and found that changes in ROS generation paralleled the entire germination dynamics of kiwifruit pollen. The AgNP treatment delayed H_2O_2 production, whereas AgNPs dramatically induced ROS overproduction at the late stage during pollen germination, leading to decreases in pollen viability and performance [110]. Moreover, Torre-Roche et al. found that AgNP exposure with concentration at 500 and 2000 mg/L caused significant increases (54–75%) in malondialdehyde (MDA) formation in soybean (*Glycine max*) [111]. MDA is a major peroxidation product under stress conditions and is indicative of the extent of lipid peroxidation [112]. Similarly, Nair and Chung reported that lipid peroxidation increased significantly after exposure to 0.2, 0.5, and 1 mg/L AgNPs in *Arabidopsis* [84]. In rice, Nair and Chung found that exposure to 0.5 and 1 mg/L AgNPs resulted in a significant increase in H_2O_2 formation and lipid peroxidation in shoots and roots; further analysis suggested that ROS production was promoted by AgNPs in a dose-dependent manner [53]. Thiruvengadam et al. reported the impact of AgNP exposure in turnip seedlings, and found that a higher concentration of AgNPs caused excessive generation of superoxide radicals and increased lipid peroxidation; H_2O_2 formation was also significantly increased after exposure to 5 and 10 mg/L AgNPs. Dichlorofluorescein (DCF) fluorescence indicated a sharp increase in ROS production in turnip seedling roots, suggesting the existence of oxidative stress in the roots after AgNP exposure. Further analysis by comet assay and terminal deoxynucleotidyl transferase-mediated dUTP nick end labeling (TUNEL) assay confirmed that DNA damage was significant, suggesting that AgNPs can induce cell death through apoptosis [113].

4.2. Silver-Specific Toxicity

It was shown that AgNPs can leach ionic silver (Ag^+) into the surroundings through the oxidation of zero-valent Ag [114]. During AgNP uptake and translocation, Ag^+ is released from AgNPs, resulting in oxidative stress through the generation of ROS and disturbing cell function, causing phytotoxicity by binding to cell components and modifying their activities [115–117]. For example, Speranza et al. analyzed the ion release kinetics of AgNPs in the pollen culture medium, and found that AgNPs rapidly dissolved into ions and reached a maximum of 11.8 wt.% ion release. The released Ag$^+$ caused a fivefold increase in H$_2$O$_2$ production over controls; moreover, the released Ag$^+$ damaged pollen membranes and inhibited germination to a greater extent than the AgNPs themselves, suggesting that Ag$^+$ may excert its impacts mostly through chemical or physicochemical interactions with nucleic acids to induce DNA damage [110]. A gene expression study by microarray in *Arabidopsis* compared gene expression profiles between AgNP and silver ion (Ag$^+$) treatments and found a significant overlap between genes differentially expressed in the two treatments, suggesting a similarity between plants' responses to AgNPs and Ag$^+$ [51]. Actually, when AgNPs oxidize in water, they can make bonds with anions and transform into the characteristics of heavy metals, which is more hazardous [43,118]. It was demonstrated that the conversion of AgNPs to a complex of anion or heavy metal could cause toxic effects on various living organisms [25,119–121].

Ag$^+$ can also affect photosynthesis through competitive substitution of Cu$^+$ in plastocyanin (Pc). Pc is a soluble copper-binding protein found in the thylakoid lumen of the chloroplast. It functions as an electron carrier to transfer electrons from cytochrome b_6/f to photosystem 1 (PS1) in the photosynthetic electron-transfer (ET) chain [122,123]. Pc contains a type 1 copper site, where the copper ion is surrounded by two histidine ligands (His87, His37) and a cysteine ligand (Cys84) [124]. Ag$^+$ can competitively replace Cu$^+$ and bind to Pc, which results in disturbance or inactivation of the photosynthetic electron transport. Sujak found that Ag-substituted Pc occupied the active Pc electron transfer site of the cytochrome f, and caused a decrease in the turnover of the cytochrome complex [125]. Similarly, Jansson and Hansson demonstrated that Ag(I)-substituted Pc competitively inhibited electron transfer between normal Cu-containing Pc and PS1 [126]. Since both AgNPs and dissolved silver can be toxic to plants, the phytotoxicity of AgNPs becomes complicated, as the plant is subjected to both silver-specific and nanoparticle-specific biological effects [51]. Therefore, it is difficult to say whether the phytotoxicity is caused by ionic sliver or by intrinsic properties of AgNPs in certain AgNP application scenarios.

4.3. AgNP-Specific Toxicity

Although the phytotoxicity of AgNPs was associated with the impact of dissolved Ag$^+$ on plants, the phytotoxicity effect could not be explained solely by the activity of the released Ag$^+$ ions. In some case, AgNPs can be even more toxic than free Ag$^+$ ions even at the same concentrations of Ag$^+$ [127]. In another study, AgNP exposure to *Cucurbita pepo* caused more reduction in biomass and transpiration when compared with bulk Ag [93]. These studies suggest that free Ag$^+$ ions contribute only partially to the phytotoxicity of AgNPs, while the intrinsic properties of AgNPs are critical for the phytotoxicity of AgNPs.

Indeed, the physical interactions between AgNPs and plant cell-transport pathways can influence the phytotoxicity of AgNPs [29]. Uptake of AgNPs into plant tissue may cause inhibition of apoplastic trafficking by clogging of pores and barriers in the cell wall or the nano-sized plasmodesmata, thereby effectively inhibit the apoplastic flow of water and nutrients [43,128].

A number of studies suggested that the effect and phytotoxicity of AgNPs are closely associated with the nature of the interactions between plants and AgNPs, which are determined by the intrinsic properties of AgNPs [100,129]. These physical and chemical properties of AgNPs, including size, shape, exposure concentration, surface coating, Ag form, and aggregation state, greatly influence the effect of AgNPs on different aspects of plant morphology, physiology, and biochemistry [130].

Among these properties, the size of AgNPs is critical for the phytotoxicity of AgNPs [129]. The smaller-sized AgNPs have a larger surface area to mass ratio, which allows better interference with cell membrane function by directly reacting with the membrane. Meanwhile, a higher proportion of the atoms of the particle on the surface can affect the interfacial reactivity and the ability to pass through physiological barriers [129,131]. It was shown that smaller AgNPs could accumulate to higher levels in plants and be more toxic than their bulk particles. Geisler-Lee et al. checked the impact of AgNPs with different sizes (20, 40, and 80 nm) on *Arabidopsis*, and found that smaller AgNPs accumulated more in seedlings than larger AgNPs (20 nm > 40 nm > 80 nm) at low concentrations. Moreover, smaller-sized AgNPs had a greater impact on root browning [39]. In another study, Wang et al. reported that smaller AgNPs (5 and 10 nm) accumulated to higher levles in poplar tissues than the larger 25-nm AgNPs when applied within the particle subinhibitory concentration range; both *Arabidopsis* and poplar showed susceptibility to the toxic effects of AgNPs, and this susceptibility increased with decreasing AgNP size [90]. Various phytotoxicity studies using different sizes of AgNPs suggested that phytotoxicity is negatively correlated with the size of AgNPs, as AgNPs with smaller size are generally more toxic to the plants than larger AgNPs [33,83,91,130]. For example, Yin et al. showed that AgNP toxicity was influenced by AgNP surface area; smaller AgNPs (6 nm) more strongly affected plant growth than larger (25 nm) AgNPs when applied with similar concentrations in *Lolium multiflorum* [59]. Similarly, another study showed that 6-nm gum arabic coated silver nanoparticles (AgNP-GA) have stronger effects on germination and growth of wetland plants than 21-nm polyvinylpyrrolidone-coated silver nanoparticles (AgNP-PVP) [91]. Abdel-Azeem and Elsayed examined the effect of different sizes of AgNPs (20, 50, and 65 nm) on *Vicia faba* and found that the effect of AgNPs on the mitotic index and chromosomal aberrations was AgNP size-dependent, as smaller-sized AgNPs caused a lower mitotic index and root growth values, confirming that smaller AgNPs are more toxic to *Vicia faba* [132].

Although these studies demonstrated that smaller AgNPs cause more phytotoxicity than larger AgNPs, this correlation between AgNP size and phytotoxicity is not always true for every AgNP exposure scenario. For example, Thuesombat et al. examined the effects of different sized AgNPs (20, 30–60, 70–120, and 150 nm) on seed germination and seedling growth in jasmine rice (*Oryza sativa* L. cv. *KDML 105*), and found that smaller AgNPs accumulated to higher levels than larger AgNPs, which is consistent with previous studies. However, both seed germination and seedling growth were decreased with increasing size; the 20-nm AgNPs treatment resulted in the less negative effects on seedling growth when compared to treatment with the larger AgNPs (150 nm), which is contrary to previous reports. Further analysis found that 20-nm AgNPs were trapped in the roots rather than transported to the leaves, thereby causing less phytotoxicity on seedling growth than 150-nm AgNPs [86].

Numerous studies on the phytotoxicity of AgNPs revealed that the phytotoxicity of AgNPs is positively correlated with the concentration of AgNPs during exposure. AgNPs can only cause negative effects on plants when applied with a concentration above a certain threshold. Mirzajani et al. showed that AgNPs were unable to change cell morphology or structure of rice root when present in low concentrations (30 μg/mL), whereas, with an increased concentration of 60 μg/mL, AgNPs not only penetrated the cell wall, but also destroyed the cell morphology and the structural features. Moreover, 30 μg/mL AgNPs even accelerated root growth, while AgNPs at 60 μg/mL restricted root growth [88]. Oukarroum et al. reported that AgNP treatment induced intracellular ROS production in the aquatic plant *Lemna gibba*; the induced oxidative stress was positively correlated with the increasing concentration of AgNPs [133]. Similarly, Thuesombat et al. showed that seed germination and subsequent seedling growth were decreased with increased concentrations of AgNPs in jasmine rice [86]. Cvjetko et al. found that AgNPs induced oxidative stress and exhibited phytotoxicity only when applied in higher concentrations in *Allium cepa* roots [67].

Engineered AgNPs are typically stabilized against aggregation through surface coating, using organic or inorganic compounds to coat the surface of AgNPs to obtain electrostatic, steric, or electrostatic repulsive forces between particles [134]. Surface coating may change AgNP properties such as optical properties, dispersion, and shape [65,135], thereby influencing the toxicity of AgNPs to plants. Cvjetko et al. compared the toxicity of three types of AgNPs with different surface coatings (citrate, polyvinylpyrrolidone (PVP), and cetyltrimethylammonium bromide (CTAB)) on *Allium cepa* roots, and found that plants treated with AgNP-CTAB had significantly higher Ag content than plants treated with AgNP-citrate and AgNP-PVP, leading to strong inhibition of root growth and oxidative damage. Among the treatments of three types of AgNPs, AgNP-CTAB caused the highest toxicity, whereas AgNP-citrate showed the weakest effects, as AgNP-citrate was much bigger in size and aggregated to larger particles. These observations suggest that the toxicity of AgNPs is correlated with the size and surface coating [67]. Similarly, Pereira et al. found that AgNP-PVP was more deleterious on the growth rate and fronds per colony than AgNP-citrate in *Lemna minor*, whereby AgNP-PVP reduced the growth rate 1.5-fold more than AgNP-citrate [65]. In another study, Liang et al. observed the responses of *Physcomitrella patens* to AgNPs with different surface coatings at the gametophyte stages, and found that AgNPs without surface coating caused the worst damage to the chlorophyll of protonemata, whereas AgNP-PVP and AgNP-citrate just displayed negligible influence, suggesting that surface coating alleviated the damage of AgNPs to the chlorophyll of protonemata. However, at the leafy gametophyte stage, exposure to AgNP-citrate led to the highest weight loss of leafy gametophytes, followed by AgNP-PVP and AgNPs without surface coating [64]. These observations suggest that the effects of AgNPs with different surface coatings on plants are complicated and are associated with the stability of AgNPs, as well as different plant systems.

In addition, the morphology of AgNPs also influences the effect of AgNPs on plants. Syu et al. studied the impacts of AgNPs with three different shapes (spherical, decahedral, and triangular) on *Arabidopsis*, and found that decahedral AgNPs induced the highest degree of root growth promotion but the lowest levels of Cu/Zn superoxide dismutase (CSD2) accumulation. Triangular AgNPs also enhanced root growth, whereas spherical AgNPs exhibited no root growth promotion, but induced the highest levels of anthocyanin and CSD2 accumulation, suggesting that different morphologies of AgNPs exhibited different levels of effects on *Arabidopsis* [85]. A schematic diagram of AgNPs-specific toxicity is shown in Figure 2.

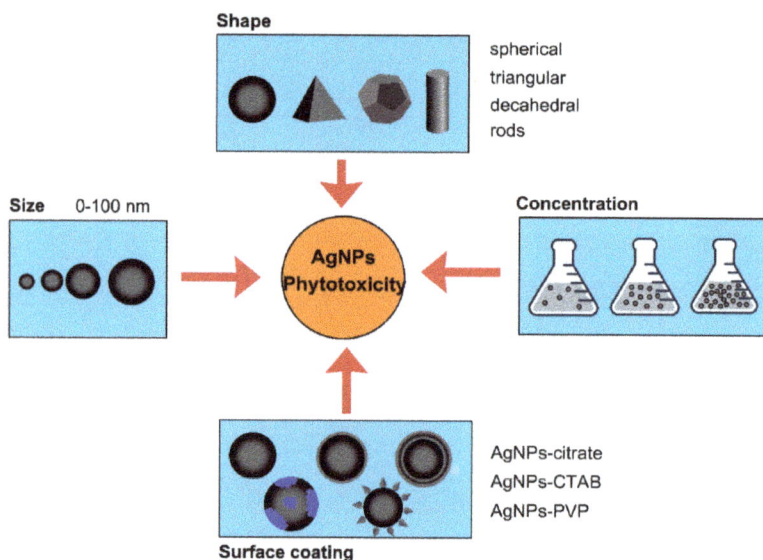

Figure 2. Schematic diagram showing AgNP-specific toxicity. The phytotoxicity of AgNPs is determined by AgNP properties, including size, shape, concentration, and surface coating of AgNPs.

Based on various studies on the phytotoxicity of AgNPs, it is evident that the interaction between plants and AgNPs is highly complicated and is not only dependent on the intrinsic properties of AgNPs, but is also influenced by plant species, developmental stages, different tissues, and sample preparation methodologies.

5. Tolerance Mechanisms

Phytotoxicity of AgNPs is highly associated with oxidative stress, which is caused by the production of excess amounts of ROS after AgNP exposure. To avoid the detrimental effects of

ROS, a set of antioxidant defense mechanisms are activated in plant cells. The defense mechanism involves the activities of enzymatic antioxidants such as superoxide dismutase (SOD), catalase (CAT), ascorbate peroxidase (APX), guaiacol peroxidase (GPX), dehydroascorbate reductase (DHAR), and glutathione reductase (GR) [100,136]. As different types of ROS have different modes of action and exhibit different effects on cellular organelles of plant cells, they can be balanced or removed by specific antioxidant enzymes [36,137]. For example, there are three types of SOD in plant cells, including Fe-SOD, Mn-SOD, and Cu-Zn-SOD, and they can rapidly convert highly toxic ROS ($O_2^{\bullet-}$) to less toxic species (H_2O_2). CAT can convert H_2O_2 to H_2O and O_2. APX is able to convert H_2O_2 to H_2O via ascorbate oxidation into monodehydroascorbate (MDA) and dehydroascorbate (DHA), both of which can be recycled to produce more ascorbate via the catalysis of MDA reductase (MADR) and DHAR [36]. Upon exposure to AgNPs, activities of these enzymatic antioxidants are elevated in plant cells to protect the cells from oxidative stress. For example, Zou et al. observed obvious oxidative damage to *Wolffia globosa* when the plants were exposed to 10 mg/L AgNPs. Meanwhile, the SOD activity was increased by 2.52 times, suggesting that the ROS-scavenging mechanism was activated [79]. Similarly, elevated SOD activity was also observed after AgNP exposure in tomatoes (*Lycopersicon esculentum*) [89]. Enhancement of peroxidase and catalase activity was also observed in *Bacopa monnieri* (Linn.) after AgNP treatment [74,83,138]. Jiang et al. found that the catalase activity in cells of *Spirodela polyrhiza* was significantly increased. Moreover, the SOD and peroxidase activity, and the antioxidant glutathione content were increased in a dose-dependent manner after exposure to 6-nm AgNPs [83]. In addition, Bagherzadeh Homaee and Ehsanpour examined the effects of AgNPs on potato (*Solanum tuberosum* L.) and observed that the activities of SOD, CAT, APX, and GR were all significantly increased in AgNP-treated plantlets [74].

Non-enzymatic antioxidants, such as anthocyanin, ascorbate, glutathione, and thiols, also contribute to the antioxidant defense mechanisms [100,136]. Anthocyanin is a kind of pigment that is implicated in tolerance to various biotic or abiotic stresses, such as herbivores and pathogens, drought, cold, ultraviolet (UV) radiation, and heavy metals [139]. Anthocyanin commonly serves as a non-enzymatic antioxidant to scavenge free radicals and chelate metals under stress conditions [36,104,106,139]. It was reported that anthocyanin accumulation was significantly induced in the spherical AgNP-treated *Arabidopsis* seedlings and was dose-dependent [85]. Similarly, anthocyanin accumulation was also significantly increased after exposure to higher concentrations of AgNPs in turnip [113]. In addition, other antioxidants such as ascorbic acid, carotenoids, and proline are also implicated in antioxidant defense responses of plants to AgNPs. Carotenoids are able to induce antioxidant activity and potentially reduce the toxic effects of ROS [140,141]. After AgNP exposure, a large increase in shoot carotenoid content was observed in rice, suggesting that plants employ carotenoid to reduce the effects of ROS caused by AgNPs [88]. An increase in ascorbic acid content was observed in *Asparagus officinalis* [99].

At the molecular level, the expression changes of genes that are associated with the response to AgNPs may underlie the antioxidant defense mechanisms of plants in response to AgNPs. Dimkpa et al. checked the transcription of a gene encoding metallothionein (MT), which is a cysteine-rich protein involved in detoxification by metal ion sequestration, and found that the expression of this gene was highly induced after AgNP treatment in wheat (*Triticum aestivum* L.) [52]. A gene expression study by microarray suggested that AgNP exposure to *Arabidopsis* led to the upregulation of genes that are associated with response to metal and oxidative stress, including genes encoding SOD, cytochrome P-450-dependent oxidase, and peroxidase, whereas AgNP exposure caused the downregulation of genes that are involved in response to pathogens and hormonal stimuli [51]. In *Arabidopsis*, the expressions of sulfur assimilation, glutathione biosynthesis, glutathione S-transferase, and glutathione reductase genes were significantly upregulated after exposure to AgNPs [84]. Sulfur metabolism in plants plays an important role in stress tolerance, especially in metal detoxification and in the maintenance of cellular redox homeostasis [117,142]. In addition, exposure of AgNPs to rice seedlings led to the differential transcription of genes associated with oxidative stress tolerance in shoots and roots, such as *FSD1, MSD1, CSD1, CSD2, CATa, CATb, CATc, APXa,* and *APXb* [53].

6. Potential Risk in Human Health Posed by AgNPs via Food Chain

Plants are producers in the ecosystem and represent the primary trophic level in the food chain. Regarding the food safety issue, most of the harvested edible tissues or organs of vegetables or cereals are consumed by livestock and humans. Since AgNPs can be taken up and accumulated in plants, they can further pose a risk to human health through invading the food chain and ultimately transferring to the human body. Actually, it was demonstrated that AgNPs could cycle in the ecosystem through various trophic levels in an aquatic or terrestrial food chain [9,100,143,144].

In aquatic ecosystems, planktonic algae as primary producers are located at the base of the aquatic food chain; therefore, algae were selected as the basic trophic level to investigate trophic transfer of AgNPs in a few studies. McTeer et al. investigated the bioavailability, toxicity, and trophic transfer of AgNPs between the alga *Chlamydomonas reinhardtii* and the grazing crustacean *Daphnia magna*, which belong to two different trophic levels. Nano Ag derived from AgNPs was accumulated into microalgae. After feeding on Ag-containing algae, *Daphnia magna* accumulated nano-derived Ag, confirming the trophic transfer of AgNPs between algae and *Daphnia magna* [145]. Similarly, Kalman et al. studied the bioaccumulation and trophic transfer of AgNPs in a simplified freshwater food chain comprising the green alga *Chlorella vulgaris* and *Daphnia magna*. After AgNPs were accumulated in algae, the Ag-contaminated algae were fed to *Daphnia magna*. Ag uptake in *Daphnia magna* was observed a few days later. Further analysis indicated that diet is the dominant pathway route of Ag uptake in *Daphnia magna* [144]. In addition, a recent study used paddy microcosm systems to estimate the trophic transfer of AgNP-citrate and AgNP-PVP among various trophic level organisms (aquatic plants, biofilms, river snails, and Chinese muddy loaches). After exposure, AgNPs rapidly coagulated and precipitated on the sediment. Stable isotope analysis indicated a close correlation between the Ag content in the prey and that in their corresponding predators, demonstrating the impact of AgNPs on ecological receptors and food chains [146].

In terrestrial food chains, studies on the potential trophic transfer of AgNPs remain scarce. However, the terrestrial trophic transfer of other metallic nanoparticles was investigated, such as AuNPs [147], CeO_2-NPs [148], and La_2O_3-NPs [149]. In a simulated terrestrial food chain, tobacco hornworm (*Manduca sexta*) caterpillars were fed tomato leaf that were surface-contaminated with AuNPs. Later, the transfer of AuNPs from tomato to tobacco hornworm was observed [147]. Hence, these studies imply a possibility that AgNPs may also be transferred in the terrestrial food chains.

Both in vivo and in vitro studies demonstrated the toxicity of AgNPs on mammalian cells. For example, AgNP exposure reduced lung function and produced inflammatory lesions in the lungs of rat [150], and resulted in the accumulation of AgNPs in the olfactory bulbs and in the brain of rats [151]. Since AgNPs can be accumulated and transferred in the food chain, they may become dangerous to humans. Indeed, AgNPs exposure to human cells can stimulate inflammatory and immunological responses, cause oxidative stress, and lead to cellular damage [152–154]. Therefore, there is an urgent need to increase our understanding of the bioaccumulation and trophic transfer of AgNPs in the food chain, which is critical for assessing and mitigating their potential harm to human health.

7. Conclusions and Perspectives

Due to the immense application of AgNPs in various fields in modern society, their dispersal and permeation into the ecosystem became inevitable. Hence, a great concern is arising related to the potential risk of destruction in the ecosystem, decline in food quality and yield, and even undermining of human health imposed by AgNPs. To this concern, understanding how AgNPs transfer through the ecosystem and exert impacts on plants is of crucial importance. During the past decade, the research communities undertook the responsibility to increase our knowledge of the impacts of AgNPs on plants, by carrying out numerous studies regarding the interactions between plants and AgNPs.

Int. J. Mol. Sci. **2019**, *20*, 1003

Most of these studies revealed the detrimental effects of AgNPs on plants in various aspects, including at morphological, physiological, cellular, and molecular levels. However, a few studies reported the positive effects of AgNPs on plant growth and development. These contradictory results indicate the complexity of the responses of plants to AgNPs, which are not only determined by the properties of AgNPs (size, concentration, shape, surface coating, Ag chemical form, etc.), but are also dependent on the plant system used (species, tissue, organ, developmental stage, etc.) and experimental methodology (medium, exposure method, exposure time, etc.)

In response to AgNPs, it is rational that multiple detoxification strategies may be activated; different plant species may employ different detoxification mechanisms to eliminate the toxic effects of AgNPs. Therefore, it is difficult to make a general conclusion on how different detoxification pathways in response to diverse AgNPs conditions are activated in different plant species. To address this issue, it is necessary to use representative species, such as the commonly used model plant *Arabidopsis*, to evaluate the phytotoxicity of AgNPs and tolerance mechanisms. Meanwhile, the establishment of a standardized methodology is required to conduct normalized AgNP exposure, thereby allowing comparisons between different species.

Although joint efforts by research communities generated essential knowledge of the impacts of AgNPs on plants, most of these experimental outcomes were based on laboratory experiments under controlled conditions that are likely far from field conditions, such as the exposure method (hydroponic vs. soil), exposure dosage, and time (acute vs. chronic). Therefore, it is hard to predict whether the phytotoxicity of AgNPs and tolerance mechanisms under laboratory conditions are the same as under field conditions. To this end, the establishment of well-designed, plant life-cycle experimental systems under environmentally realistic conditions is required to accurately evaluate the impacts of AgNPs on plants and to generate environmentally relevant implications.

In addition, most studies performed during the last decade focused on the impacts of AgNPs on plants at the morphological and physiological levels; however, the profound impacts of AgNPs at the molecular level did not draw enough attention. Benefits from the development of systems biology and multiple omics methodologies, such as transcriptomics, proteomics, and metabonomics, can be employed in future studies to comprehensively assess the phytotoxicity mechanism of AgNPs and tolerance mechanisms in plants.

Author Contributions: Z.C. conceived and designed the topic and content. A.Y. and Z.C. wrote the article.

Acknowledgments: The authors thank funding supports from the NIE AcRF grant (RI 8/16 CZ) and the Singapore National Parks Board grant (NParks-Phytoremediation).

Conflicts of Interest: The authors declare no conflict of interest.

References

1. Vance, M.E.; Kuiken, T.; Vejerano, E.P.; McGinnis, S.P.; Hochella, M.F., Jr.; Rejeski, D.; Hull, M.S. Nanotechnology in the real world: Redeveloping the nanomaterial consumer products inventory. *Beilstein J. Nanotechnol.* **2015**, *6*, 1769–1780. [CrossRef] [PubMed]
2. Ahamed, M.; Posgai, R.; Gorey, T.J.; Nielsen, M.; Hussain, S.M.; Rowe, J.J. Silver nanoparticles induced heat shock protein 70, oxidative stress and apoptosis in *Drosophila melanogaster*. *Toxicol. Appl. Pharm.* **2010**, *242*, 263–269. [CrossRef] [PubMed]
3. Durán, N.; Marcato, P.D.; De Souza, G.I.; Alves, O.L.; Esposito, E. Antibacterial effect of silver nanoparticles produced by fungal process on textile fabrics and their effluent treatment. *J. Biomed. Nanotechnol.* **2007**, *3*, 203–208. [CrossRef]
4. Pandian, A.M.K.; Karthikeyan, C.; Rajasimman, M.; Dinesh, M.G. Synthesis of silver nanoparticle and its application. *Ecotoxicol. Environ. Saf.* **2015**, *121*, 211–217. [CrossRef] [PubMed]
5. Rai, M.; Yadav, A.; Gade, A. Silver nanoparticles as a new generation of antimicrobials. *Biotechnol. Adv.* **2009**, *27*, 76–83. [CrossRef] [PubMed]
6. Quang Huy, T.; Van Quy, N.; Anh-Tuan, L. Silver nanoparticles: Synthesis, properties, toxicology, applications and perspectives. *Adv. Nat. Sci. Nanosci. Nanotechnol.* **2013**, *4*, 033001. [CrossRef]

7. Zhang, C.; Hu, Z.; Li, P.; Gajaraj, S. Governing factors affecting the impacts of silver nanoparticles on wastewater treatment. *Sci. Total Environ.* **2016**, *572*, 852–873. [CrossRef] [PubMed]

8. Steinitz, B.; Bilavendran, A.D. Thiosulfate stimulates growth and alleviates silver and copper toxicity in tomato root cultures. *Plant Cell Tissue Organ Cult.* **2011**, *107*, 355–363. [CrossRef]

9. Monica, R.C.; Cremonini, R. Nanoparticles and higher plants. *Caryologia* **2009**, *62*, 161–165. [CrossRef]

10. Alavi, S.; Dehpour, A. Evaluation of the nanosilver colloidal solution in comparison with the registered fungicide to control greenhouse cucumber downy mildew disease in the north of Iran. In Proceedings of the VI International Postharvest Symposium, Antalya, Turkey, 11 November 2010; pp. 1643–1646.

11. Sah, S.; Sorooshzadeh, A.; Rezazadeh, H.; Naghdibadi, H. Effect of nano silver and silver nitrate on seed yield of borage. *J. Med. Plants Res.* **2011**, *5*, 706–710.

12. Vinković, T.; Novák, O.; Strnad, M.; Goessler, W.; Jurašin, D.D.; Parađiković, N.; Vrček, I.V. Cytokinin response in pepper plants (*Capsicum annuum* L.) exposed to silver nanoparticles. *Environ. Res.* **2017**, *156*, 10–18. [CrossRef]

13. Hedberg, J.; Skoglund, S.; Karlsson, M.-E.; Wold, S.; Odnevall Wallinder, I.; Hedberg, Y. Sequential studies of silver released from silver nanoparticles in aqueous media simulating sweat, laundry detergent solutions and surface water. *Environ. Sci. Technol.* **2014**, *48*, 7314–7322. [CrossRef] [PubMed]

14. Künniger, T.; Gerecke, A.C.; Ulrich, A.; Huch, A.; Vonbank, R.; Heeb, M.; Wichser, A.; Haag, R.; Kunz, P.; Faller, M. Release and environmental impact of silver nanoparticles and conventional organic biocides from coated wooden façades. *Environ. Pollut.* **2014**, *184*, 464–471. [CrossRef] [PubMed]

15. Lombi, E.; Donner, E.; Scheckel, K.G.; Sekine, R.; Lorenz, C.; Goetz, N.V.; Nowack, B. Silver speciation and release in commercial antimicrobial textiles as influenced by washing. *Chemosphere* **2014**, *111*, 352–358. [CrossRef] [PubMed]

16. Sun, T.Y.; Gottschalk, F.; Hungerbühler, K.; Nowack, B. Comprehensive probabilistic modelling of environmental emissions of engineered nanomaterials. *Environ. Pollut.* **2014**, *185*, 69–76. [CrossRef] [PubMed]

17. Gottschalk, F.; Sonderer, T.; Scholz, R.W.; Nowack, B. Modeled environmental concentrations of engineered nanomaterials (TiO$_2$, ZnO, Ag, CNT, Fullerenes) for different regions. *Environ. Sci. Technol.* **2009**, *43*, 9216–9222. [CrossRef] [PubMed]

18. Mueller, N.C.; Nowack, B. Exposure modeling of engineered nanoparticles in the environment. *Environ. Sci. Technol.* **2008**, *42*, 4447–4453. [CrossRef] [PubMed]

19. Benn, T.M.; Westerhoff, P. Nanoparticle silver released into water from commercially available sock fabrics. *Environ. Sci. Technol.* **2008**, *42*, 4133–4139. [CrossRef] [PubMed]

20. Kaegi, R.; Sinnet, B.; Zuleeg, S.; Hagendorfer, H.; Mueller, E.; Vonbank, R.; Boller, M.; Burkhardt, M. Release of silver nanoparticles from outdoor facades. *Environ. Pollut.* **2010**, *158*, 2900–2905. [CrossRef] [PubMed]

21. Gottschalk, F.; Nowack, B. The release of engineered nanomaterials to the environment. *J. Environ. Monit.* **2011**, *13*, 1145–1155. [CrossRef] [PubMed]

22. Hoque, M.E.; Khosravi, K.; Newman, K.; Metcalfe, C.D. Detection and characterization of silver nanoparticles in aqueous matrices using asymmetric-flow field flow fractionation with inductively coupled plasma mass spectrometry. *J. Chromatogr. A* **2012**, *1233*, 109–115. [CrossRef] [PubMed]

23. Blaser, S.A.; Scheringer, M.; MacLeod, M.; Hungerbühler, K. Estimation of cumulative aquatic exposure and risk due to silver: Contribution of nano-functionalized plastics and textiles. *Sci. Total Environ.* **2008**, *390*, 396–409. [CrossRef] [PubMed]

24. Gottschalk, F.; Sonderer, T.; Scholz, R.W.; Nowack, B. Possibilities and limitations of modeling environmental exposure to engineered nanomaterials by probabilistic material flow analysis. *Environ. Toxicol. Chem.* **2010**, *29*, 1036–1048. [CrossRef] [PubMed]

25. Fabrega, J.; Luoma, S.N.; Tyler, C.R.; Galloway, T.S.; Lead, J.R. Silver nanoparticles: Behaviour and effects in the aquatic environment. *Environ. Int.* **2011**, *37*, 517–531. [CrossRef] [PubMed]

26. Kaegi, R.; Voegelin, A.; Sinnet, B.; Zuleeg, S.; Hagendorfer, H.; Burkhardt, M.; Siegrist, H. Behavior of metallic silver nanoparticles in a pilot wastewater treatment plant. *Environ. Sci. Technol.* **2011**, *45*, 3902–3908. [CrossRef] [PubMed]

27. Dietz, K.-J.; Herth, S. Plant nanotoxicology. *Trends Plant Sci.* **2011**, *16*, 582–589. [CrossRef] [PubMed]

28. Lazareva, A.; Keller, A.A. Estimating potential life cycle releases of engineered nanomaterials from wastewater treatment plants. *ACS Sustain. Chem. Eng.* **2014**, *2*, 1656–1665. [CrossRef]

29. Ma, X.; Geiser-Lee, J.; Deng, Y.; Kolmakov, A. Interactions between engineered nanoparticles (ENPs) and plants: Phytotoxicity, uptake and accumulation. *Sci. Total Environ.* **2010**, *408*, 3053–3061. [CrossRef] [PubMed]

30. Maynard, A.D.; Warheit, D.B.; Philbert, M.A. The new toxicology of sophisticated materials: Nanotoxicology and beyond. *Toxicol. Sci.* **2011**, *120*, S109–S129. [CrossRef] [PubMed]

31. Beer, C.; Foldbjerg, R.; Hayashi, Y.; Sutherland, D.S.; Autrup, H. Toxicity of silver nanoparticles-Nanoparticle or silver ion? *Toxicol. Lett.* **2012**, *208*, 286–292. [CrossRef] [PubMed]

32. Colman, B.P.; Arnaout, C.L.; Anciaux, S.; Gunsch, C.K.; Hochella, M.F., Jr.; Kim, B.; Lowry, G.V.; McGill, B.M.; Reinsch, B.C.; Richardson, C.J.; et al. Low concentrations of silver nanoparticles in biosolids cause adverse ecosystem responses under realistic field scenario. *PLoS ONE* **2013**, *8*, e57189. [CrossRef] [PubMed]

33. Cvjetko, P.; Zovko, M.; Štefanić, P.P.; Biba, R.; Tkalec, M.; Domijan, A.-M.; Vrček, I.V.; Letofsky-Papst, I.; Šikić, S.; Balen, B. Phytotoxic effects of silver nanoparticles in tobacco plants. *Environ. Sci. Pollut. Res.* **2018**, *25*, 5590–5602. [CrossRef] [PubMed]

34. Moreno-Garrido, I.; Pérez, S.; Blasco, J. Toxicity of silver and gold nanoparticles on marine microalgae. *Mar. Environ. Res.* **2015**, *111*, 60–73. [CrossRef] [PubMed]

35. Ratte, H.T. Bioaccumulation and toxicity of silver compounds: A review. *Environ. Toxicol. Chem.* **1999**, *18*, 89–108. [CrossRef]

36. Ma, C.; White, J.C.; Dhankher, O.P.; Xing, B. Metal-based nanotoxicity and detoxification pathways in higher plants. *Environ. Sci. Technol.* **2015**, *49*, 7109–7122. [CrossRef] [PubMed]

37. Maurer-Jones, M.A.; Gunsolus, I.L.; Murphy, C.J.; Haynes, C.L. Toxicity of engineered nanoparticles in the environment. *Anal. Chem.* **2013**, *85*, 3036–3049. [CrossRef] [PubMed]

38. Gardea-Torresdey, J.L.; Rico, C.M.; White, J.C. Trophic transfer, transformation, and impact of engineered nanomaterials in terrestrial environments. *Environ. Sci. Technol.* **2014**, *48*, 2526–2540. [CrossRef] [PubMed]

39. Geisler-Lee, J.; Wang, Q.; Yao, Y.; Zhang, W.; Geisler, M.; Li, K.; Huang, Y.; Chen, Y.; Kolmakov, A.; Ma, X. Phytotoxicity, accumulation and transport of silver nanoparticles by *Arabidopsis thaliana*. *Nanotoxicology* **2013**, *7*, 323–337. [CrossRef] [PubMed]

40. Geisler-Lee, J.; Brooks, M.; Gerfen, J.R.; Wang, Q.; Fotis, C.; Sparer, A.; Ma, X.; Berg, R.H.; Geisler, M. Reproductive toxicity and life history study of silver nanoparticle effect, uptake and transport in *Arabidopsis thaliana*. *Nanomaterials* **2014**, *4*, 301–318. [CrossRef] [PubMed]

41. Miralles, P.; Church, T.L.; Harris, A.T. Toxicity, uptake, and translocation of engineered nanomaterials in vascular plants. *Environ. Sci. Technol.* **2012**, *46*, 9224–9239. [CrossRef] [PubMed]

42. Aslani, F.; Bagheri, S.; Muhd Julkapli, N.; Juraimi, A.S.; Hashemi, F.S.G.; Baghdadi, A. Effects of engineered nanomaterials on plants growth: An overview. *Sci. World J.* **2014**, *2014*. [CrossRef] [PubMed]

43. Tripathi, D.K.; Tripathi, A.; Singh, S.; Singh, Y.; Vishwakarma, K.; Yadav, G.; Sharma, S.; Singh, V.K.; Mishra, R.K.; Upadhyay, R.G.; et al. Uptake, accumulation and toxicity of silver nanoparticle in autotrophic plants, and heterotrophic microbes: A concentric review. *Front. Microbiol.* **2017**, *8*. [CrossRef] [PubMed]

44. Navarro, E.; Baun, A.; Behra, R.; Hartmann, N.B.; Filser, J.; Miao, A.-J.; Quigg, A.; Santschi, P.H.; Sigg, L. Environmental behavior and ecotoxicity of engineered nanoparticles to algae, plants, and fungi. *Ecotoxicology* **2008**, *17*, 372–386. [CrossRef] [PubMed]

45. Carpita, N.C.; Gibeaut, D.M. Structural models of primary cell walls in flowering plants: Consistency of molecular structure with the physical properties of the walls during growth. *Plant J.* **1993**, *3*, 1–30. [CrossRef] [PubMed]

46. Heinlein, M.; Epel, B.L. Macromolecular transport and signaling through plasmodesmata. *Int. Rev. Cytol.* **2004**. [CrossRef]

47. Lucas, W.J.; Lee, J.-Y. Plasmodesmata as a supracellular control network in plants. *Nat. Rev. Mol. Cell Biol.* **2004**, *5*, 712. [CrossRef] [PubMed]

48. Larue, C.; Castillo-Michel, H.; Sobanska, S.; Cécillon, L.; Bureau, S.; Barthès, V.; Ouerdane, L.; Carrière, M.; Sarret, G. Foliar exposure of the crop *Lactuca sativa* to silver nanoparticles: Evidence for internalization and changes in Ag speciation. *J. Hazard. Mater.* **2014**, *264*, 98–106. [CrossRef] [PubMed]

49. Li, C.-C.; Dang, F.; Li, M.; Zhu, M.; Zhong, H.; Hintelmann, H.; Zhou, D.-M. Effects of exposure pathways on the accumulation and phytotoxicity of silver nanoparticles in soybean and rice. *Nanotoxicology* **2017**, *11*, 699–709. [CrossRef] [PubMed]

50. Jiang, H.-S.; Li, M.; Chang, F.-Y.; Li, W.; Yin, L.-Y. Physiological analysis of silver nanoparticles and AgNO₃ toxicity to *Spirodela polyrhiza*. *Environ. Toxicol. Chem.* **2012**, *31*, 1880–1886. [CrossRef] [PubMed]
51. Kaveh, R.; Li, Y.-S.; Ranjbar, S.; Tehrani, R.; Brueck, C.L.; Van Aken, B. Changes in *Arabidopsis thaliana* gene expression in response to silver nanoparticles and silver ions. *Environ. Sci. Technol.* **2013**, *47*, 10637–10644. [CrossRef] [PubMed]
52. Dimkpa, C.O.; McLean, J.E.; Martineau, N.; Britt, D.W.; Haverkamp, R.; Anderson, A.J. Silver nanoparticles disrupt wheat (*Triticum aestivum* L.) growth in a sand matrix. *Environ. Sci. Technol.* **2013**, *47*, 1082–1090. [CrossRef] [PubMed]
53. Nair, P.M.G.; Chung, I.M. Physiological and molecular level effects of silver nanoparticles exposure in rice (*Oryza sativa* L.) seedlings. *Chemosphere* **2014**, *112*, 105–113. [CrossRef] [PubMed]
54. Stampoulis, D.; Sinha, S.K.; White, J.C. Assay-dependent phytotoxicity of nanoparticles to plants. *Environ. Sci. Technol.* **2009**, *43*, 9473–9479. [CrossRef] [PubMed]
55. Qian, H.; Peng, X.; Han, X.; Ren, J.; Sun, L.; Fu, Z. Comparison of the toxicity of silver nanoparticles and silver ions on the growth of terrestrial plant model *Arabidopsis thaliana*. *J. Environ. Sci.* **2013**, *25*, 1947–1956. [CrossRef]
56. Amooaghaie, R.; Tabatabaei, F.; Ahadi, A.-M. Role of hematin and sodium nitroprusside in regulating *Brassica nigra* seed germination under nanosilver and silver nitrate stresses. *Ecotoxicol. Environ. Saf.* **2015**, *113*, 259–270. [CrossRef] [PubMed]
57. Gubbins, E.J.; Batty, L.C.; Lead, J.R. Phytotoxicity of silver nanoparticles to *Lemna minor* L. *Environ. Pollut.* **2011**, *159*, 1551–1559. [CrossRef] [PubMed]
58. Lee, W.-M.; Kwak, J.I.; An, Y.-J. Effect of silver nanoparticles in crop plants *Phaseolus radiatus* and *Sorghum bicolor*: Media effect on phytotoxicity. *Chemosphere* **2012**, *86*, 491–499. [CrossRef] [PubMed]
59. Yin, L.; Cheng, Y.; Espinasse, B.; Colman, B.P.; Auffan, M.; Wiesner, M.; Rose, J.; Liu, J.; Bernhardt, E.S. More than the Ions: The effects of silver nanoparticles on *Lolium multiflorum*. *Environ. Sci. Technol.* **2011**, *45*, 2360–2367. [CrossRef] [PubMed]
60. Ejaz, M.; Raja, N.I.; Ahmad, M.S.; Hussain, M.; Iqbal, M. Effect of silver nanoparticles and silver nitrate on growth of rice under biotic stress. *IET Nanobiotechnol.* **2018**. [CrossRef] [PubMed]
61. Yang, J.; Jiang, F.; Ma, C.; Rui, Y.; Rui, M.; Adeel, M.; Cao, W.; Xing, B. Alteration of crop yield and quality of wheat upon exposure to silver nanoparticles in a life cycle study. *J. Agric. Food Chem.* **2018**, *66*, 2589–2597. [CrossRef] [PubMed]
62. Al-Huqail, A.A.; Hatata, M.M.; Al-Huqail, A.A.; Ibrahim, M.M. Preparation, characterization of silver phyto nanoparticles and their impact on growth potential of *Lupinus termis* L. seedlings. *Saudi J. Biol. Sci.* **2018**, *25*, 313–319. [CrossRef] [PubMed]
63. Abdelsalam, N.R.; Abdel-Megeed, A.; Ali, H.M.; Salem, M.Z.M.; Al-Hayali, M.F.A.; Elshikh, M.S. Genotoxicity effects of silver nanoparticles on wheat (*Triticum aestivum* L.) root tip cells. *Ecotoxicol. Environ. Saf.* **2018**, *155*, 76–85. [CrossRef] [PubMed]
64. Liang, L.; Tang, H.; Deng, Z.; Liu, Y.; Chen, X.; Wang, H. Ag nanoparticles inhibit the growth of the bryophyte, *Physcomitrella patens*. *Ecotoxicol. Environ. Saf.* **2018**, *164*, 739–748. [CrossRef] [PubMed]
65. Pereira, S.P.P.; Jesus, F.; Aguiar, S.; de Oliveira, R.; Fernandes, M.; Ranville, J.; Nogueira, A.J.A. Phytotoxicity of silver nanoparticles to *Lemna minor*: Surface coating and exposure period-related effects. *Sci. Total Environ.* **2018**, *618*, 1389–1399. [CrossRef] [PubMed]
66. Tripathi, D.K.; Singh, S.; Singh, S.; Srivastava, P.K.; Singh, V.P.; Singh, S.; Prasad, S.M.; Singh, P.K.; Dubey, N.K.; Pandey, A.C.; et al. Nitric oxide alleviates silver nanoparticles (AgNPs)-induced phytotoxicity in *Pisum sativum* seedlings. *Plant Physiol. Biochem.* **2017**, *110*, 167–177. [CrossRef] [PubMed]
67. Cvjetko, P.; Milošić, A.; Domijan, A.-M.; Vinković Vrček, I.; Tolić, S.; Peharec Štefanić, P.; Letofsky-Papst, I.; Tkalec, M.; Balen, B. Toxicity of silver ions and differently coated silver nanoparticles in *Allium cepa* roots. *Ecotoxicol. Environ. Saf.* **2017**, *137*, 18–28. [CrossRef] [PubMed]
68. Saha, N.; Dutta Gupta, S. Low-dose toxicity of biogenic silver nanoparticles fabricated by *Swertia chirata* on root tips and flower buds of *Allium cepa*. *J. Hazard. Mater.* **2017**, *330*, 18–28. [CrossRef] [PubMed]
69. Jasim, B.; Thomas, R.; Mathew, J.; Radhakrishnan, E.K. Plant growth and diosgenin enhancement effect of silver nanoparticles in Fenugreek (*Trigonella foenum-graecum* L.). *Saudi Pharm. J.* **2017**, *25*, 443–447. [CrossRef] [PubMed]

70. Sun, J.; Wang, L.; Li, S.; Yin, L.; Huang, J.; Chen, C. Toxicity of silver nanoparticles to *Arabidopsis*: Inhibition of root gravitropism by interfering with auxin pathway. *Environ. Toxicol. Chem.* **2017**, *36*, 2773–2780. [CrossRef] [PubMed]

71. Vishwakarma, K.; Upadhyay, N.; Singh, J.; Liu, S.; Singh, V.P.; Prasad, S.M.; Chauhan, D.K.; Tripathi, D.K.; Sharma, S. Differential phytotoxic impact of plant mediated silver nanoparticles (AgNPs) and silver nitrate (AgNO$_3$) on *Brassica* sp. *Front. Plant Sci.* **2017**, *8*, 1501. [CrossRef] [PubMed]

72. Wang, P.; Lombi, E.; Sun, S.; Scheckel, K.G.; Malysheva, A.; McKenna, B.A.; Menzies, N.W.; Zhao, F.-J.; Kopittke, P.M. Characterizing the uptake, accumulation and toxicity of silver sulfide nanoparticles in plants. *Environ. Sci.* **2017**, *4*, 448–460. [CrossRef]

73. Abd-Alla, M.H.; Nafady, N.A.; Khalaf, D.M. Assessment of silver nanoparticles contamination on faba bean-*Rhizobium leguminosarum* bv. *viciae-Glomus aggregatum* symbiosis: Implications for induction of autophagy process in root nodule. *Agric. Ecosyst. Environ.* **2016**, *218*, 163–177. [CrossRef]

74. Bagherzadeh Homaee, M.; Ehsanpour, A.A. Silver nanoparticles and silver ions: Oxidative stress responses and toxicity in potato (*Solanum tuberosum* L.) grown in vitro. *Hortic. Environ. Biotechnol.* **2016**, *57*, 544–553. [CrossRef]

75. Mehta, C.M.; Srivastava, R.; Arora, S.; Sharma, A.K. Impact assessment of silver nanoparticles on plant growth and soil bacterial diversity. *3 Biotech* **2016**, *6*, 254. [CrossRef]

76. Zuverza-Mena, N.; Armendariz, R.; Peralta-Videa, J.R.; Gardea-Torresdey, J.L. Effects of silver nanoparticles on radish sprouts: Root growth reduction and modifications in the nutritional value. *Front. Plant Sci.* **2016**, *7*, 90. [CrossRef] [PubMed]

77. Sosan, A.; Svistunenko, D.; Straltsova, D.; Tsiurkina, K.; Smolich, I.; Lawson, T.; Subramaniam, S.; Golovko, V.; Anderson, D.; Sokolik, A.; et al. Engineered silver nanoparticles are sensed at the plasma membrane and dramatically modify the physiology of *Arabidopsis thaliana* plants. *Plant J.* **2016**, *85*, 245–257. [CrossRef] [PubMed]

78. Wen, Y.; Zhang, L.; Chen, Z.; Sheng, X.; Qiu, J.; Xu, D. Co-exposure of silver nanoparticles and chiral herbicide imazethapyr to *Arabidopsis thaliana*: Enantioselective effects. *Chemosphere* **2016**, *145*, 207–214. [CrossRef] [PubMed]

79. Zou, X.; Li, P.; Huang, Q.; Zhang, H. The different response mechanisms of *Wolffia globosa*: Light-induced silver nanoparticle toxicity. *Aquat. Toxicol.* **2016**, *176*, 97–105. [CrossRef] [PubMed]

80. Kohan-Baghkheirati, E.; Geisler-Lee, J. Gene expression, protein function and pathways of *Arabidopsis thaliana* responding to silver nanoparticles in comparison to silver ions, cold, salt, drought, and heat. *Nanomaterials* **2015**, *5*, 436–467. [CrossRef] [PubMed]

81. García-Sánchez, S.; Bernales, I.; Cristobal, S. Early response to nanoparticles in the *Arabidopsis* transcriptome compromises plant defence and root-hair development through salicylic acid signalling. *BMC Genom.* **2015**, *16*, 341. [CrossRef] [PubMed]

82. Wang, P.; Menzies, N.W.; Lombi, E.; Sekine, R.; Blamey, F.P.C.; Hernandez-Soriano, M.C.; Cheng, M.; Kappen, P.; Peijnenburg, W.J.G.M.; Tang, C.; et al. Silver sulfide nanoparticles (Ag$_2$S-NPs) are taken up by plants and are phytotoxic. *Nanotoxicology* **2015**, *9*, 1041–1049. [CrossRef] [PubMed]

83. Jiang, H.-S.; Qiu, X.-N.; Li, G.-B.; Li, W.; Yin, L.-Y. Silver nanoparticles induced accumulation of reactive oxygen species and alteration of antioxidant systems in the aquatic plant *Spirodela polyrhiza*. *Environ. Toxicol. Chem.* **2014**, *33*, 1398–1405. [CrossRef] [PubMed]

84. Nair, P.M.G.; Chung, I.M. Assessment of silver nanoparticle-induced physiological and molecular changes in *Arabidopsis thaliana*. *Environ. Sci. Pollut. Res.* **2014**, *21*, 8858–8869. [CrossRef] [PubMed]

85. Syu, Y.-Y.; Hung, J.-H.; Chen, J.-C.; Chuang, H.-W. Impacts of size and shape of silver nanoparticles on *Arabidopsis* plant growth and gene expression. *Plant Physiol. Biochem.* **2014**, *83*, 57–64. [CrossRef] [PubMed]

86. Thuesombat, P.; Hannongbua, S.; Akasit, S.; Chadchawan, S. Effect of silver nanoparticles on rice (*Oryza sativa* L. cv. *KDML 105*) seed germination and seedling growth. *Ecotoxicol. Environ. Saf.* **2014**, *104*, 302–309. [CrossRef] [PubMed]

87. Pokhrel, L.R.; Dubey, B. Evaluation of developmental responses of two crop plants exposed to silver and zinc oxide nanoparticles. *Sci. Total Environ.* **2013**, *452–453*, 321–332. [CrossRef] [PubMed]

88. Mirzajani, F.; Askari, H.; Hamzelou, S.; Farzaneh, M.; Ghassempour, A. Effect of silver nanoparticles on *Oryza sativa* L. and its rhizosphere bacteria. *Ecotoxicol. Environ. Saf.* **2013**, *88*, 48–54. [CrossRef] [PubMed]

89. Song, U.; Jun, H.; Waldman, B.; Roh, J.; Kim, Y.; Yi, J.; Lee, E.J. Functional analyses of nanoparticle toxicity: A comparative study of the effects of TiO$_2$ and Ag on tomatoes (*Lycopersicon esculentum*). *Ecotoxicol. Environ. Saf.* **2013**, *93*, 60–67. [CrossRef] [PubMed]

90. Wang, J.; Koo, Y.; Alexander, A.; Yang, Y.; Westerhof, S.; Zhang, Q.; Schnoor, J.L.; Colvin, V.L.; Braam, J.; Alvarez, P.J.J. Phytostimulation of Poplars and Arabidopsis exposed to silver nanoparticles and Ag$^+$ at sublethal concentrations. *Environ. Sci. Technol.* **2013**, *47*, 5442–5449. [CrossRef] [PubMed]

91. Yin, L.; Colman, B.P.; McGill, B.M.; Wright, J.P.; Bernhardt, E.S. Effects of silver nanoparticle exposure on germination and early growth of eleven wetland plants. *PLoS ONE* **2012**, *7*, e47674. [CrossRef] [PubMed]

92. Hawthorne, J.; Musante, C.; Sinha, S.K.; White, J.C. Accumulation and phytotoxicity of engineered nanoparticles to *Cucurbita Pepo*. *Int. J. Phytoremediat.* **2012**, *14*, 429–442. [CrossRef] [PubMed]

93. Musante, C.; White, J.C. Toxicity of silver and copper to *Cucurbita pepo*: Differential effects of nano and bulk-size particles. *Environ. Toxicol.* **2012**, *27*, 510–517. [CrossRef] [PubMed]

94. Patlolla, A.K.; Berry, A.; May, L.; Tchounwou, P.B. Genotoxicity of silver nanoparticles in *Vicia faba*: A pilot study on the environmental monitoring of nanoparticles. *Int. J. Environ. Res. Public Health* **2012**, *9*, 1649. [CrossRef] [PubMed]

95. Sharma, P.; Bhatt, D.; Zaidi, M.G.H.; Saradhi, P.P.; Khanna, P.K.; Arora, S. Silver nanoparticle-mediated enhancement in growth and antioxidant status of *Brassica juncea*. *Appl. Biochem. Biotech.* **2012**, *167*, 2225–2233. [CrossRef] [PubMed]

96. Mazumdar, H.; Ahmed, G. Phytotoxicity effect of silver nanoparticles on *Oryza sativa*. *IJ Chemtech. Res.* **2011**, *3*, 1494–1500.

97. Panda, K.K.; Achary, V.M.M.; Krishnaveni, R.; Padhi, B.K.; Sarangi, S.N.; Sahu, S.N.; Panda, B.B. In vitro biosynthesis and genotoxicity bioassay of silver nanoparticles using plants. *Toxicol. In Vitro* **2011**, *25*, 1097–1105. [CrossRef] [PubMed]

98. Kumari, M.; Mukherjee, A.; Chandrasekaran, N. Genotoxicity of silver nanoparticles in *Allium cepa*. *Sci. Total Environ.* **2009**, *407*, 5243–5246. [CrossRef] [PubMed]

99. An, J.; Zhang, M.; Wang, S.; Tang, J. Physical, chemical and microbiological changes in stored green asparagus spears as affected by coating of silver nanoparticles-PVP. *LWT Food Sci. Technol.* **2008**, *41*, 1100–1107. [CrossRef]

100. Tripathi, D.K.; Singh, S.; Singh, S.; Pandey, R.; Singh, V.P.; Sharma, N.C.; Prasad, S.M.; Dubey, N.K.; Chauhan, D.K. An overview on manufactured nanoparticles in plants: Uptake, translocation, accumulation and phytotoxicity. *Plant Physiol. Biochem.* **2017**, *110*, 2–12. [CrossRef] [PubMed]

101. Mazumdar, H. Comparative assessment of the adverse effect of silver nanoparticles to *Vigna radiata* and *Brassica campestris* crop plants. *Int. J. Eng. Res. Appl.* **2014**, *4*, 118–124.

102. Yan, A.; Chen, Z. Detection methods of nanoparticles in plant tissues. In *New Visions in Plant Science*; IntechOpen: London, UK, 2018.

103. Nair, R.; Varghese, S.H.; Nair, B.G.; Maekawa, T.; Yoshida, Y.; Kumar, D.S. Nanoparticulate material delivery to plants. *Plant Sci.* **2010**, *179*, 154–163. [CrossRef]

104. Mourato, M.; Reis, R.; Martins, L.L. Characterization of plant antioxidative system in response to abiotic stresses: A focus on heavy metal toxicity. In *Advances in Selected Plant Physiology Aspects*; IntechOpen: London, UK, 2012.

105. Møller, I.M.; Jensen, P.E.; Hansson, A. Oxidative modifications to cellular components in plants. *Annu. Rev. Plant Biol.* **2007**, *58*, 459–481. [CrossRef] [PubMed]

106. Carocho, M.; Ferreira, I.C.F.R. A review on antioxidants, prooxidants and related controversy: Natural and synthetic compounds, screening and analysis methodologies and future perspectives. *Food Chem. Toxicol.* **2013**, *51*, 15–25. [CrossRef] [PubMed]

107. Capaldi Arruda, S.C.; Diniz Silva, A.L.; Moretto Galazzi, R.; Antunes Azevedo, R.; Zezzi Arruda, M.A. Nanoparticles applied to plant science: A review. *Talanta* **2015**, *131*, 693–705. [CrossRef] [PubMed]

108. Sen Raychaudhuri, S.; Deng, X.W. The role of superoxide dismutase in combating oxidative stress in higher plants. *Bot. Rev.* **2000**, *66*, 89–98. [CrossRef]

109. Yuan, L.; Richardson, C.J.; Ho, M.; Willis, C.W.; Colman, B.P.; Wiesner, M.R. Stress responses of aquatic plants to silver nanoparticles. *Environ. Sci. Technol.* **2018**, *52*, 2558–2565. [CrossRef] [PubMed]

110. Speranza, A.; Crinelli, R.; Scoccianti, V.; Taddei, A.R.; Iacobucci, M.; Bhattacharya, P.; Ke, P.C. In vitro toxicity of silver nanoparticles to kiwifruit pollen exhibits peculiar traits beyond the cause of silver ion release. *Environ. Pollut.* **2013**, *179*, 258–267. [CrossRef] [PubMed]

111. De La Torre-Roche, R.; Hawthorne, J.; Musante, C.; Xing, B.; Newman, L.A.; Ma, X.; White, J.C. Impact of Ag nanoparticle exposure on p,p'-DDE bioaccumulation by *Cucurbita pepo* (Zucchini) and *Glycine max* (Soybean). *Environ. Sci. Technol.* **2013**, *47*, 718–725. [CrossRef] [PubMed]

112. Lin, T.-H.; Huang, Y.-L.; Huang, S.-F. Lipid peroxidation in liver of rats administrated with methyl mercuric chloride. *Biol. Trace Elem. Res.* **1996**, *54*, 33–41. [CrossRef] [PubMed]

113. Thiruvengadam, M.; Gurunathan, S.; Chung, I.-M. Physiological, metabolic, and transcriptional effects of biologically-synthesized silver nanoparticles in turnip (*Brassica rapa* ssp. *rapa* L.). *Protoplasma* **2015**, *252*, 1031–1046. [CrossRef] [PubMed]

114. Dobias, J.; Bernier-Latmani, R. Silver release from silver nanoparticles in natural waters. *Environ. Sci. Technol.* **2013**, *47*, 4140–4146. [CrossRef] [PubMed]

115. Park, H.-J.; Kim, J.Y.; Kim, J.; Lee, J.-H.; Hahn, J.-S.; Gu, M.B.; Yoon, J. Silver-ion-mediated reactive oxygen species generation affecting bactericidal activity. *Water Res.* **2009**, *43*, 1027–1032. [CrossRef] [PubMed]

116. Tripathi, A.; Liu, S.; Singh, P.K.; Kumar, N.; Pandey, A.C.; Tripathi, D.K.; Chauhan, D.K.; Sahi, S. Differential phytotoxic responses of silver nitrate (AgNO$_3$) and silver nanoparticle (AgNPs) in *Cucumis sativus* L. *Plant Gene* **2017**, *11*, 255–264. [CrossRef]

117. Montes, A.; Bisson, M.A.; Gardella, J.A.; Aga, D.S. Uptake and transformations of engineered nanomaterials: Critical responses observed in terrestrial plants and the model plant *Arabidopsis thaliana*. *Sci. Total Environ.* **2017**, *607–608*, 1497–1516. [CrossRef] [PubMed]

118. Mura, S.; Greppi, G.; Irudayaraj, J. Latest developments of nanotoxicology in plants. In *Nanotechnology and Plant Sciences: Nanoparticles and Their Impact on Plants*; Siddiqui, M.H., Al-Whaibi, M.H., Mohammad, F., Eds.; Springer: Cham, Switzerland, 2015; pp. 125–151.

119. Chen, X.; Schluesener, H.J. Nanosilver: A nanoproduct in medical application. *Toxicol. Lett.* **2008**, *176*, 1–12. [CrossRef] [PubMed]

120. Wijnhoven, S.W.P.; Peijnenburg, W.J.G.M.; Herberts, C.A.; Hagens, W.I.; Oomen, A.G.; Heugens, E.H.W.; Roszek, B.; Bisschops, J.; Gosens, I.; Van De Meent, D.; et al. Nano-silver—A review of available data and knowledge gaps in human and environmental risk assessment. *Nanotoxicology* **2009**, *3*, 109–138. [CrossRef]

121. Anjum, N.A.; Gill, S.S.; Duarte, A.C.; Pereira, E.; Ahmad, I. Silver nanoparticles in soil-plant systems. *J. Nanopart. Res.* **2013**, *15*, 1896. [CrossRef]

122. Gross, E.L. Plastocyanin: Structure and function. *Photosynth. Res.* **1993**, *37*, 103–116. [CrossRef] [PubMed]

123. Sigfridsson, K. Plastocyanin, an electron-transfer protein. *Photosynth. Res.* **1998**, *57*, 1–28. [CrossRef]

124. Sas, K.N.; Haldrup, A.; Hemmingsen, L.; Danielsen, E.; Øgendal, L.H. pH-dependent structural change of reduced spinach plastocyanin studied by perturbed angular correlation of γ-rays and dynamic light scattering. *JBIC J. Biol. Inorg. Chem.* **2006**, *11*, 409. [CrossRef] [PubMed]

125. Sujak, A. Interaction between cadmium, zinc and silver-substituted plastocyanin and cytochrome b6f complex—Heavy metals toxicity towards photosynthetic apparatus. *Acta Physiol. Plant* **2005**, *27*, 61–69. [CrossRef]

126. Jansson, H.; Hansson, Ö. Competitive inhibition of electron donation to photosystem 1 by metal-substituted plastocyanin. *Biochim. Biophys. Acta* **2008**, *1777*, 1116–1121. [CrossRef] [PubMed]

127. Navarro, E.; Piccapietra, F.; Wagner, B.; Marconi, F.; Kaegi, R.; Odzak, N.; Sigg, L.; Behra, R. Toxicity of silver nanoparticles to *Chlamydomonas reinhardtii*. *Environ. Sci. Technol.* **2008**, *42*, 8959–8964. [CrossRef] [PubMed]

128. Ruotolo, R.; Maestri, E.; Pagano, L.; Marmiroli, M.; White, J.C.; Marmiroli, N. Plant response to metal-containing engineered nanomaterials: An omics-based perspective. *Environ. Sci. Technol.* **2018**, *52*, 2451–2467. [CrossRef] [PubMed]

129. Wang, P.; Lombi, E.; Zhao, F.-J.; Kopittke, P.M. Nanotechnology: A new opportunity in plant sciences. *Trends Plant Sci.* **2016**, *21*, 699–712. [CrossRef] [PubMed]

130. Rastogi, A.; Zivcak, M.; Sytar, O.; Kalaji, H.M.; He, X.; Mbarki, S.; Brestic, M. Impact of metal and metal oxide nanoparticles on plant: A critical review. *Front. Chem.* **2017**, *5*, 78. [CrossRef] [PubMed]

131. Nel, A.; Xia, T.; Mädler, L.; Li, N. Toxic potential of materials at the nanolevel. *Science* **2006**, *311*, 622–627. [CrossRef] [PubMed]

132. Abdel-Azeem, E.A.; Elsayed, B.A. Phytotoxicity of silver nanoparticles on *Vicia faba* seedlings. *N. Y. Sci. J.* **2013**, *6*, 148–156.

133. Oukarroum, A.; Barhoumi, L.; Pirastru, L.; Dewez, D. Silver nanoparticle toxicity effect on growth and cellular viability of the aquatic plant *Lemna gibba*. *Environ. Toxicol. Chem.* **2013**, *32*, 902–907. [CrossRef] [PubMed]

134. Levard, C.; Hotze, E.M.; Lowry, G.V.; Brown, G.E. Environmental transformations of silver nanoparticles: Impact on stability and toxicity. *Environ. Sci. Technol.* **2012**, *46*, 6900–6914. [CrossRef] [PubMed]

135. Tejamaya, M.; Römer, I.; Merrifield, R.C.; Lead, J.R. Stability of citrate, PVP, and PEG coated silver nanoparticles in ecotoxicology media. *Environ. Sci. Technol.* **2012**, *46*, 7011–7017. [CrossRef] [PubMed]

136. Rico, C.M.; Peralta-Videa, J.R.; Gardea-Torresdey, J.L. Chemistry, biochemistry of nanoparticles, and their role in antioxidant defense system in plants. In *Nanotechnology and Plant Sciences: Nanoparticles and Their Impact on Plants*; Siddiqui, M.H., Al-Whaibi, M.H., Mohammad, F., Eds.; Springer: Cham, Switzerland, 2015; pp. 1–17.

137. Apel, K.; Hirt, H. Reactive oxygen species: Metabolism, oxidative stress, and signal transduction. *Annu. Rev. Plant Biol.* **2004**, *55*, 373–399. [CrossRef] [PubMed]

138. Krishnaraj, C.; Jagan, E.G.; Ramachandran, R.; Abirami, S.M.; Mohan, N.; Kalaichelvan, P.T. Effect of biologically synthesized silver nanoparticles on *Bacopa monnieri* (Linn.) Wettst. plant growth metabolism. *Process Biochem.* **2012**, *47*, 651–658. [CrossRef]

139. Gould, K.S. Nature's Swiss army knife: The diverse protective roles of anthocyanins in leaves. *Biomed Res. Int.* **2004**, *2004*, 314–320. [CrossRef] [PubMed]

140. He, D.; Jones, A.M.; Garg, S.; Pham, A.N.; Waite, T.D. Silver nanoparticle–reactive oxygen species interactions: Application of a charging-discharging model. *J. Phys. Chem. C* **2011**, *115*, 5461–5468. [CrossRef]

141. Chew, B.P.; Park, J.S. Carotenoid action on the immune response. *J. Nutr.* **2004**, *134*, 257S–261S. [CrossRef] [PubMed]

142. Zechmann, B.; Müller, M.; Zellnig, G. Modified levels of cysteine affect glutathione metabolism in plant cells. In *Sulfur Assimilation and Abiotic Stress in Plants*; Khan, N.A., Singh, S., Umar, S., Eds.; Springer: Berlin/Heidelberg, Germany, 2008; pp. 193–206.

143. Tangaa, S.R.; Selck, H.; Winther-Nielsen, M.; Khan, F.R. Trophic transfer of metal-based nanoparticles in aquatic environments: A review and recommendations for future research focus. *Environ. Sci. Nano* **2016**, *3*, 966–981. [CrossRef]

144. Kalman, J.; Paul, K.B.; Khan, F.R.; Stone, V.; Fernandes, T.F. Characterisation of bioaccumulation dynamics of three differently coated silver nanoparticles and aqueous silver in a simple freshwater food chain. *Environ. Chem.* **2015**, *12*, 662–672. [CrossRef]

145. McTeer, J.; Dean, A.P.; White, K.N.; Pittman, J.K. Bioaccumulation of silver nanoparticles into *Daphnia magna* from a freshwater algal diet and the impact of phosphate availability. *Nanotoxicology* **2014**, *8*, 305–316. [CrossRef] [PubMed]

146. Park, H.-G.; Kim, J.I.; Chang, K.-H.; Lee, B.-C.; Eom, I.-C.; Kim, P.; Nam, D.-H.; Yeo, M.-K. Trophic transfer of citrate, PVP coated silver nanomaterials, and silver ions in a paddy microcosm. *Environ. Pollut.* **2018**, *235*, 435–445. [CrossRef] [PubMed]

147. Judy, J.D.; Unrine, J.M.; Rao, W.; Bertsch, P.M. Bioaccumulation of gold nanomaterials by *Manduca sexta* through dietary uptake of surface contaminated plant tissue. *Environ. Sci. Technol.* **2012**, *46*, 12672–12678. [CrossRef] [PubMed]

148. Hawthorne, J.; De la Torre Roche, R.; Xing, B.; Newman, L.A.; Ma, X.; Majumdar, S.; Gardea-Torresdey, J.; White, J.C. Particle-size dependent accumulation and trophic transfer of cerium oxide through a terrestrial food chain. *Environ. Sci. Technol.* **2014**, *48*, 13102–13109. [CrossRef] [PubMed]

149. De la Torre Roche, R.; Servin, A.; Hawthorne, J.; Xing, B.; Newman, L.A.; Ma, X.; Chen, G.; White, J.C. Terrestrial trophic transfer of bulk and nanoparticle La_2O_3 does not depend on particle size. *Environ. Sci. Technol.* **2015**, *49*, 11866–11874. [CrossRef] [PubMed]

150. Sung, J.H.; Ji, J.H.; Yoon, J.U.; Kim, D.S.; Song, M.Y.; Jeong, J.; Han, B.S.; Han, J.H.; Chung, Y.H.; Kim, J.; et al. Lung function changes in Sprague-Dawley rats after prolonged inhalation exposure to silver nanoparticles. *Inhal. Toxicol.* **2008**, *20*, 567–574. [CrossRef] [PubMed]

Int. J. Mol. Sci. **2019**, *20*, 1003

151. Kim, Y.S.; Kim, J.S.; Cho, H.S.; Rha, D.S.; Kim, J.M.; Park, J.D.; Choi, B.S.; Lim, R.; Chang, H.K.; Chung, Y.H.; et al. Twenty-eight-day oral toxicity, genotoxicity, and gender-related tissue distribution of silver nanoparticles in Sprague-Dawley rats. *Inhal. Toxicol.* **2008**, *20*, 575–583. [CrossRef] [PubMed]

152. Luo, Y.-H.; Chang, L.W.; Lin, P. Metal-based nanoparticles and the immune system: Activation, inflammation, and potential applications. *Biomed Res. Int.* **2015**, *2015*, 143720. [CrossRef] [PubMed]

153. Jang, J.; Lim, D.-H.; Choi, I.-H. The impact of nanomaterials in immune system. *Immune Netw.* **2010**, *10*, 85–91. [CrossRef] [PubMed]

154. Arora, S.; Jain, J.; Rajwade, J.M.; Paknikar, K.M. Cellular responses induced by silver nanoparticles: In vitro studies. *Toxicol. Lett.* **2008**, *179*, 93–100. [CrossRef] [PubMed]

International Journal of
Molecular Sciences

MDPI

Review

Silver Nanoparticles Based Ink with Moderate Sintering in Flexible and Printed Electronics

Lixin Mo [1,*], Zhenxin Guo [1], Li Yang [2], Qingqing Zhang [1], Yi Fang [1], Zhiqing Xin [1], Zheng Chen [3], Kun Hu [1], Lu Han [1] and Luhai Li [1,*]

[1]   Beijing Engineering Research Center of Printed Electronics, Beijing Institute of Graphic Communication, Beijing 102600, China; beiyinguozhenxin@163.com (Z.G.); zqq15201169516@163.com (Q.Z.); fangyi@bigc.edu.cn (Y.F.); zhiqingxin@bigc.edu.cn (Z.X.); hukun@bigc.edu.cn (K.H.); hanlu@bigc.edu.cn (L.H.)
[2]   Research Institutes of Sweden (RISE), RISE Bioeconomy, Drottning Kristinas väg 61, 11428 Stockholm, Sweden; li.yang@ri.se
[3]   Shine Optoelectronics (Kunshan) Co., Ltd., Shenzhou Industrial Park, No. 33 Yuanfeng Rd, Kunshan 215300, China; zchen2015@163.com
*   Correspondence: molixin@bigc.edu.cn (L.M.); liluhai@bigc.edu.cn (L.L.)

Received: 9 March 2019; Accepted: 7 April 2019; Published: 29 April 2019

Abstract: Printed electronics on flexible substrates has attracted tremendous research interest research thanks its low cost, large area production capability and environmentally friendly advantages. Optimal characteristics of silver nanoparticles (Ag NPs) based inks are crucial for ink rheology, printing, post-print treatment, and performance of the printed electronics devices. In this review, the methods and mechanisms for obtaining Ag NPs based inks that are highly conductive under moderate sintering conditions are summarized. These characteristics are particularly important when printed on temperature sensitive substrates that cannot withstand sintering of high temperature. Strategies to tailor the protective agents capping on the surface of Ag NPs, in order to optimize the sizes and shapes of Ag NPs as well as to modify the substrate surface, are presented. Different (emerging) sintering technologies are also discussed, including photonic sintering, electrical sintering, plasma sintering, microwave sintering, etc. Finally, applications of the Ag NPs based ink in transparent conductive film (TCF), thin film transistor (TFT), biosensor, radio frequency identification (RFID) antenna, stretchable electronics and their perspectives on flexible and printed electronics are presented.

Keywords: silver nanoparticles; flexible and printed electronics; moderate sintering; protective agent; substrate modification; photonic sintering; transparent conductive film; biosensor

---

1. Introduction

Over the past few decades, silver nanoparticles (Ag NPs) have made a substantial impact on various fields, such as biomedical [1–3], optoelectronics [4,5], catalysis [6–9], imaging [10–12], etc., due to their superior physical, chemical and biological characteristics compared to their macroscale counterparts. For instance, Ag NPs have made great progresses in the development of novel antimicrobial agents [13–16], drug-delivery formulations [17–19], detection and diagnosis platforms [20–22], performance-enhanced biomaterial and medical devices [23,24], etc. In the emerging and fast growing multidisciplinary research field, flexible and printed electronics (FPE), Ag NPs have also been a key component of conductive ink [25–27]. FPE refers to the application of printing technologies for the fabrication of electronic circuits and devices on flexible substrates [28,29]. It differs from the traditional manufacturing technologies of electronic devices, e.g., photolithography, vacuum deposition and electroless plating process. The traditional technologies involve multiple steps, require high cost equipment and production environment (clean room), and the use of environmentally undesirable chemicals, which result usually in the formation of large amounts of waste. In contrast, FPE may be viewed as an additive manufacture method that brings about the possibility of preparing relatively high-resolution devices in a much simpler, faster and more cost-effective way.

Like other emerging science and technologies, advances in materials [30–35] have been a major driving force for FPE, including printable organic and inorganic materials: conductive, semi-conductive and insulative. Among the conductive materials, Ag NPs hold a unique position when making high performance conductive ink because of their high electric conductivity and good oxidation resistance. For Ag NPs based printed electronics, there are two major factors that dominate the conductivity of the printed device, e.g., packability of Ag NPs and sintering. The morphology and size distribution of Ag NPs are responsible for packability. A good packability means a dense Ag NPs based film structure, which is essential for good conductivity. After Ag NPs based conductive ink was printed on the substrate, the sintering process is often needed to remove or decompose the protective agents from the surfaces of Ag NPs, enabling direct physical contacts between Ag NPs, and to establish a dense and conductive network throughout the printed feature. As the devices are usually printed on heat sensitive flexible substrates, it is crucial to keep the sintering in a moderate condition. Thus, obtaining Ag NPs based ink, which only requires for moderate sintering and high conductivity, is of the utmost important for the development of FPE.

In this review, recent developments in Ag NPs based conductive inks with moderate sintering and their applications in the FPE are summarized, with particular emphasis on the methods and mechanisms to achieve highly conductive Ag NPs based ink under moderate sintering. The review describes the relevant strategies in Section 2, including tailoring the protective agents capping on the surfaces of Ag NPs, optimizing the sizes and shapes of the Ag NPs, and substrates modification. Some emerging sintering technologies, e.g., infra-red sintering, intense pulsed light sintering, laser sintering, electrical sintering, plasma sintering and microwave sintering, are also included. Applications of the Ag NPs based ink for FPE devices are presented in Section 3, including the transparent conductive film, thin film transistor, biosensor, stretchable electronics and radio frequency identification antenna. Finally, we conclude this review with a summary and discussions on the perspectives and challenges of the Ag NPs based ink and the related sintering techniques in FPE areas in Section 4.

2. Strategies of Achieving Highly Conductive Ag NPs Based Ink under Moderate Sintering

For Ag NPs based ink, sintering means that the Ag NPs begin to make physical contact with each other and form a continuous percolating network in the printed pattern. To achieve a high conductivity, further sintering is required to transform the initially very small contact areas into thicker necks and, eventually, to a dense layer. In the initial stage of sintering, the driving forces are mainly surface energy reduction due to the Ag NPs' large surface-to-volume ratio, a process known as Ostwald ripening [35]. Ostwald ripening triggers surface and grain boundary diffusion within the coalesced Ag NPs. Grain boundary diffusion allows for neck formation and neck radii increase, which is diminished by the energy required for grain boundary creation. As the sintering develops into a deep level, the relative density of the printed Ag NPs based film increase and the electric conductivity increase too. In this section, we focus the attention on the strategies of obtaining highly conductive Ag NPs based ink under moderate sintering and their mechanisms. The key influential factors related to the moderate sintering of Ag NPs based ink, such as protective agents, Ag NPs size and shapes, substrate modification as well as the emerging selective sintering techniques, are discussed in the following.

2.1. Protective Agents

Protective agents are commonly used to improve the stability of the metallic nanoparticles suspension. It is well known that the protective agents could be adsorbed onto the surface of the nanoparticles thus controlling their nucleation and growth rates as well as preventing agglomeration and sedimentation of the prepared nanoparticles [36–38]. Meanwhile, the adsorbed protective agents, even though as thin as a few nanometers or only in a mono molecular layer, are found to prevent electrons from moving between the metallic nanoparticles and decrease the conductivity of the printed film [39,40]. Thus, post-treatment is usually employed to reduce the protective agents covering and to sinter the metallic nanoparticles, both resulting in improved conductivity. Therefore, a better understanding of the sintering process as well as the effects of the protective agents on the conductivity of the printed Ag NPs based pattern is needed. Usually, two kinds of protective agents are commonly used in Ag NPs based inks: first, the polymers bearing carboxylate, amino or hydroxyl functional groups, such as poly(acrylic acid) (PAA) [41–44], poly(vinyl pyrolidone) (PVP) [45–49] and poly(vinyl alcohol) (PVA) [50,51]; second, the small molecular compounds with a long alkyl chain and polar head, such as alkanethiols [52–54], alkylamines [53,55,56] and carboxylic acids [57,58]. Through investigating the behavior of protective agent in sintering, some efforts have been made to improve the conductivity of the Ag NPs based ink under moderate sintering.

Magdassi et al. [41], Grouchko et al. [42] and Tang et al. [46] realized room temperature sintering of the Ag NPs capped polymer protective agents by adding the destabilizing agents, oppositely charged Cl⁻ containing electrolyte, into the ink to promote the Ag NPs aggregation and coalescence in the drying processes. The optimized electric conductivities achieved were 20%, 41% and 40%, respectively, of that of bulk silver. The destabilizing agents, which contain Cl^- ions, cause detachment of the anchoring groups of the protective agents from the surface of Ag NPs and thus enable their sintering (Figure 1). Further study showed that this sintering is dependent on coalescence and Ostwald ripening spontaneous behaviors of Ag NPs after they have been destabilized. In addition, these two behaviors could be extremely affected by the size of the Ag NPs [46]. On this basis, Layani et al. [59] reported a rapid and simple process to obtain high conductive printed patterns, above 30% of bulk silver, by sequential printing of the Ag NPs based ink and solutions of electrolyte such as NaCl and $MgCl_2$ (Figure 2).

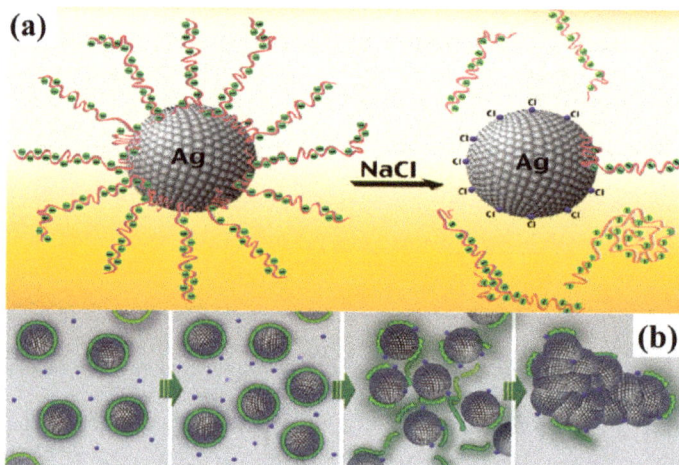

Figure 1. (a) schematic illustration of the Ag NPs before (left) and after (right) the addition of NaCl, and (b) schematic illustration of the protective agents detachment, which leads to the Ag NPs sintering (the green lines represent the polymeric stabilizer; the blue spheres represent the sintering agent). Reproduced with permission from [42]; Copyright 2011 American Chemical Society.

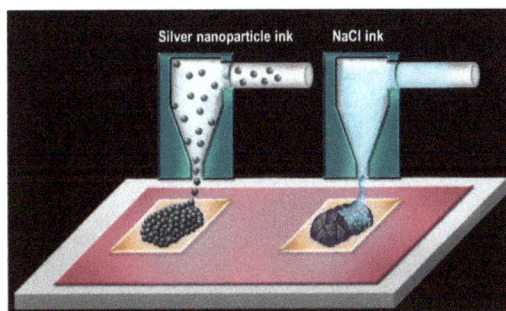

Figure 2. Scheme of the double printing process. First, a pattern of Ag NPs based ink is printed, followed by printing a salt solution on top of the silver pattern. Reproduced with permission from [59]; Copyright 2012 Royal Society of Chemistry.

The influence and behavior of small molecules protective agents on the conductivity and sintering of the Ag NPs based ink were also investigated. Previously, we prepared Ag NPs, with dodecylamine (DDA) and dodecanethiol (DDT) as the protective agent, and studied the effect of protective agents on the properties of the Ag NPs based film in the post-treatment [53]. The results showed that the molecular structure of the DDA and DDT as well as the bonding strength between the protective agents and the Ag NPs surface affect the conductivity, sintering temperature and morphology of the Ag NPs based film significantly. The bonding energy of Ag-S being higher than that of Ag-N and a higher alkyl chain ordering of capping DDT molecules lead to a stronger interaction between the alkyl chains than that of capping DDA molecules. Thus, Ag-DDA film requires a lower treatment temperature to convert it into conductive than that of Ag-DDT film. The results showed that the printed Ag-DDA NPs based film even could transfer from insulative into conductive with an electric resistivity as low as 15.1 $\mu\Omega\cdot$cm after air storage at room temperature for less than seven days. In addition, the electric resistivity of the Ag-DDA NPs based ink after 60 min heat-treatment at 140 °C reached 2.9 $\mu\Omega\cdot$cm, which is 1.8 times the bulk Ag resistivity. Jung et al. [60] achieved low temperature sintering and highly conductive Ag NPs based ink by ligand exchange and ligand reduction using an acetic acid (AA) immersion treatment. The original surface capping agent of oleylamine (OA) was replaced by AA through the ligand exchange, simultaneously resulting in the capping ligand weight reduction by 10 wt.%. The ligand exchange was explained by the difference in adsorption energy of the two ligands, as estimated by density functional theory (DFT) calculation. The relative energy difference between the state of OA being adsorbed and the state of AA being adsorbed is approximately −1.98 ev. Thus, AA adsorption is energetically much more favorable than the OA adsorption. Both the reduced ligand weight and relatively lower bonding energy between Ag NPs and ligand contributed to the lower sintering temperature of the Ag NPs based ink compared to its counterpart before ligand exchange.

2.2. Ag NPs Sizes and Shapes

It is well known that nanomaterials usually exhibit novel specific properties that may be significantly different from that of bulk materials in mechanical, optical, electrical, thermal and magnetic properties. For instance, according to the phenomenological model and the experimental observations presented by Buffat and coworker (Figure 3a) [61], the melting temperature of gold particles significantly drops when the diameter is smaller than 5–7 nm. This size dependent melting temperature decrease is also observed and investigated in Ag NPs [62]. The Ag NPs approximately 2 nm show melting behavior at significantly low temperatures (≈150 °C) compared to the melting temperature of bulk Ag (960 °C), as illustrated in Figure 3b. This huge melting temperature depression is not only very interesting from a fundamental research perspective, but indicates that the atomic diffusion becomes very active in nanoparticles near the surface which is very important for the flexible and printed electronics applications requiring low temperature processing. On the other hand, the shape and size distribution of the nano-Ag fillers in conductive ink could affect the packing density, filler interconnect and morphology of the printed film during post treatment process, which have a significant impact on the conductivity and sintering of the Ag NPs based ink [63].

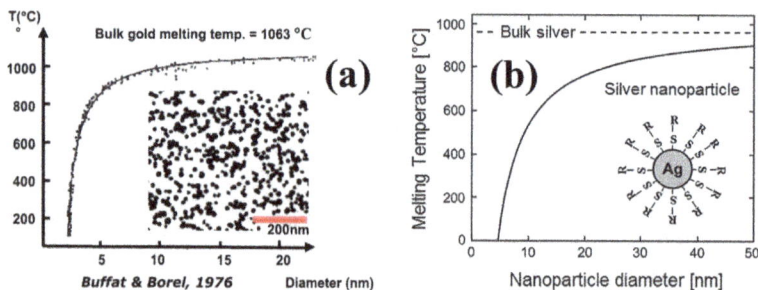

Figure 3. Experimental and theoretical values of the melting point temperature of (**a**) gold particles. Reproduced with permission from [61]. Copyright 1986 American Physical Society; (**b**) silver particles as the function of decreasing particle size; reproduced with permission from [62]; Copyright 2006 Springer.

Balantrapu et al. [64] and Ding et al. [65] studied the relationship between the size distribution and electrical properties of the printed Ag NPs based film. The results showed that the electric resistivity and sintering of the printed pattern are highly dependent on the Ag NPs size distribution. The Ag NPs based ink with bimodal distribution or relatively broad size distribution is more favorable to form extensive conductive 3D network in the printed pattern during sintering by forming a large number of contact points in different sized Ag NPs. In addition, the voids caused by volumetric shrinkage of the relatively large Ag NPs during sintering could be filled with relatively small Ag NPs, resulting in a compact morphology and high conductivity of the printed film. The optimal electrical resistivity values of ~6.7 $\mu\Omega\cdot$cm and ~3.83 $\mu\Omega\cdot$cm were achieved by Balantrapu et al. and Ding et al. at 200 °C and 160 °C, respectively. Seo et al. [66] focused their research on the effects of both the Ag NPs size and the type of protective agents on the conductivity and morphology of the Ag NPs based film during the sintering process. It was found that the size of the Ag NPs was the main factor influencing the initial decrease in the resistivity because of the neck formation between Ag NPs and the type of protective agents was the most important factor for determining the final resistivity of the conductive films due to interconnections of the Ag NPs via extended neck formation. The lowest resistivity (2.2 $\mu\Omega\cdot$cm) was obtained for the film that was prepared using 3.4 nm Ag NPs, hexylamine as a stabilizer, and sintered at 220 °C. Han et al. [67], Yang et al. [68] and Lee et al. [69] investigated the shape influence on the electrical property of the nano-Ag based film by using the Ag NPs (spherical shape), nanorods, nanoplates and their mixtures as the conductive fillers. It was found that, when combining the Ag NPs with Ag nanoplates or nanorods at a certain ratio as the conductive filler, the different shapes of nano-Ag mixture based ink demonstrate a higher conductivity at a relatively low temperature compared to that of single Ag NPs based ink. The conductive mechanism research shows that the small sized Ag NPs provide sufficient energy to motivate the grain and lattice transport to facilitate strong bonding and the large sized Ag nanoplates or nanorods stack densely to reduce the porous space in the pattern. Specifically, Han et al. obtained the resistivity of 10.3 $\mu\Omega\cdot$cm at 100 °C for 30 min which was only 6.5 times of the bulk Ag by mixing Ag NPs and nanoplates with the weight ratio of 1:1.

2.3. Substrate Facilitated Sintering

In the above sections, we have discussed that the sintering and electrical property of the Ag NPs based ink could be tailored by controlling the property of protective agents and optimizing the shape and size distribution of Ag NPs. In this section, we pay attention to another key component of the flexible and printed electronics: the substrate. It is well known that the requirements when printing for electronics are totally different from those for printing graphic arts. Graphic printing needs images or text with a good visual impression, whereas electronic applications require continuous and homogeneous patterns with restrictions on the layer thickness, roughness, and print resolution. Therefore, the substrates, whether plastics or papers, must be able to offer some or most of the following properties: thermal stability, dimensional stability, barrier properties, solvent resistance, low coefficient of thermal expansion, a smooth surface and optical clarity for display purposes. MacDonald et al. [70] reported the issues associated with the selection of a plastic film with the required property set for development and the leading candidate materials for plastic-based flexible electronics. In addition, Tobjörk et al. [71] reviewed recent progress in the development of electronic devices on paper substrates.

A recent research provides an extremely interesting approach, where substrate modification leads to the spontaneous coalescence and sintering of Ag NPs at a relatively low temperature. This substrate facilitating sintering of Ag NPs is attributed to two aspects' reasons, which are mainly related to the superficial physical and chemical properties of the substrates, respectively. The superficial physical properties include the surface roughness, solvent wettability, solvent absorption rate and mechanical stability, etc. The chemical modification of the substrates is intended to provide chemical removal of the protective agents from the surface of the Ag NPs, which is in accordance with Refs. [70–73] in Section 2.1. While the main distinguishing factor of the chemical related substrate facilitated sintering compared to the methods mentioned in Refs. [70–73] of Section 2.1 is that the sintering agent is added in the paper coating during manufacturing and do not need any post treatment of the printed Ag NPs pattern. This is significant for the large-scale production and high speed roll to roll printed electronics.

Lee et al. [72] characterized the commercial available photo-papers with respect to their superficial physical and chemical properties to obtain highly conductive Ag NPs based printed patterns at a relatively low sintering temperature. The results showed that chloride ions on the paper's surface when they are under a certain value could activate the decomposition of polymer protective agent and sintering between the Ag NPs. On the other hand, the surface roughness and pore size of the paper were inversely related to the conductivity of the Ag NPs pattern.

Öhlund et al. [73] incorporated the sintering agent of chloride as an ingredient of the mesoporous paper coating to achieve chemical sintering and investigated the effect of the variations in the pore size of paper coating and precoating type on the sintering of Ag NPs. Figure 4 shows that the Cl$^-$ migrate into the Ag NPs film when Ag NPs deposit in the printing process and react with the Ag NPs matrix to assist the low temperature sintering. Meanwhile, the sintering is impaired by increasing the pore size of the paper coating, but greatly enhanced by using a porous CaCO$_3$ precoating.

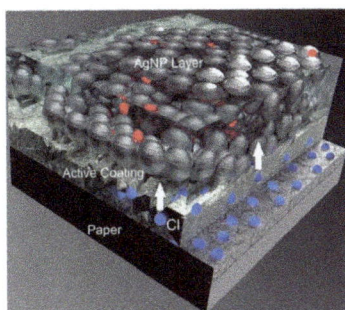

Figure 4. Schematic image showing the principle of the active papers. A small amount of chloride is contained in the coating as a sintering agent. During the deposition of the Ag NPs dispersion and absorption of the carrier fluid, Cl ions migrate into the Ag NPs film and react with the Ag NPs matrix to assist the sintering. Reproduced with permission from [73]; Copyright 2015 Royal Society of Chemistry.

Allen et al. [74] and Andersson et al. [75] also found that, by choosing the type of ink receptive coating, it is possible to manufacture printed Ag NPs based pattern without the need for, or at least to reduce the need for, post print sintering. Allen et al. [74] demonstrated that the room temperature sintering of Ag NPs could be achieved on the substrates with the ink receptive coating that contains silanol groups. The silanol groups could dissolve the protective agent of PVP on the Ag NPs surface by providing enhanced water absorption in the substrate coating layer as well as providing strong binding sites so that it is energetically favorable to detach the protective agent from the Ag NPs. Andersson et al. [75] observed an extreme difference in electric resistivity for tracks printed on paper substrates with aluminum oxide based coatings compared to silica based coatings. Nge et al. [76] paid attention to obtain the superficial nanostructured paper and studied its influence on the electrical property of the inkjet printed Ag NPs patterns. They introduced a direct sheet casting method to prepare cellulose nanofibers (CNF) based paper, with unique surface features including a nanoporous network structure and low surface roughness. The CNF based paper shows a shorter sintering time at a low temperature and a less pronounced coffee ring effect compared to the commonly used paper and plastic because of the permeation of the ink vehicles through the nanopores and absorption along the nanofibrils that compete with the initial spreading and the final evaporation process.

2.4. Photonic Sintering Method

Recently, various emerging sintering techniques have been used to obtain highly conductive printed patterns based on Ag NPs ink under moderate condition. In this section, the photonic sintering, which is the most popular method in this related field, is presented. The sintering of metallic NPs based inks via electromagnetic (EM) irradiation ranging between the ultra-violet (UV) and infra-red (IR) is called photonic sintering. Frequently reported bands are in the infra-red (IR), ultra-violet (UV) and visible region, which is called intense pulsed light (IPL) or photonic flash sintering. Since the absorption of metallic NPs based inks (plasmon resonance) is in the visible region (Figure 5a), UV irradiation (ranging from 100 to 400 nm) is not suitable for the selective heating of these materials but mainly applied to metal organic compounds (MOD) inks, which is not in the discussion scope of this review. In addition, a special form of irradiation is laser sintering, where the emission of the laser can be tuned in a narrow wavelength window or even a single wavelength to match the absorption spectrum of the respective ink formulation. Rather than heating the entire system indiscriminately, photonic sintering enables targeting specific components selectively, leaving the substrate that tends to absorb only in the UV range (Figure 5b, the polyimide substrate is the exception because of its brown color) unaffected.

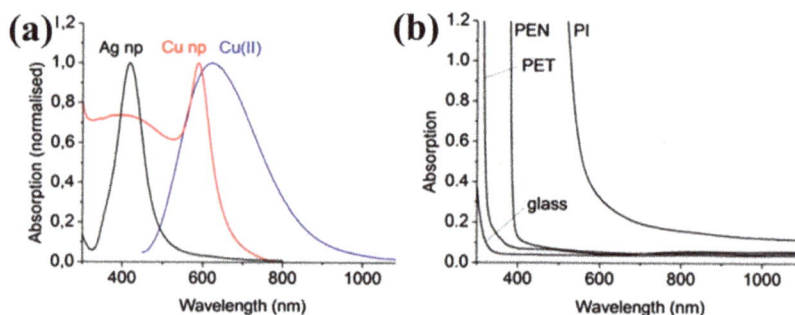

Figure 5. (**a**) UV-Vis absorption spectra of commonly used metallic NPs dispersion of conductive inks and (**b**) substrates for printed electronic applications. Reproduced with permission from [77]; Copyright 2014 Royal Society of Chemistry.

2.4.1. Infra-Red (IR) Sintering

IR technology using irradiation in the range of the NIR to MIR region (700 to 15,000 nm) facilitates the contact-less and selective drying and sintering of printed metallic NPs based layers within a very short time. Denneulin et al. [78] used an IR lamp operating at wavelengths of 8 to 15 μm to sinter the inkjet printed pattern of Ag NPs. A similar level of electric resistance was obtained by IR sintering within a relatively short time of 3 min compared to that by conventional heating at 200 °C for 5 min. while the high wavelength of the using IR also caused a fast temperature increasing of the substrate to 180 °C–210 °C, which limits its application on the temperature sensitive substrate. A more selective approach of IR sintering was performed by Cherrington et al. [79], who used irradiation in the near-IR (NIR) region to sinter the slot-die coated Ag NPs pattern on Polyethylene terephthalate (PET) substrate within 2 s yielding a conductivity of about 16% of bulk Ag. Irradiation in the NIR is shown to be less absorbed by the used PET, enabling a selective sintering of the metal ink without substrate deformation. The NIR irradiation was also used by Tobjörk et al. [52] and Gu et al. [80] to sinter printed Ag NPs inks on paper and plastic substrates. An optimal sintering result can be achieved by carefully adjusting settings like power output, distance between lamp and sample and treatment time. The resistivity of 2.78 μΩ·cm was achieved after only 8 s exposure to NIR irradiation with no damage to the substrate, which was only 1.7 fold higher than that of bulk Ag. Figure 6 shows the electrical resistivity and morphology evolution of the printed Ag NPs based film during sintering process [80].

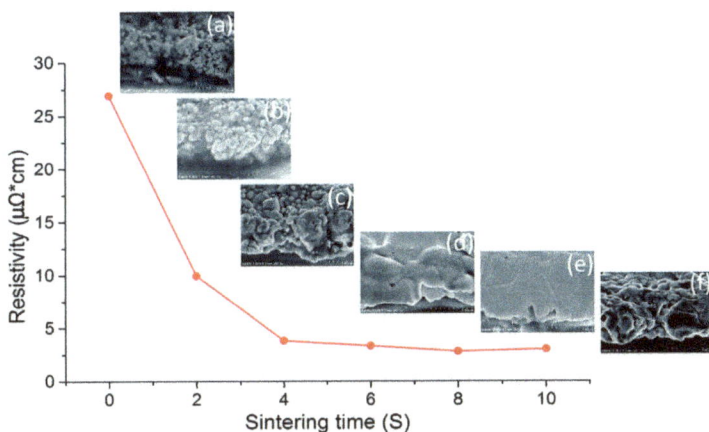

Figure 6. The resistivity of Ag NPs film sintered by NIR with power of 360 kW·m2 over 10 s and SEM images of the sintered film at (**a**) 0 s; (**b**) 2 s; (**c**) 4 s; (**d**) 6 s; (**e**) 8 s,;and (**f**) 10 s. Reproduced with permission from [80]; Copyright 2018 Royal Society of Chemistry.

Sowade et al. [81] reported a roll to roll (R2R) NIR drying and sintering process for inkjet printed Ag NPs layers on Polyethylene naphthalate (PEN) substrate (Figure 7). Relevant process conditions, e.g., intensity of IR radiation, duration of exposure, velocity of moving substrate, usage of IR reflectors, the distance between IR emitters and printed Ag NPs layers, were varied to evaluate the effects on the morphology and conductivity of sintered Ag NPs layer. The optimized electric conductivity up to 15% of Ag bulk was achieved at high web velocities up to 1 m/s with an exposure time of less than 0.5 s. Basically, IR sintering is a very fast (in the order of seconds) method to sinter Ag NPs based inks to obtain conductivity values in the range of 10%–35% of the Ag bulk. Considering heat dissipation from the printed Ag NPs coating into the substrate happened also very fast, the sintering parameters should be carefully optimized and the paper substrate with high diffuse reflectance, relatively high thermal stability and low thermal conductivity is especially suitable.

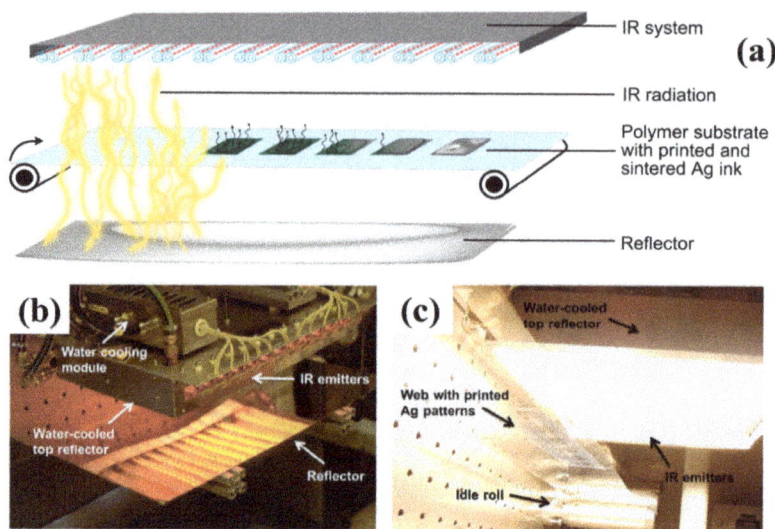

Figure 7. (**a**) scheme of the experimental setup of roll to roll (R2R) IR drying and sintering of inkjet-printed Ag NPs layers on Polyethylene naphthalate (PEN) substrates. The R2R sintering instrument in (**b**) top view and (**c**) from below with activated IR radiation. Reproduced with permission from [81]; Copyright 2015 Royal Society of Chemistry.

2.4.2. Intense Pulsed Light (IPL) Sintering

Intense pulsed light (IPL) or photonic flash sintering is essentially a thermal technique which employs the heat generated by the absorption of visible light in the target materials to achieve the necessary temperature increase. In contrast to conventional thermal sintering, where the sample is exposed continuously to a high temperature, IPL irradiates the sample with multiple short flashes, each with a pulse length in the range of a few micro-to milliseconds. The most commonly used light source for IPL sintering is a xenon stroboscope lamp, which emits radiation in the range between roughly 200 and 1200 nm, encompassing the entire visible spectrum. Figure 8 gives the schematic of IPL sintering of Ag NPs based film [82].

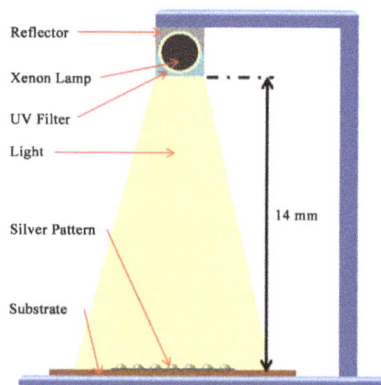

Figure 8. Schematic of intense pulsed light (IPL) sintering for Ag NPs based film using xenon flash irradiation. Reproduced with permission from [82]; Copyright 2011 Springer.

Although, in most cases, the IPL was used to sinter Cu NPs based ink because of its superiority in the reduction of the oxide layer on the surface of Cu NPs, a number of reports concerning Ag NPs have appeared about the influence of various IPL parameters on sintering time, final conductivity, film morphology and substrate damage. Chung et al. [83] obtained the optimal IPL sintering conditions for the gravure offset printed Ag NPs film on PET substrate by in situ monitoring of the IPL sintering process. The optimized IPL process reduced the sheet resistance of Ag NPs based film to below that of thermally sintering without damaging the PET substrate or allowing interfacial delamination between the Ag NPs film and PET. Kang et al. [82], Abbel et al. [84], Lee et al. [85] and Sarkar et al. [86] investigated the effect of the IPL parameters such as flashing frequency, intensity, pulse duration and number on the electrical property and morphology of the Ag NPs based film. The results showed that variation of the IPL sintering parameters offers a wide range of conditions for process optimization. In addition, the ink composition and type of substrate also have a decisive influence on the IPL sintering of Ag NPs based ink. According to the investigation of Lee et al. [85], the protective agent and organic additives play a critical role in the microstructure formation inside IPL sintered film, which affects the final electric resistivity. The vaporization induced from the thermal decomposition of the protective agent and organic additives could result in film swelling during the re-melting stage of the surface Ag NPs layer. Weise et al. [87] presented and analyzed the application of IPL sintering on inkjet printed Ag NPs based patterns on various flexible substrates, like PEN, PET, Polyimide (PI) and paper. A high dependency of the electrical and structural properties of the printed Ag NPs layer on the substrate was observed. This observation was explained as resulting from the different surface roughness, solvent absorbing rate and thermal conductivity of the substrates.

2.4.3. Laser Sintering

Laser sintering has shown great promises to achieve high-quality sintering locally through controlling the heat penetration to preserve the substrates' integrity. The printed Ag NPs based layer absorbs the laser irradiation in the affected area followed by heating up and sintering the Ag NPs due to the photothermal effect shown in Figure 9. The generated temperature inside of the Ag NPs based layer has to be controlled and kept as low as possible to avoid heat dissipation into the substrate material. Thus, a careful adaption of sintering parameters like power output, writing velocity, wavelength and operation mode (continuous wave or pulsed) should be carried out allowing a reduction of the processing temperatures on the substrate. Balliu et al. [88] investigated laser sintering of inkjet printed Ag NPs inks on papers. High conductivity of 1.63×10^7 s·m^{-1}, nearly 26% of the bulk Ag, was achieved where a special care was taken in sintering parameters to prevent the substrates from damage by intense laser light. Yeo et al. [89] sintered the R2R printed Ag NPs layer by laser to a conductivity up to 20% of bulk Ag on PET substrates. Bolduc et al. [90] indicated that controlling the incident laser pulse's energy distribution in the time-domain was paramount to optimizing sintering process in Ag NPs based ink. A multi-step microsecond-pulsed laser process and a time-domain pulse-shaping modulation sintering caused a uniform and high conductive printed Ag trace on polymer substrates.

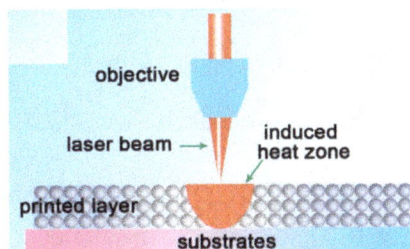

Figure 9. Scheme of laser sintering principle: a focused beam locally heats the printed Ag NPs layer. Reproduced with permission from [32]; Copyright 2017 Royal Society of Chemistry.

In the other hand, the spot size of the laser and its heat affected zone is far more smaller than the minimum trace of printing technologies, which makes laser sintering a suitable tool for high resolution and lithography free manufacturing [91]. Figure 10 shows the selective laser sintering process of metallic NPs based ink. Hong et al. [92] fabricated a metallic grid transparent conductor on PET and glass substrates using selective laser sintering of Ag NPs based ink (Figure 11). Such the transparent conductor with high transmittance (85%) and low sheet resistance (30 Ω/sq) could be produced at a large scale without any vacuum or high temperature environment.

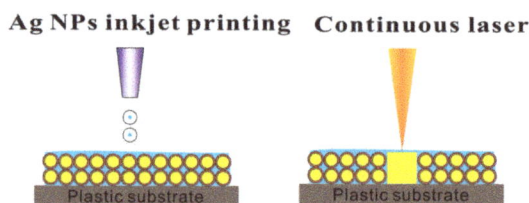

Ag NPs inkjet printing   Continuous laser

Figure 10. Selective laser sintering process of inkjet printed metallic NPs on a polymer substrate. The circles represent metallic NPs with protective agents and the square block indicates a conductor pattern of sintered metallic NPs. Unsintered NPs are simply washed away in an organic solvent.

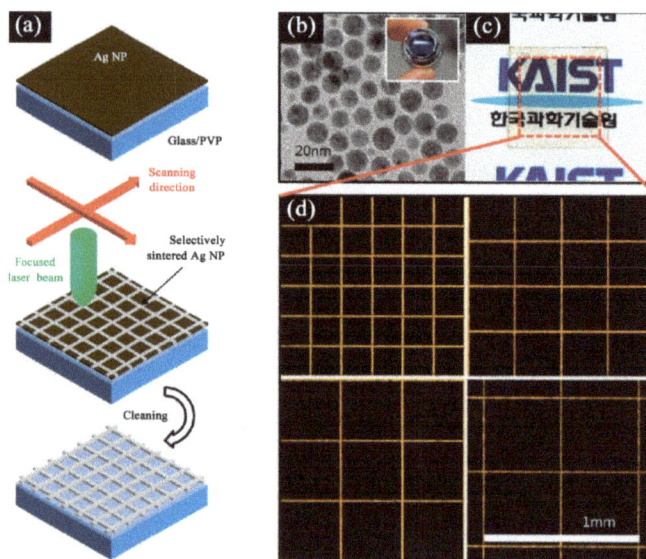

Figure 11. (**a**) schematic diagram of selective laser sintering of Ag NPs for the fabrication of a transparent conductor; (**b**) TEM: Transmission electron microscope (TEM) image of synthesized Ag NPs (inset: optical photograph of Ag NP ink); (**c**) photograph of a transparent conductor on a glass substrate (metallic grid in the red-boxed region); (**d**) optical stereoscope images of square-metallic grids at different grid sizes (200 to 500 μm, increment 100 μm). Reproduced with permission from [92]; Copyright 2013 American Chemical Society.

2.5. Other Emerging Sintering Methods

Electrical sintering describes the application of a current to printed Ag NPs based inks causing local heating within the ink, which is due to its highly resistive nature before sintering. This process occurs on a timescale of a few milliseconds to seconds and is called rapid electrical sintering (RES). RES is demonstrated on printed Ag NPs structures by applying direct current (DC) voltage as well as via a near-field coupled alternating current (AC) electric field [93]. Figure 12a illustrates a sintering setup, where sintering electrodes are in contact with the Ag NPs layer. When a voltage U is coupled between the sintering electrodes, a non-zero current flow (indicated by arrows in Figure 12a) causes local heating in the layer. This initiates the sintering process and the structure undergoes a rapid transition in conductivity. The series resistor Rs limits the maximum current once the structure is sintered. Contact-mode electrical sintering has been applied using DC voltage. However, the requirement of directly contacting the printed pattern during sintering demonstrates an obstacle in large quantity fabrication. Therefore, contactless electrical sintering using AC current was developed. This is accomplished by applying sintering electrodes above the sample, which couple to the printed layer (Figure 12b). Allen et al. [94] obtained excellent conductivity up to 60% of bulk Ag in very short time of 2 µs using DC current with the power density of at least 100 nW/µm^3.

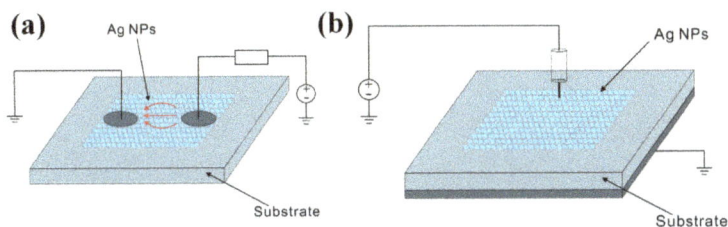

Figure 12. (a) schematic illustration of a contact-mode Direct current (DC) electrical sintering setup and (b) contactless Alternating current (AC) sintering between a probe above the NPs layer and a ground plate beneath the printing substrate.

Plasma sintering is usually performed by exposure of printed patterns to low pressure Ar plasma. During plasma exposure of the Ag NPs based ink, the plasma inherent active species decompose the protective agents on the surface of Ag NPs due to chain scission, which results in the sintering of the Ag NPs. The sintering process shows a clear evolution starting from the top layer into the bulk. Reinhold et al. [95] used a low pressure argon plasma in order to sinter Ag NP inks on glass, PC and PET to a conductivity of up to 30% of bulk Ag. Recently, Wolf et al. [96] reported that low pressure Ar plasma sintering resulted in the conductivity of printed patterns equal to 11% of bulk silver after only 1 min of exposure, and 40% of bulk silver after 60 min of exposure, while the processing temperature was below 70 °C. To avoid the need for sophisticated equipment for low pressure plasma sintering, Ar plasma sintering at atmospheric pressure and room temperature was developed by Wünscher et al. [97,98]. With this technique or combining with a mild heating of the substrate less than 110 °C, relatively high conductivity of the printed Ag NPs based trace was obtained in a short sintering time without substrate damage. This approach enables sintering of patterns printed onto plastic substrates and can be utilized in R2R processes. Ma et al. [99] sintered the Ag NPs film on glass substrate by applying the Ar plasma and studied the effects of plasma conditions on the morphology, composition and electrical property of the sintered Ag NPs film. The optimized resistivity of the sintered Ag NPs film was about five times higher than bulk Ag.

Microwave radiation also can be used as an alternative and selective sintering technique [100]. Typically, Ag NPs based film have a penetration depth of 1–2 µm at a microwave frequency of 2.45 GHz. While since the Ag NPs is a good thermal conductor, the printed pattern will be heated uniformly by thermal conductance enabling the microwave radiation to be applied to sintering patterns with thickness exceeding the penetration depth. In contrast to the relatively strong microwave absorption by the Ag NPs, the polarization of dipoles in thermoplastic polymers below Tg is limited, which makes the polymer substrate transparent to microwave radiation. Perelaer and his colleges focused their research on the microwave sintering of the Ag NPs based ink for many years. Their research results showed that the exposure of inkjet printed Ag NPs to microwaves decreased the sintering time by a factor of 20 with the conductivity value of 5% compared to the bulk Ag [100]. Further decreasing the sintering time to only a few seconds with the conductivity up to 10% to 34% of the bulk Ag have been achieved by placing conductive antennae structure around the Ag NPs based pattern [101]. This process can be implemented into R2R production. Meanwhile, combining microwave and other sintering techniques have been proved to be an effective way to improve the sintering performance. Combining photonic and microwave flash treatments enabled obtaining 40% of bulk silver conductivity in less than 15 s [102]. Even higher conductivity, 60% of bulk Ag, was obtained in less than 10 min by combining low-pressure Ar plasma and microwave sintering of printed Ag NPs on PEN foil without damage of the polymeric substrate [43].

3. Applications of the Ag NPs Based Ink

In this section, we will discuss several applications of Ag NPs based ink and their perspectives in FPE. This will include fabrication and properties of transparent conductive film (TCF), which are essential features nowadays for many optoelectronic devices, printed thin film transistor (TFT), biosensor, radio frequency identification (RFID) antenna as well as emerging stretchable and wearable electronics.

3.1. Transparent Conductive Films

Indium tin oxide (ITO) with both excellent transparence and conductivity has been the most widely used transparent conductive film (TCF) in decades. However, an ITO film also has a number of unavoidable disadvantages and weaknesses, such as the relatively high cost and poor flexibility. As the development of the large area and flexible devices such as solar cells, touch panel, light-emitting device and display, extensive efforts have been made to obtain alternatives to ITO [103–108]. Among alternative materials and approaches, patterned Ag NPs grids are a promising candidate for high performance TCF. For instance, we prepared a high-performance ITO-free TCF by combining high-resolution flexography printed Ag NPs grids with a carbon nanotubes (CNTs) coating [45]. The Ag NPs grids/CNTs hybrid TCF with a 20 μm grid width at an interval of 400 μm exhibits excellent overall performances, with a typical sheet resistance of 14.8 Ω/sq and 82.6% light transmittance at room temperature as well as good mechanical flexibility. Magdassi et al. [103] produced TCF by inkjet printing of diluted Ag NPs based inks to form overlapping metallic rings, forming in spontaneous self-assembly of Ag NPs during solvent evaporation. The resulting array Ag NPs based ring with rims <10 μm in width and <300 nm in height has a transparency of 95% and sheet resistance of 4 Ω/sq. Ahn et al. [104] produced the TCF with high transparency of 94.1% by direct writing of concentrated Ag NPs based ink. Deganello et al. [105] obtained TCF with a transparency of 81.4% and sheet resistance of 1.26 Ω/sq by patterned micro-scale Ag NPs based grids using roll-to-roll flexographic printing. Kahng et al. [106] obtained highly conductive flexible TCF with a sheet resistance of 12 Ω/sq and 73% transparency at 550 nm by combining ink-jet printed Ag NPs grids with graphene film. In addition, Jeong et al. [107] obtained an Ag NPs grid/ITO hybrid TCF by inkjet printing. The hybrid TCF has a sandwich structure with the Ag NPs grids in the middle of two ITO layers, showing a sheet resistance of 2.86 Ω/sq and transparency of 74.06%.

Ag NPs grids based TCF exhibits excellent performance in terms of optical transparency, electrical conductance, and mechanical flexibility. Thus, they have found applications in many optoelectronic devices such as displays, touch screens, organic light emitting diodes (OLEDs) and solar cells. Many proof-of-concept devices such as solar cells and OLEDs with incorporated TCF have been demonstrated. Li et al. [108] reported using Ag NPs grids based TCF to fabricate ITO-free flexible organic solar cell. The Ag NPs grids' TCF has very fine honeycomb structure with the width of around 3 μm and the diagonal length of 130 μm, showing low sheet resistance less than 5 Ω/sq and high transparency of 85%. A layer of highly conductive PEDOT:PSS(3,4-ethylenedioxythiophene): poly(styrenesulfonate) on the top of Ag NPs grids is added to increase the charge collection efficiency. Flexible organic solar cell using this TCF as electrode, and P3HT:PCBM as the photoactive layer, achieved a PCE of 1.36%. When using high performance conjugated polymer, PTB7, as the donor, the highest power conversion efficiency of 5.85% was achieved for a large area flexible polymer solar cell [109]. Cai et al. [110] reported a novel electrochromo-supercapacitor based on an Ag NPs/PEDOT:PSS hybrid transparent electrode. The bifunctional device performs as a regular energy storage device and simultaneously monitors the level of stored energy with rapid and reversible color variation, even in high current charge/discharge conditions.

3.2. Thin Film Transistor

Thin film transistor (TFT) is an electronic device widely used in applications including display back plane, sensor and logic circuit, etc [111–113]. In addition, it is considered as a model device to investigate the intrinsic physics of semiconductor film and metallic electrode contact, which also are significant for other devices, such as photovoltaic cell and organic light emitting diode. TFTs are constructed with four parts at least: an active semiconductor film usually named as channel layer, a dielectric layer used as gate insulator, a gate electrode and a couple of source/drain electrodes. Figure 13 shows a schematic of four types of TFTs [114].

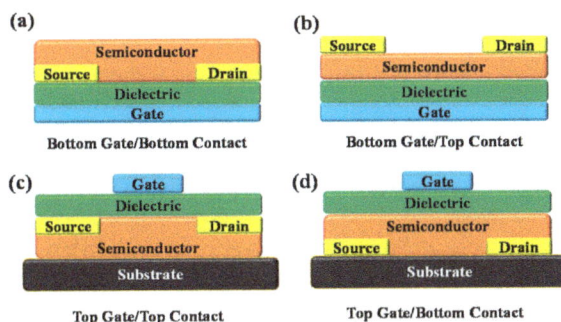

Figure 13. Schematic diagram of four types of TFTs (TFTs: Thin film transistors); (**a**) bottom gate/bottom contact structure; (**b**) bottom gate/top contact structure; (**c**) top gate/top contact structure; (**d**) top gate/bottom contact structure.

The printing fabrication of electrodes avoids the conventional vacuum deposition of metallic films, and is expected for full-printed devices. More importantly, the etching inevitable during the photolithograph process for electrode patterns is excluded. The etching process for source/drain electrodes would damage the underlying semiconductor active layer, resulting in degradation of device performance. Ag NPs are considered the primary candidate material to printing electrodes, due to its high conductivity and low temperature treatment in atmospheric ambient [115]. Printed Ag NPs based films have been employed as the source/drain and gate electrodes in some TFTs [115–117].

The requirement of printed Ag NPs based electrodes is not only low resistance but also good contact with semiconductor and good morphology. Printed Ag NPs based source/drain electrodes have been commonly reported in p-type organic and carbon-nanotube TFT [118–123], though the intrinsic work function of Ag is as low as 4.26 eV [124]. It may be due to the formation of Ag_2O on a surface that is a high-doped degenerate p-type semiconductor with high work function and moderate electric conductivity. Furthermore, a bottom contact device structure was usually used in these devices, where the source-drain electrodes were printed prior to channel layer. The surface of printed Ag NPs electrodes could be modified to increase work function and improve electrode contact with channel layer, so as to improve device performance [125–128].

Vacuum-deposited Ag source/drain electrodes have been confirmed very appropriate to contact with n-type oxide semiconductor with a low specific contact resistance, due to its inherent low work function of 4.26 eV [124]. However, for a long time, the oxide thin film transistors were reported, exhibiting low mobility of device (less than 0.5 cm^2 V^{-1} s^{-1}), as printed Ag NPs conductors were utilized as source/drain electrodes [129–131]. It suggests that there are some specific issues involving the use of Ag NPs ink in the fabrication process.

Metallic Ag NPs undergo undesirable electrical/thermal migration [115,132], which would result in unpredictable degradation of the device performance [115,122]. Hong et al. reported high-performance oxide thin-film transistors, by surrounding Ag nanoparticles firmly with oleic acid to suppress the migration of Ag inside adjacent oxide semiconductors. The devices exhibited comparable performances to their counterparts of vacuum-deposited metal electrodes. It was also found that the suppressed formation of Ag_2O and the reduced incorporation of organic components inside the electrodes all play critical roles in facilitating the realization of high performance, oxide thin-film transistors employing printed Ag NPs based source/drain electrodes [115].

Ueoka et al. analyzed printed Ag NPs source/drain electrodes on amorphous indium gallium zinc oxide (IGZO), and found that carbon and hydrogen seriously affect the TFT characteristic. The carbon and hydrogen were abscised from the printed Ag NPs during annealing, generating additional carriers and electron traps [131]. Ning's group also found carbon at the interface between a-IGZO and printed Ag electrodes. They suggested that the presence of carbon adversely impacted on contact, whereas the diffusion of silver into IGZO semiconductor layer resulted in a better contact at the interface [116]. Recently, the same group reported IGZO TFT with printed Ag source/drain electrodes. The devices show high performance comparable with the analogous devices with sputtered electrodes: a maximum saturation mobility of 8.73 cm^2 V^{-1} s^{-1} and an average saturation mobility of 6.97 cm^2 V^{-1} s^{-1}, Ion/Ioff ratio more than 107 and subthreshold swing of 0.28 V/decade [133].

To improve electrode contact, the interfaces of printed Ag electrodes with n-type oxide semiconductors were modified. A universal method was reported to produce low-work function electrodes for electronics devices, with surface modifiers based on polymers containing simple aliphatic amine groups [134]. The method was instantiated to achieve high performance solution processed oxide TFTs with inkjet printed Ag source/drain electrodes recently [135]. In other research work, graphene was embedded insert IGZO thin film and printed Ag source-drain electrodes. High-performance IGZO TFTs were achieved with an electron mobility of ~6 cm^2 V^{-1} s^{-1} and I_{on}/I_{off} ratio of ~10^5 [136].

The morphology of printed Ag line is critical in some cases. More specially, the thickness and profiles of printed Ag line must be well-controlled when it acts as bottom gate electrode in device, to achieve low resistance and low leakage current simultaneously. Usually, a thick film is good for low resistance but results in high leakage current. In addition, smooth surface is required to void electric breakdown. Guo et al. prepared all ink-jet printed low-voltage organic field-effect transistors with Ag NPs based ink, one kind of metal-organic precursor type ink. The printed convex Ag lines have a small thickness of 30 nm and root-mean-square (RMS) roughness of about 1.8 nm [137].

Short channel length of thin film transistors is expected to achieve high operating frequency. There are important technological challenges in printing methods to achieve high-resolution electrodes patterning [111]. Some high-resolution printing equipment has been used to achieve channel length around 1 μm with printed Ag source/drain electrodes, including sub-femtoliter nozzle [138], EHD printer [139], and printing plate with high resolution pattern [140,141]. Recently, Ning's group reported a short channel length of printed Ag electrodes with a common printer [142,143]. Silver electrodes with 2.4 μm channel length were printed by piezoelectric inkjet printing of 10 pL nozzle, without an extra process. It was attributed to the difference in the retraction velocities on both sides of an ink droplet during the printing process [142].

3.3. Biosensors

Ag NPs could be also used as electrode materials for developing electrochemical biosensors. Compared to other metallic NPs, Ag NPs have a relatively low price and superior conductivity, which are undoubtedly the best option for reducing the cost of the biosensors. Han et al. [144] fabricated a label-free electrochemical immunosensor for prostate specific antigen (PSA) detection using rGO/Ag NPs composites as electrode materials on a screen-printed three-electrode system. The rGO/Ag NPs possessed superior electrical conductivity compared to rGO because the small Ag NPs, stabilized by sodium citrate were anchored onto the rGO sheets. The electrochemical immunosensors (EIs) demonstrated a wide linear response range (1.0 to 1000 ng/mL) and low detection limit (0.01 ng/mL). Figure 14 gives the preparing process of rGO and rGO/Ag NPs materials, and the schematic illustration of fabricated electrochemical immunosensor for PSA detection. Ag NPs can speed up the transfer of electrons between the enzyme and the electrode in biosensor, thereby speeding up the reaction and shortening the reaction time. For instance, Rad et al. [145] prepared a hydrogen peroxide sensor having high electron transport efficiency by modifying an electrode with Ag NPs. Wang et al. [146] showed that the gold electrode modified with Ag NPs based nanocomposite demonstrated relatively high sensitivity, fast response time and low detection limit compared to those of their counterpart. In addition, Ag NPs, exhibiting strong localized surface plasmon response (LSPR) absorption in the visible region, have potential as the optical biosensors [147]. Currently printed electrodes for biosensor possess a lot of advantages, such as low price, low sample volume requirement, high sensitivity, high and rapid volume production, portability and easy handling. The Ag NPs based ink supplies a very good opportunity pushing the newly and emerging printed biosensor into the market. It is expected that, in the near future, these printed biosensors will be widely commercialized, and we can use them to find solutions to health monitoring and disease control issues, especially in remote areas.

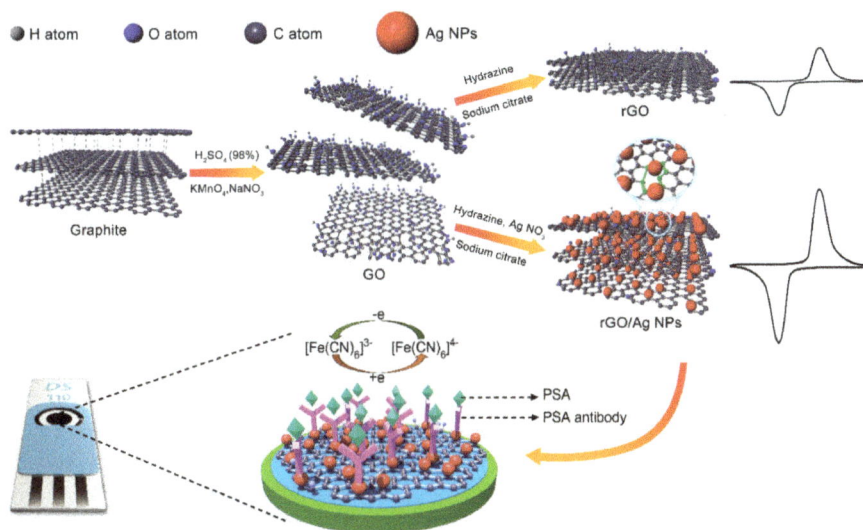

Figure 14. Preparing process of rGO (reduced graphene oxide) and rGO/Ag NPs as well as schematic illustration of fabricated electrochemical immunosensor for PSA (prostate specific antigen) detection. Reproduced with permission from [144]; Copyright 2017 Elsevier.

3.4. RFID

RFID (Radio Frequency Identification) tag is a device that provides storing and remote reading of data from items equipped with such tags. The main elements of an RFID tag are a microchip and an antenna that provide power to the tag and are responsible for communication with a reading device. Direct printing of antennas on plastic and paper substrates with the use of Ag NPs inks is a promising approach to the production of RFID tags [148–153]. Inkjet printing, screen printing and gravure printing have been used to fabricate Ag NPs based RFID antenna on paper and plastic substrates. The printing parameters, printability of the Ag NPs based ink, properties of the substrates and the sintering methods were investigated to reveal their relationship to the property of RFID tags. Particularly, Jung et al. [151] reported a practical way to provide all-printed and R2R-printable antenna, rectifiers, and ring oscillators on plastic foils and demonstrated 13.56 MHz operated 1-bit RF tags. The whole process used three different printing technologies, R2R gravure, inkjet and pad printing. This is the first report that not only fabricates the antenna but the whole RFID tag by printing. Sanchez-Romaguera et al. [152,153] reported an inkjet and screen printed low cost passive UHF RFID based on Ag NPs, which can be transferred from the tattoo paper to skin. This is significant for the development of wearable electronics and E-skin.

In spite of the printed RFID antenna based on Ag NPs based inks, this is relatively mature in its technique and has shown its obvious advantages in high production efficiency and lower environmental impact; its price is still a little bit high in most cases compared to that of a counterpart fabricated with traditional methods and using aluminum and copper as conductive materials. To overcome this difficulty, the development of conductive inks based on low cost nanometals or carbon is a choice, such as copper, aluminum, nickel, carbon nanotube, graphene, etc. However, such nanometals are easy to oxidize and the attempts to make carbon-based materials into ink and printable have so far not been very successful. On the contrary, to further improve the electric conductivity of the Ag NPs based ink under moderate sintering and sequentially to decrease their dosage in the devices with a relatively thin and uniform printed, coating can be a promising way.

3.5. Stretchable Electronics

Fabrication of large-area stretchable electronic devices is necessary for future applications in wearables, healthcare and robotics, etc. For integrating stretchable electronics, stretchable wiring is the most important component. Obtaining reliable conductance against strain could be achieved by the use of intrinsically stretchable materials, such as liquid metals, conducting polymers, and ionic conductors. Another approach is to fabricate conductive pathways using micro-structures, which can be obtained by mainly two methods. One is a metallization of artificially made microstructures, including serpentine, micro- or nano-meshes, or accordion motifs. The other is to develop a conductive nanocomposites mix with conductive fillers and elastomer matrix, which is advantageous in terms of large-area, low-cost and high-throughput fabrication. Among various conductive fillers, including carbon nanotubes, graphene and metallic nanowires, the nano- or micro-Ag with different shapes as well as their composites with other nanomaterials have attracted much attention in recent years.

The use of Ag NPs has yielded a relatively high conductivity at strains larger than 100% [154–159]. Matsuhisa et al. [154] reported a high performance stretchable and printable elastic conductor, with the conductivity of 935 S/cm at 400% strain, by the in situ formation of Ag NPs which were created via printing and heating an ink comprising Ag flakes, fluorine rubber, fluorine surfactant and methylisobutylketone (MIBK) as solvents. The results showed that even a small fraction of Ag NPs could reduce the percolation threshold of the composite, increasing the conductivity significantly. Park et al. [155] introduced a conductive and stretchable mat compositing of Ag NPs and rubber fibres. Percolation of the Ag NPs inside the fibres led to a high conductivity of 2200 S/cm at 100% strain for a 150 μm thick mat. Chung et al. [156] introduced a stretchable electrode on wave structured elastomeric substrate by ink jet printed Ag NPs based ink. The printed Ag NPs based electrode showed a relatively good adhesion and conductive stability in the stretching test.

Ag with different shapes and size distributions also have their applications in the stretchable electronics. Ag flakes and fractal structure were used as conductive fillers for their larger contact area compared to that of Ag NPs [160–165]. Matsuhisa et al. [160] demonstrated a printable and stretchable conductor that had a conductivity of 182 S/cm and stretchability at a strain of 215% using a nanocomposite material composed of Ag flakes, fluorine rubber and a fluorine surfactant. The fluorine surfactant constitutes a key component that directs the formation of surface-localized conductive networks in the printed elastic conductor, leading to a high conductivity and stretchability. Zhang et al. [165] used fractal structure Ag particles as conductive fillers in PDMS (polydimethylsiloxane) to fabricate a flexible and stretchable conductor, which could stretch up to 100% and twist up to 180° and possessed good mechanical and electronic stability. In addition, a stretchable conductor with Ag nanowires (Ag NWs) embedded in elastomer has been suggested [166–170]. The high aspect ratio of Ag NWs contributed to the high conductivity at a relatively high stain.

4. Conclusions

Ag NPs are favored materials for high performance FPE applications because of high electric conductivity, good oxidation resistance and easiness for large-scale preparation. When FPE devices are manufactured, a moderate sintering condition is often required in order to be compatible with heat sensitive substrates and various functional materials. Thus, highly conductive Ag NPs based inks with moderate sintering are essential for FPE applications. The comprehensive research regarding the strategies of achieving highly conductive Ag NPs based ink under moderate sintering has been covered in this review. A better understanding of the relationship of the sintering condition and conductive property of Ag NPs based film with respect to the protective agents, Ag NPs size distribution and shapes, and superficial characterization of the substrates is given. The mechanisms of the various emerging mild and selective sintering technologies are also highlighted. In addition, diverse applications of the Ag NPs based inks and their perspectives in FPE were also presented and discussed, including the transparent conductive film, thin film transistor, biosensor, stretchable electronics and RFID antenna.

Although there has been remarkable progress for Ag NPs based inks in FPE applications, there are still challenges that call for further development in order to gain widely industrial acceptance and in significant quantities. The current high price of the commercial available Ag NPs based inks impedes their wide use for large area flexible and printed electronics. Therefore, research should be focused on the development of new Ag NPs based inks with higher electric conductivity, which could decrease the ink dosage and printed film thickness to achieve the electric performance requirement of the device. For instance, the comprehensive cost of the ultrahigh frequency Ag NPs based RFID antenna fabricated by flexography printing method has already been lower than their counterpart, which uses aluminum as conductive materials and etching technology as the production method. The thickness of the flexography printed RFID antenna is only 0.5–0.8 μm, causing the dosage of Ag NPs to be relatively small and the cost to also be reduced. Meanwhile, for Ag NPs based ink, reduction in sintering temperature could be achieved by lowering the amount of a protective agent or decreasing the bonding energy between Ag and protective agent, which are presented in Section 2.1. However, these processes could adversely affect the stability and printability of the ink, and have unwelcome implications for mechanical integrity and adhesion. Thus, how to make a balance between the electric property, stability and the printability of the Ag NPs based ink is also an urgent problem that needs to be solved. In addition, the underlying physics and chemistry mechanisms governing the sintering process and the interaction between the influence factors of the sintering and electric conductivity in Ag NPs based ink should be further strengthened.

On the other hand, the increasing growth of application in FPE is calling for fast and reliable mild sintering technologies that can perform in high-throughput and R2R manufacturing. The research on emerging sintering techniques presented in this review has created possibilities to a different extent. Among all emerging sintering techniques, IPL sintering is the most promising one. The white-light flashes used in this technique can selectively heat the Ag NPs based ink without damaging the substrate. This sintering technique is also well understood and compatible with the R2R process. In fact, it has already been employed in a pilot production line for fabricating different electrodes and antenna structures. Further strengthening the effort to establish the relationship between different sintering parameters and ink/substrate combination will be likely to push it into the industry as a mainstream technique. One more important perspective in the application of Ag NPs based ink with moderate sintering is 3D printing of conductive patterns. Nowadays, this field is at its very early stages of research and development, and the search for new Ag NPs based ink as well as suitable 3D fabrication tools, is a stimulating challenge for materials scientists. It is likely that a combination of highly conductive Ag NPs based ink with fast and R2R compatible sintering techniques will realize more superior performance of Ag NPs based ink in a wider field of FPE.

Author Contributions: L.X.M. conceived the idea and planned for the review article and performed all literature surveys. All authors prepared and reviewed the manuscript.

Funding: This research was funded by NSFC project (61474144), the 2018 Beijing Municipal Commission of Education project (KM201810015004), the Beijing Municipal Commission of Education 2011 Collaborative Innovation Centre, the 2018 Beijing University Talents Cross Training Plan (Shipei Plan), the 2017 Beijing Municipal Commission of Education Outstanding Young Scholars (CIT&TCD201704051), the Research and Development Program of BIGC (Ea201803) and the Beijing Municipal Commission of the Education Foundation (PXM2017_014223_000036).

Conflicts of Interest: The funders had no role in the design of the study; in the collection, analyses, or interpretation of data; in the writing of the manuscript, and in the decision to publish the results.

Abbreviations

Ag NPs	Silver nanoparticles
TCF	Transparent conductive film
TFTs	Thin film transistors
FPE	Flexible and printed electronics
PAA	Poly(acrylic acid)
PVP	Poly(vinyl pyrolidone)
PVA	Poly(vinyl alcohol)
DDA	Dodecylamine
DDT	Dodecanethiol
AA	Acetic acid
OA	Oleylamine
CNF	Cellulose nanofibers
EM	Electromagnetic
UV	Ultra-violet
IR	Infra-red
IPL	Intense pulsed light
MOD	Metal organic compounds
NIR	Near-IR
R2R	Roll to roll
RFID	Radio frequency identification
TEM	Transmission electron microscope
PET	Polyethylene terephthalate
rGO	Reduced graphene oxide
PSA	Prostate specific antigen
PDMS	Polydimethylsiloxane
PI	Polyimide
PEN	Polyethylene naphthalate
RES	Rapid electrical sintering
DC	Direct current
AC	Alternating current
ITO	Indium tin oxide
CNTs	Carbon nanotubes
OLEDs	Organic light emitting diodes
IGZO	Indium gallium zinc oxide
RMS	Root-mean-square
PSA	Prostate specific antigen
EIs	Electrochemical immunosensors
LSPR	Localized surface plasmon response
RFID	Radio frequency identification
MIBK	Methylisobutylketone
Ag NWs	Ag nanowires

References

1. Lee, S.H.; Jun, B.H. Silver Nanoparticles: Synthesis and Application for Nanomedicine. *Int. J. Mol. Sci.* **2019**, *20*, 865. [CrossRef] [PubMed]

2. Burdusel, A.C.; Gherasim, O.; Grumezescu, A.M.; Mogoanta, L.; Ficai, A.; Andronescu, E. Biomedical Applications of Silver Nanoparticles: An Up-to-Date Overview. *Nanomaterials* **2018**, *8*, 681. [CrossRef] [PubMed]

3. Jain, P.K.; Huang, X.; El-Sayed, I.H.; El-Sayed, M.A. Noble Metals on the Nanoscale: Optical and Photothermal Properties and Some Applications in Imaging, Sensing, Biology, and Medicine. *Acc. Chem. Res.* **2008**, *41*, 1578–1586. [CrossRef]

4. Zhang, D.; Tang, Y.; Jiang, F.; Han, Z.; Chen, J. Electrodeposition of silver nanoparticle arrays on transparent conductive oxides. *Appl. Surf. Sci.* **2016**, *369*, 178–182. [CrossRef]

5. Sarina, S.; Waclawik, E.R.; Zhu, H. Photocatalysis on supported gold and silver nanoparticles under ultraviolet and visible light irradiation. *Green Chem.* **2013**, *15*, 1814. [CrossRef]

6. Zhang, P.; Shao, C.; Zhang, Z.; Zhang, M.; Mu, J.; Guo, Z.; Liu, Y. In situ assembly of well-dispersed Ag nanoparticles (AgNPs) on electrospun carbon nanofibers (CNFs) for catalytic reduction of 4-nitrophenol. *Nanoscale* **2011**, *3*, 3357–3363. [CrossRef]

7. Dhakshinamoorthy, A.; Garcia, H. Catalysis by metal nanoparticles embedded on metal-organic frameworks. *Chem. Soc. Rev.* **2012**, *41*, 5262–5284. [CrossRef]

8. Kim, C.; Jeon, H.S.; Eom, T.; Jee, M.S.; Kim, H.; Friend, C.M.; Min, B.K.; Hwang, Y.J. Achieving Selective and Efficient Electrocatalytic Activity for CO_2 Reduction Using Immobilized Silver Nanoparticles. *J. Am. Chem. Soc.* **2015**, *137*, 13844–13850. [CrossRef]

9. Chopra, R.; Sharma, K.; Kumar, M.; Bhalla, V. Pentacenequinone-Stabilized Silver Nanoparticles: A Reusable Catalyst for the Diels-Alder [4 + 2] Cycloaddition Reactions. *J. Org. Chem.* **2016**, *81*, 1039–1046. [CrossRef] [PubMed]

10. Stranahan, S.M.; Titus, E.J.; Willets, K.A. SERS Orientational Imaging of Silver Nanoparticle Dimers. *J. Phys. Chem. Lett.* **2011**, *2*, 2711–2715. [CrossRef]

11. Yan, C.C.; Zhang, D.H.; Li, D.D. Spherical metallic nanoparticle arrays for super-resolution imaging. *J. Appl. Phys.* **2011**, *109*, 063105. [CrossRef]

12. Hsiao, H.H.; Yeh, P.C.; Wang, H.H.; Cheng, T.Y.; Chang, H.C.; Wang, Y.L.; Wang, J.K. Enhancing bright-field image of microorganisms by local plasmon of Ag nanoparticle array. *Opt. Lett.* **2014**, *39*, 1173–1176. [CrossRef]

13. Chernousova, S.; Epple, M. Silver as antibacterial agent: Ion, nanoparticle, and metal. *Angew. Chem. Int. Ed.* **2013**, *52*, 1636–1653. [CrossRef]

14. Ahmed, S.; Ahmad, M.; Swami, B.L.; Ikram, S. A review on plants extract mediated synthesis of silver nanoparticles for antimicrobial applications: A green expertise. *J. Adv. Res.* **2016**, *7*, 17–28. [CrossRef]

15. Hajipour, M.J.; Fromm, K.M.; Ashkarran, A.A.; Jimenez de Aberasturi, D.; de Larramendi, I.R.; Rojo, T.; Serpooshan, V.; Parak, W.J.; Mahmoudi, M. Antibacterial properties of nanoparticles. *Trends Biotechnol.* **2012**, *30*, 499–511. [CrossRef]

16. Dizaj, S.M.; Lotfipour, F.; Barzegar-Jalali, M.; Zarrintan, M.H.; Adibkia, K. Antimicrobial activity of the metals and metal oxide nanoparticles. *Mater. Sci. Eng. C Mater.* **2014**, *44*, 278–284. [CrossRef]

17. Anandhakumar, S.; Mahalakshmi, V.; Raichur, A.M. Silver nanoparticles modified nanocapsules for ultrasonically activated drug delivery. *Mater. Sci. Eng. C* **2012**, *32*, 2349–2355. [CrossRef]

18. Rai, M.; Ingle, A.P.; Gupta, I.; Brandelli, A. Bioactivity of noble metal nanoparticles decorated with biopolymers and their application in drug delivery. *Int. J. Pharm.* **2015**, *496*, 159–172. [CrossRef]

19. Benyettou, F.; Rezgui, R.; Ravaux, F.; Jaber, T.; Blumer, K.; Jouiad, M.; Motte, L.; Olsen, J.C.; Platas-Iglesias, C.; Magzoub, M.; et al. Synthesis of silver nanoparticles for the dual delivery of doxorubicin and alendronate to cancer cells. *J. Mater. Chem. B* **2015**, *3*, 7237–7245. [CrossRef]

20. Nantaphol, S.; Chailapakul, O.; Siangproh, W. A novel paper-based device coupled with a silver nanoparticle-modified boron-doped diamond electrode for cholesterol detection. *Anal. Chim. Acta* **2015**, *891*, 136–143. [CrossRef]

21. Kim, W.; Kim, Y.H.; Park, H.K.; Choi, S. Facile Fabrication of a Silver Nanoparticle Immersed, Surface-Enhanced Raman Scattering Imposed Paper Platform through Successive Ionic Layer Absorption and Reaction for On-Site Bioassays. *ACS. Appl. Mater. Int.* **2015**, *7*, 27910–27917. [CrossRef]

22. Ye, Y.D.; Xia, L.; Xu, D.D.; Xing, X.J.; Pang, D.W.; Tang, H.W. DNA-stabilized silver nanoclusters and carbon nanoparticles oxide: A sensitive platform for label-free fluorescence turn-on detection of HIV-DNA sequences. *Biosens. Bioelectron.* **2016**, *85*, 837–843. [CrossRef]

23. Ge, L.; Li, Q.; Wang, M.; Ouyang, J.; Li, X.; Xing, M.M. Nanosilver particles in medical applications: Synthesis, performance, and toxicity. *Int. J. Nanomed.* **2014**, *9*, 2399–2407. [CrossRef]

24. Taheri, S.; Cavallaro, A.; Christo, S.N.; Smith, L.E.; Majewski, P.; Barton, M.; Hayball, J.D.; Vasilev, K. Substrate independent silver nanoparticle based antibacterial coatings. *Biomaterials* **2014**, *35*, 4601–4609. [CrossRef]

25. Kamyshny, A.; Steinke, J.; Magdassi, S. Metal-based Inkjet Inks for Printed Electronics. *Open Appl. Phys. J.* **2011**, *4*, 19–36. [CrossRef]

26. Karthik, P.S.; Singh, S.P. Conductive silver inks and their applications in printed and flexible electronics. *RSC Adv.* **2015**, *5*, 77760–77790. [CrossRef]

27. Raut, N.C.; Al-Shamery, K. Inkjet printing metals on flexible materials for plastic and paper electronics. *J. Mater. Chem. C* **2018**, *6*, 1618–1641. [CrossRef]

28. Magliulo, M.; Mulla, M.Y.; Singh, M.; Macchia, E.; Tiwari, A.; Torsi, L.; Manoli, K. Printable and flexible electronics: From TFTs to bioelectronic devices. *J. Mater. Chem. C* **2015**, *3*, 12347–12363. [CrossRef]

29. Tran, T.S.; Dutta, N.K.; Choudhury, N.R. Graphene inks for printed flexible electronics: Graphene dispersions, ink formulations, printing techniques and applications. *Adv. Colloid Interface Sci.* **2018**, *261*, 41–61. [CrossRef]

30. Søndergaard, R.R.; Hösel, M.; Krebs, F.C. Roll-to-Roll fabrication of large area functional organic materials. *J. Polym. Sci. Part B Polym. Phys.* **2013**, *51*, 16–34. [CrossRef]

31. Rim, Y.S.; Bae, S.H.; Chen, H.; De Marco, N.; Yang, Y. Recent Progress in Materials and Devices toward Printable and Flexible Sensors. *Adv. Mater.* **2016**, *28*, 4415–4440. [CrossRef] [PubMed]

32. Wu, W. Inorganic nanomaterials for printed electronics: A review. *Nanoscale* **2017**, *9*, 7342–7372. [CrossRef] [PubMed]

33. Garlapati, S.K.; Divya, M.; Breitung, B.; Kruk, R.; Hahn, H.; Dasgupta, S. Printed Electronics Based on Inorganic Semiconductors: From Processes and Materials to Devices. *Adv. Mater.* **2018**, *30*, 1707600. [CrossRef]

34. Kamyshny, A.; Magdassi, S. Conductive Nanomaterials for Printed Electronics. *Small* **2014**, *10*, 3515–3535. [CrossRef]

35. Perelaer, J.; Smith, P.J.; Mager, D.; Soltman, D.; Volkman, S.K.; Subramanian, V.; Korvink, J.G.; Schubert, U.S. Printed electronics: The challenges involved in printing devices, interconnects, and contacts based on inorganic materials. *J. Mater. Chem.* **2010**, *20*, 8446. [CrossRef]

36. Soukupová, J.; Kvítek, L.; Panáček, A.; Nevěčná, T.; Zbořil, R. Comprehensive study on surfactant role on silver nanoparticles (NPs) prepared via modified Tollens process. *Mater. Chem. Phys.* **2008**, *111*, 77–81. [CrossRef]

37. Zhang, Z.; Zhao, B.; Hu, L. PVP Protective Mechanism of Ultrafine Silver Powder Synthesized by Chemical Reduction Processes. *J. Solid State Chem.* **1996**, *121*, 105–110. [CrossRef]

38. Chen, M.; Feng, Y.G.; Wang, X.; Li, T.C.; Zhang, J.Y.; Qian, D.J. Silver nanoparticles capped by oleylamine: Formation, growth, and self-organization. *Langmuir* **2007**, *23*, 5296–5304. [CrossRef]

39. Lovinger, A.J. Development of Electrical Conduction in Silver-filled Epoxy Adhesives. *J. Adhesion* **1979**, *10*, 1–15. [CrossRef]

40. Ruschau, G.R.; Yoshikawa, S.; Newnham, R.E. Resistivities of conductive composites. *J. Appl. Phys.* **1992**, *72*, 953–959. [CrossRef]

41. Shlomo, M.; Michael, G.; Oleg, B.; Alexander, K. Triggering the sintering of silver nanoparticles at room temperature. *ACS Nano* **2010**, *4*, 1943–1948. [CrossRef]

42. Grouchko, M.; Kamyshny, A.; Mihailescu, C.F.; Anghel, D.F.; Magdassi, S. Conductive Inks with a "Built-In" Mechanism That Enables Sintering at Room Temperature. *ACS Nano* **2011**, *5*, 3354–3359. [CrossRef] [PubMed]

43. Perelaer, J.; Jani, R.; Grouchko, M.; Kamyshny, A.; Magdassi, S.; Schubert, U.S. Plasma and microwave flash sintering of a tailored silver nanoparticle ink, yielding 60% bulk conductivity on cost-effective polymer foils. *Adv. Mater.* **2012**, *24*, 3993–3998. [CrossRef] [PubMed]

44. Lee, S.; Kim, J.H.; Wajahat, M.; Jeong, H.; Chang, W.S.; Cho, S.H.; Kim, J.T.; Seol, S.K. Three-dimensional Printing of Silver Microarchitectures Using Newtonian Nanoparticle Inks. *ACS Appl. Mater. Interface* **2017**, *9*, 18918–18924. [CrossRef] [PubMed]

45. Mo, L.; Ran, J.; Yang, L.; Fang, Y.; Zhai, Q.; Li, L. Flexible transparent conductive films combining flexographic printed silver grids with CNT coating. *Nanotechnology* **2016**, *27*, 065202. [CrossRef] [PubMed]

46. Tang, Y.; He, W.; Wang, S.; Tao, Z.; Cheng, L. New insight into the size-controlled synthesis of silver nanoparticles and its superiority in room temperature sintering. *CrystEngComm* **2014**, *16*, 4431–4440. [CrossRef]

47. Khalil, A.M.; Hassan, M.L.; Ward, A.A. Novel nanofibrillated cellulose/polyvinylpyrrolidone/silver nanoparticles films with electrical conductivity properties. *Carbohyd. Polym.* **2017**, *157*, 503–511. [CrossRef]

48. Polavarapu, L.; Manga, K.K.; Cao, H.D.; Loh, K.P.; Xu, Q.-H. Preparation of Conductive Silver Films at Mild Temperatures for Printable Organic Electronics. *Chem. Mater.* **2011**, *23*, 3273–3276. [CrossRef]

49. Kim, D.; Jeong, S.; Shin, H.; Xia, Y.; Moon, J. Heterogeneous Interfacial Properties of Ink-Jet-Printed Silver Nanoparticulate Electrode and Organic Semiconductor. *Adv. Mater.* **2008**, *20*, 3084–3089. [CrossRef]

50. Liu, P.; Chen, W.; Bai, S.; Liu, Y.; Wang, Q. Fabrication of an ultralight flame-induced high conductivity hybrid sponge based on poly (vinyl alcohol)/silver nitrate composite. *Mater. Des.* **2018**, *139*, 96–103. [CrossRef]

51. Liang, K.-L.; Wang, Y.-C.; Lin, W.-L.; Lin, J.-J. Polymer-assisted self-assembly of silver nanoparticles into interconnected morphology and enhanced surface electric conductivity. *RSC Adv.* **2014**, *4*, 15098. [CrossRef]

52. Tobjörk, D.; Aarnio, H.; Pulkkinen, P.; Bollström, R.; Määttänen, A.; Ihalainen, P.; Mäkelä, T.; Peltonen, J.; Toivakka, M.; Tenhu, H.; et al. IR-sintering of ink-jet printed metal-nanoparticles on paper. *Thin Solid Films* **2012**, *520*, 2949–2955. [CrossRef]

53. Mo, L.; Liu, D.; Li, W.; Li, L.; Wang, L.; Zhou, X. Effects of dodecylamine and dodecanethiol on the conductive properties of nano-Ag films. *Appl. Surf. Sci.* **2011**, *257*, 5746–5753. [CrossRef]

54. Volkman, S.K.; Yin, S.; Bakhishev, T.; Puntambekar, K.; Subramanian, V.; Toney, M.F. Mechanistic Studies on Sintering of Silver Nanoparticles. *Chem. Mater.* **2011**, *23*, 4634–4640. [CrossRef]

55. Yuning, L.; Yiliang, W.; Ong, B.S. Facile synthesis of silver nanoparticles useful for fabrication of high-conductivity elements for printed electronics. *J. Am. Chem. Soc.* **2005**, *127*, 3266–3267. [CrossRef]

56. Polavarapu, L.; Manga, K.K.; Yu, K.; Ang, P.K.; Cao, H.D.; Balapanuru, J.; Loh, K.P.; Xu, Q.H. Alkylamine capped metal nanoparticle "inks" for printable SERS substrates, electronics and broadband photodetectors. *Nanoscale* **2011**, *3*, 2268–2274. [CrossRef]

57. Ankireddy, K.; Vunnam, S.; Kellar, J.; Cross, W. Highly conductive short chain carboxylic acid encapsulated silver nanoparticle based inks for direct write technology applications. *J. Mater. Chem. C* **2013**, *1*, 572–579. [CrossRef]

58. Yiliang, W.; Yuning, L.; Ong, B.S. Printed silver ohmic contacts for high-mobility organic thin-film transistors. *J. Am. Chem. Soc.* **2006**, *128*, 4202–4203. [CrossRef]

59. Layani, M.; Grouchko, M.; Shemesh, S.; Magdassi, S. Conductive patterns on plastic substrates by sequential inkjet printing of silver nanoparticles and electrolyte sintering solutions. *J. Mater. Chem.* **2012**, *22*, 14349. [CrossRef]

60. Jung, I.; Shin, K.; Kim, N.R.; Lee, H.M. Synthesis of low-temperature-processable and highly conductive Ag ink by a simple ligand modification: The role of adsorption energy. *J. Mater. Chem. C* **2013**, *1*, 1855. [CrossRef]

61. Buffat, P.; Borel, J.P. Size effect on the melting temperature of gold particles. *Phys. Rev. A* **1976**, *13*, 2287–2298. [CrossRef]

62. Shyjumon, I.; Gopinadhan, M.; Ivanova, O.; Quaas, M.; Wulff, H.; Helm, C.A.; Hippler, R. Structural deformation, melting point and lattice parameter studies of size selected silver clusters. *Eur. Phys. J. D* **2005**, *37*, 409–415. [CrossRef]

63. Durairaj, R.; Man, L.W. Effect of epoxy and filler concentrations on curing behaviour of isotropic conductive adhesives. *J. Therm. Anal. Calorim.* **2011**, *105*, 151–155. [CrossRef]

64. Balantrapu, K.; McMurran, M.; Goia, D.V. Inkjet printable silver dispersions: Effect of bimodal particle-size distribution on film formation and electrical conductivity. *J. Mater. Res.* **2010**, *25*, 821–827. [CrossRef]

65. Ding, J.; Liu, J.; Tian, Q.; Wu, Z.; Yao, W.; Dai, Z.; Liu, L.; Wu, W. Preparing of Highly Conductive Patterns on Flexible Substrates by Screen Printing of Silver Nanoparticles with Different Size Distribution. *Nanoscale Res. Lett.* **2016**, *11*, 412. [CrossRef] [PubMed]

66. Seo, M.; Kim, J.S.; Lee, J.G.; Kim, S.B.; Koo, S.M. The effect of silver particle size and organic stabilizers on the conductivity of silver particulate films in thermal sintering processes. *Thin Solid Films* **2016**, *616*, 366–374. [CrossRef]

67. Han, Y.D.; Zhang, S.M.; Jing, H.Y.; Wei, J.; Bu, F.H.; Zhao, L.; Lv, X.Q.; Xu, L.Y. The fabrication of highly conductive and flexible Ag patterning through baking Ag nanosphere–nanoplate hybrid ink at a low temperature of 100 °C. *Nanotechnology* **2018**, *29*, 135301. [CrossRef]

68. Yang, X.; He, W.; Wang, S.; Zhou, G.; Tang, Y.; Yang, J. Effect of the different shapes of silver particles in conductive ink on electrical performance and microstructure of the conductive tracks. *J. Mater. Sci. Mater. Electron.* **2012**, *23*, 1980–1986. [CrossRef]

69. Lee, C.-L.; Chang, K.-C.; Syu, C.-M. Silver nanoplates as inkjet ink particles for metallization at a low baking temperature of 100 °C. *Colloids Surf. A Physicochem. Eng. Asp.* **2011**, *381*, 85–91. [CrossRef]

70. MacDonald, W.A. Engineered films for display technologies. *J. Mater. Chem.* **2004**, *14*, 4. [CrossRef]

71. Tobjork, D.; Osterbacka, R. Paper electronics. *Adv. Mater.* **2011**, *23*, 1935–1961. [CrossRef]

72. Lee, J.; Kim, J.; Park, J.; Lee, C. Characterization of in situ sintering of silver nanoparticles on commercial photo papers in inkjet printing. *Flex. Print. Electron.* **2018**, *3*, 025001. [CrossRef]

73. Öhlund, T.; Schuppert, A.; Andres, B.; Andersson, H.; Forsberg, S.; Schmidt, W.; Nilsson, H.-E.; Andersson, M.; Zhang, R.; Olin, H. Assisted sintering of silver nanoparticle inkjet ink on paper with active coatings. *RSC Adv.* **2015**, *5*, 64841–64849. [CrossRef]

74. Allen, M.; Leppaniemi, J.; Vilkman, M.; Alastalo, A.; Mattila, T. Substrate-facilitated nanoparticle sintering and component interconnection procedure. *Nanotechnology* **2010**, *21*, 475204. [CrossRef]

75. Andersson, H.; Manuilskiy, A.; Lidenmark, C.; Gao, J.; Ohlund, T.; Forsberg, S.; Ortegren, J.; Schmidt, W.; Nilsson, H.E. The influence of paper coating content on room temperature sintering of silver nanoparticle ink. *Nanotechnology* **2013**, *24*, 455203. [CrossRef] [PubMed]

76. Nge, T.T.; Nogi, M.; Suganuma, K. Electrical functionality of inkjet-printed silver nanoparticle conductive tracks on nanostructured paper compared with those on plastic substrates. *J. Mater. Chem. C* **2013**, *1*, 5235. [CrossRef]

77. Wünscher, S.; Abbel, R.; Perelaer, J.; Schubert, U.S. Progress of alternative sintering approaches of inkjet-printed metal inks and their application for manufacturing of flexible electronic devices. *J. Mater. Chem. C* **2014**, *2*, 10232–10261. [CrossRef]

78. Denneulin, A.; Blayo, A.; Neuman, C.; Bras, J. Infra-red assisted sintering of inkjet printed silver tracks on paper substrates. *J. Nanoparticle Res.* **2011**, *13*, 3815–3823. [CrossRef]

79. Cherrington, M.; Claypole, T.C.; Deganello, D.; Mabbett, I.; Watson, T.; Worsley, D. Ultrafast near-infrared sintering of a slot-die coated nano-silver conducting ink. *J. Mater. Chem.* **2011**, *21*, 7562. [CrossRef]

80. Gu, W.; Yuan, W.; Zhong, T.; Wu, X.; Zhou, C.; Lin, J.; Cui, Z. Fast near infrared sintering of silver nanoparticle ink and applications for flexible hybrid circuits. *RSC Adv.* **2018**, *8*, 30215–30222. [CrossRef]

81. Sowade, E.; Kang, H.; Mitra, K.Y.; Weiß, O.J.; Weber, J.; Baumann, R.R. Roll-to-roll infrared (IR) drying and sintering of an inkjet-printed silver nanoparticle ink within 1 second. *J. Mater. Chem. C* **2015**, *3*, 11815–11826. [CrossRef]

82. Kang, J.S.; Ryu, J.; Kim, H.S.; Hahn, H.T. Sintering of Inkjet-Printed Silver Nanoparticles at Room Temperature Using Intense Pulsed Light. *J. Electron. Mater.* **2011**, *40*, 2268–2277. [CrossRef]

83. Chung, W.H.; Hwang, H.J.; Lee, S.H.; Kim, H.S. In situ monitoring of a flash light sintering process using silver nano-ink for producing flexible electronics. *Nanotechnology* **2013**, *24*, 035202. [CrossRef]

84. Abbel, R.; van Lammeren, T.; Hendriks, R.; Ploegmakers, J.; Rubingh, E.J.; Meinders, E.R.; Groen, W.A. Photonic flash sintering of silver nanoparticle inks: A fast and convenient method for the preparation of highly conductive structures on foil. *MRS Commun.* **2012**, *2*, 145–150. [CrossRef]

85. Lee, D.J.; Park, S.H.; Jang, S.; Kim, H.S.; Oh, J.H.; Song, Y.W. Pulsed light sintering characteristics of inkjet-printed nanosilver films on a polymer substrate. *J. Micromech. Microeng.* **2011**, *21*, 125023. [CrossRef]

86. Sarkar, S.K.; Gupta, H.; Gupta, D. Flash Light Sintering of Silver Nanoink for Inkjet-Printed Thin-Film Transistor on Flexible Substrate. *IEEE Trans. Nanotechnol.* **2017**, *16*, 375–382. [CrossRef]

87. Weise, D.; Mitra, K.Y.; Sowade, E.; Baumann, R.R. Intense Pulsed Light Sintering of Inkjet Printed Silver Nanoparticle Ink: Influence of Flashing Parameters and Substrate. *Proc. MRS* **2015**, *1761*. [CrossRef]

88. Balliu, E.; Andersson, H.; Engholm, M.; Ohlund, T.; Nilsson, H.E.; Olin, H. Selective laser sintering of inkjet-printed silver nanoparticle inks on paper substrates to achieve highly conductive patterns. *Sci. Rep.* **2018**, *8*, 10408. [CrossRef]

89. Yeo, J.; Kim, G.; Hong, S.; Kim, M.S.; Kim, D.; Lee, J.; Lee, H.B.; Kwon, J.; Suh, Y.D.; Kang, H.W.; et al. Flexible supercapacitor fabrication by room temperature rapid laser processing of roll-to-roll printed metal nanoparticle ink for wearable electronics application. *J. Power Sources* **2014**, *246*, 562–568. [CrossRef]

90. Bolduc, M.; Trudeau, C.; Beaupre, P.; Cloutier, S.G.; Galarneau, P. Thermal Dynamics Effects using Pulse-Shaping Laser Sintering of Printed Silver Inks. *Sci. Rep.* **2018**, *8*, 1418. [CrossRef]

91. Ko, S.H.; Pan, H.; Grigoropoulos, C.P.; Luscombe, C.K.; Fréchet, J.M.J.; Poulikakos, D. All-inkjet-printed flexible electronics fabrication on a polymer substrate by low-temperature high-resolution selective laser sintering of metal nanoparticles. *Nanotechnology* **2007**, *18*, 345202. [CrossRef]

92. Hong, S.; Yeo, J.; Kim, G.; Kim, D.; Lee, H.; Kwon, J.; Lee, H.; Lee, P.; Ko, S.H. Nonvacuum, Maskless Fabrication of a Flexible Metal Grid Transparent Conductor by Low-Temperature Selective Laser Sintering of Nanoparticle Ink. *ACS Nano.* **2013**, *7*, 5024–5031. [CrossRef]

93. Allen, M.L. *Nanoparticle Sintering Methods and Applications for Printed Electronics*; Aalto University: Espoo/Helsinki, Finland, 2011.

94. Allen, M.L.; Aronniemi, M.; Mattila, T.; Alastalo, A.; Ojanpera, K.; Suhonen, M.; Seppa, H. Electrical sintering of nanoparticle structures. *Nanotechnology* **2008**, *19*, 175201. [CrossRef] [PubMed]

95. Reinhold, I.; Hendriks, C.E.; Eckardt, R.; Kranenburg, J.M.; Perelaer, J.; Baumann, R.R.; Schubert, U.S. Argon plasma sintering of inkjet printed silver tracks on polymer substrates. *J. Mater. Chem.* **2009**, *19*, 3384. [CrossRef]

96. Wolf, F.M.; Perelaer, J.; Stumpf, S.; Bollen, D.; Kriebel, F.; Schubert, U.S. Rapid low-pressure plasma sintering of inkjet-printed silver nanoparticles for RFID antennas. *J. Mater. Res.* **2013**, *28*, 1254–1261. [CrossRef]

97. Wünscher, S.; Stumpf, S.; Teichler, A.; Pabst, O.; Perelaer, J.; Beckert, E.; Schubert, U.S. Localized atmospheric plasma sintering of inkjet printed silver nanoparticles. *J. Mater. Chem.* **2012**, *22*, 24569. [CrossRef]

98. Wünscher, S.; Stumpf, S.; Perelaer, J.; Schubert, U.S. Towards single-pass plasma sintering: Temperature influence of atmospheric pressure plasma sintering of silver nanoparticle ink. *J. Mater. Chem. C* **2014**, *2*, 1642. [CrossRef]

99. Ma, S.; Bromberg, V.; Liu, L.; Egitto, F.D.; Chiarot, P.R.; Singler, T.J. Low temperature plasma sintering of silver nanoparticles. *Appl. Surf. Sci.* **2014**, *293*, 207–215. [CrossRef]

100. Perelaer, J.; de Gans, B.J.; Schubert, U.S. Ink-jet Printing and Microwave Sintering of Conductive Silver Tracks. *Adv. Mater.* **2006**, *18*, 2101–2104. [CrossRef]

101. Perelaer, J.; Klokkenburg, M.; Hendriks, C.E.; Schubert, U.S. Microwave flash sintering of inkjet-printed silver tracks on polymer substrates. *Adv. Mater.* **2009**, *21*, 4830–4834. [CrossRef]

102. Perelaer, J.; Abbel, R.; Wunscher, S.; Jani, R.; van Lammeren, T.; Schubert, U.S. Roll-to-roll compatible sintering of inkjet printed features by photonic and microwave exposure: From non-conductive ink to 40% bulk silver conductivity in less than 15 seconds. *Adv. Mater.* **2012**, *24*, 2620–2625. [CrossRef]

103. Layani, M.; Gruchko, M.; Milo, O.; Balberg, I.; Azulay, D.; Magdassi, S. Transparent Conductive Coatings by Printing coffee Ring Arrays Obtained at Room Temperature. *ACS Nano* **2009**, *3*, 3537–3542. [CrossRef]

104. Ahn, B.Y.; Lorang, D.J.; Lewis, J.A. Transparent conductive grids via direct writing of silver nanoparticle inks. *Nanoscale* **2011**, *3*, 2700–2702. [CrossRef]

105. Deganello, D.; Cherry, J.A.; Gethin, D.T.; Claypole, T.C. Patterning of micro-scale conductive networks using reel-to-reel flexographic printing. *Thin Solid Films* **2010**, *518*, 6113–6116. [CrossRef]

106. Kahng, Y.H.; Kim, M.-K.; Lee, J.-H.; Kim, Y.J.; Kim, N.; Park, D.-W.; Lee, K. Highly conductive flexible transparent electrodes fabricated by combining graphene films and inkjet-printed silver grids. *Sol. Energy Mater. Sol. C* **2014**, *124*, 86–91. [CrossRef]

107. Jeong, J.-A.; Kim, J.; Kim, H.-K. Ag grid/ITO hybrid transparent electrodes prepared by inkjet printing. *Sol. Energy Mater. Sol. C* **2011**, *95*, 1974–1978. [CrossRef]

108. Li, Y.; Mao, L.; Gao, Y.; Zhang, P.; Li, C.; Ma, C.; Tu, Y.; Cui, Z.; Chen, L. ITO-free photovoltaic cell utilizing a high-resolution silver grid current collecting layer. *Sol. Energy Mater Sol. C* **2013**, *113*, 85–89. [CrossRef]

109. Mao, L.; Chen, Q.; Li, Y.; Li, Y.; Cai, J.; Su, W.; Bai, S.; Jin, Y.; Ma, C.-Q.; Cui, Z.; et al. Flexible silver grid/PEDOT:PSS hybrid electrodes for large area inverted polymer solar cells. *Nano Energy* **2014**, *10*, 259–267. [CrossRef]

110. Cai, G.; Darmawan, P.; Cui, M.; Wang, J.; Chen, J.; Magdassi, S.; Lee, P.S. Highly Stable Transparent Conductive Silver Grid/PEDOT:PSS Electrodes for Integrated Bifunctional Flexible Electrochromic Supercapacitors. *Adv. Energy Mater.* **2016**, *6*, 1501882. [CrossRef]

111. Fukuda, K.; Someya, T. Recent Progress in the Development of Printed Thin-Film Transistors and Circuits with High-Resolution Printing Technology. *Adv. Mater.* **2017**, *29*, 1602736. [CrossRef]

112. Wu, X.; Zhou, J.; Huang, J. Integration of Biomaterials into Sensors Based on Organic Thin-Film Transistors. *Macromol. Rapid Commun.* **2018**, *39*, 1800084. [CrossRef] [PubMed]

113. Liang, X.; Xia, J.; Dong, G.; Tian, B.; Peng, L. Carbon Nanotube Thin Film Transistors for Flat Panel Display Application. *Top. Curr. Chem.* **2016**, *374*, 80. [CrossRef]

114. Tong, S.C.; Sun, J.; Yang, J.L. Printed Thin-Film Transistors: Research from China. *ACS Appl. Mater. Interface* **2018**, *10*, 25902–25924. [CrossRef] [PubMed]

115. Hong, G.R.; Lee, S.S.; Park, H.J.; Jo, Y.; Kim, J.Y.; Lee, H.S.; Kang, Y.C.; Ryu, B.H.; Song, A.; Chung, K.B.; et al. Unraveling the Issue of Ag Migration in Printable Source/Drain Electrodes Compatible with Versatile Solution-Processed Oxide Semiconductors for Printed Thin-Film Transistor Applications. *ACS Appl. Mater. Interface* **2017**, *9*, 14058–14066. [CrossRef] [PubMed]

116. Ning, H.; Chen, J.; Fang, Z.; Tao, R.; Cai, W.; Yao, R.; Hu, S.; Zhu, Z.; Zhou, Y.; Yang, C.; et al. Direct Inkjet Printing of Silver Source/Drain Electrodes on an Amorphous InGaZnO Layer for Thin-Film Transistors. *Materials* **2017**, *10*, 51. [CrossRef] [PubMed]

117. Feng, L.R.; Jiang, C.; Ma, H.B.; Guo, X.J.; Nathan, A. All ink-jet printed low-voltage organic field-effect transistors on flexible substrate. *Org. Electron.* **2016**, *38*, 186–192. [CrossRef]

118. Castro, H.F.; Sowade, E.; Rocha, J.G.; Alpuim, P.; Lanceros-Mendez, S.; Baumann, R.R. All-Inkjet-Printed Bottom-Gate Thin-Film Transistors Using UV Curable Dielectric for Well-Defined Source-Drain Electrodes. *J. Electron. Mater.* **2014**, *43*, 2631–2636. [CrossRef]

119. Fukuda, K.; Sekine, T.; Kumaki, D.; Tokito, S. Profile Control of Inkjet Printed Silver Electrodes and Their Application to Organic Transistors. *ACS Appl. Mater. Interface* **2013**, *5*, 3916–3920. [CrossRef]

120. Liu, T.T.; Zhao, J.W.; Xu, W.W.; Dou, J.Y.; Zhao, X.L.; Deng, W.; Wei, C.T.; Xu, W.Y.; Guo, W.R.; Su, W.M.; et al. Flexible integrated diode-transistor logic (DTL) driving circuits based on printed carbon nanotube thin film transistors with low operation voltage. *Nanoscale* **2018**, *10*, 614–622. [CrossRef]

121. Tu, L.; Yuan, S.J.; Zhang, H.T.; Wang, P.F.; Cui, X.L.; Wang, J.; Zhan, Y.Q.; Zheng, L.R. Aerosol jet printed silver nanowire transparent electrode for flexible electronic application. *J. Appl. Phys.* **2018**, *123*, 174905. [CrossRef]

122. Cardenas, J.A.; Upshaw, S.; Williams, N.X.; Catenacci, M.J.; Wiley, B.J.; Franklin, A.D. Impact of Morphology on Printed Contact Performance in Carbon Nanotube Thin-Film Transistors. *Adv. Funct. Mater.* **2019**, *29*, 1805727. [CrossRef]

123. Xu, W.W.; Liu, Z.; Zhao, J.W.; Xu, W.Y.; Gu, W.B.; Zhang, X.; Qian, L.; Cui, Z. Flexible logic circuits based on top-gate thin film transistors with printed semiconductor carbon nanotubes and top electrodes. *Nanoscale* **2014**, *6*, 14891–14897. [CrossRef]

124. Shimura, Y.; Nomura, K.; Yanagi, H.; Kamiya, T.; Hirano, M.; Hosono, H. Specific contact resistances between amorphous oxide semiconductor In-Ga-Zn-O and metallic electrodes. *Thin Solid Films* **2008**, *516*, 5899–5902. [CrossRef]

125. Aoshima, K.; Arai, S.; Fukuhara, K.; Yamada, T.; Hasegawa, T. Surface modification of printed silver electrodes for efficient carrier injection in organic thin-film transistors. *Org. Electron.* **2017**, *41*, 137–142. [CrossRef]

126. Ji, D.Y.; Jiang, L.; Dong, H.L.; Meng, Q.; Zhen, Y.G.; Hu, W.P. Silver mirror reaction for organic electronics: Towards high-performance organic field-effect transistors and circuits. *J. Mater. Chem. C* **2014**, *2*, 4142–4146. [CrossRef]

127. Choi, S.; Larrain, F.A.; Wang, C.Y.; Fuentes-Hernandez, C.; Chou, W.F.; Kippelen, B. Self-forming electrode modification in organic field-effect transistors. *J. Mater. Chem. C* **2016**, *4*, 8297–8303. [CrossRef]

128. Fukuda, K.; Minamiki, T.; Minami, T.; Watanabe, M.; Fukuda, T.; Kumaki, D.; Tokito, S. Printed Organic Transistors with Uniform Electrical Performance and Their Application to Amplifiers in Biosensors. *Adv. Electron. Mater.* **2015**, *1*, 1400052. [CrossRef]

129. Park, S.K.; Kim, Y.H.; Han, J.I. All solution-processed high-resolution bottom-contact transparent metal-oxide thin film transistors. *J. Phys. D Appl. Phys.* **2009**, *42*, 125102. [CrossRef]

130. Hwang, J.K.; Cho, S.; Dang, J.M.; Kwak, E.B.; Song, K.; Moon, J.; Sung, M.M. Direct nanoprinting by liquid-bridge-mediated nanotransfer moulding. *Nat. Nanotechnol.* **2010**, *5*, 742–748. [CrossRef]

131. Ueoka, Y.; Nishibayashi, T.; Ishikawa, Y.; Yamazaki, H.; Osada, Y.; Horita, M.; Uraoka, Y. Analysis of printed silver electrode on amorphous indium gallium zinc oxide. *Jpn. J. Appl. Phys.* **2014**, *53*, 125102. [CrossRef]

132. Ando, E.; Miyazaki, M. Moisture degradation mechanism of silver-based low-emissivity coatings. *Thin Solid Films* **1999**, *351*, 308–312. [CrossRef]

133. Chen, J.Q.; Ning, H.L.; Fang, Z.Q.; Tao, R.Q.; Yang, C.G.; Zhou, Y.C.; Yao, R.H.; Xu, M.; Wang, L.; Peng, J.B. Reduced contact resistance of a-IGZO thin film transistors with inkjet-printed silver electrodes. *J. Phys. D Appl. Phys.* **2018**, *51*, 165103. [CrossRef]

134. Zhou, Y.H.; Fuentes-Hernandez, C.; Shim, J.; Meyer, J.; Giordano, A.J.; Li, H.; Winget, P.; Papadopoulos, T.; Cheun, H.; Kim, J.; et al. A Universal Method to Produce Low-Work Function Electrodes for Organic Electronics. *Science* **2012**, *336*, 327–332. [CrossRef]

135. Gillan, L.; Leppaniemi, J.; Eiroma, K.; Majumdar, H.; Alastalo, A. High performance solution processed oxide thin-film transistors with inkjet printed Ag source-drain electrodes. *J. Mater. Chem. C* **2018**, *6*, 3220–3225. [CrossRef]

136. Secor, E.B.; Smith, J.; Marks, T.J.; Hersam, M.C. High-Performance Inkjet-Printed Indium-Gallium-Zinc-Oxide Transistors Enabled by Embedded, Chemically Stable Graphene Electrodes. *ACS Appl. Mater. Interface* **2016**, *8*, 17428–17434. [CrossRef]

137. Tang, W.; Feng, L.R.; Zhao, J.Q.; Cui, Q.Y.; Chen, S.J.; Guo, X.J. Inkjet printed fine silver electrodes for all-solution-processed low-voltage organic thin film transistors. *J. Mater. Chem. C* **2014**, *2*, 1995–2000. [CrossRef]

138. Sekitani, T.; Noguchi, Y.; Zschieschang, U.; Klauk, H.; Someya, T. Organic transistors manufactured using inkjet technology with subfemtoliter accuracy. *Proc. Natl. Acad. Sci. USA* **2008**, *105*, 4976–4980. [CrossRef] [PubMed]

139. Park, J.U.; Hardy, M.; Kang, S.J.; Barton, K.; Adair, K.; Mukhopadhyay, D.K.; Lee, C.Y.; Strano, M.S.; Alleyne, A.G.; Georgiadis, J.G.; et al. High-resolution electrohydrodynamic jet printing. *Nat. Mater.* **2007**, *6*, 782–789. [CrossRef]

140. Zhang, H.; Ramm, A.; Lim, S.M.; Xie, W.; Ahn, B.Y.; Xu, W.C.; Mahajan, A.; Suszynski, W.J.; Kim, C.; Lewis, J.A.; et al. Wettability Contrast Gravure Printing. *Adv. Mater.* **2015**, *27*, 7420–7425. [CrossRef]

141. Fukuda, K.; Yoshimura, Y.; Okamoto, T.; Takeda, Y.; Kumaki, D.; Katayama, Y.; Tokito, S. Reverse-Offset Printing Optimized for Scalable Organic Thin-Film Transistors with Submicrometer Channel Lengths. *Adv. Electron. Mater.* **2015**, *1*, 1500145. [CrossRef]

142. Ning, H.L.; Tao, R.Q.; Fang, Z.Q.; Cai, W.; Chen, J.Q.; Zhou, Y.C.; Zhu, Z.A.; Zheng, Z.K.; Yao, R.H.; Xu, M.; et al. Direct patterning of silver electrodes with 2.4 mu m channel length by piezoelectric inkjet printing. *J. Colloids Interface Sci.* **2017**, *487*, 68–72. [CrossRef]

143. Tao, R.Q.; Fang, Z.Q.; Zhang, J.H.; Ning, H.L.; Chen, J.Q.; Yang, C.G.; Zhou, Y.C.; Yao, R.H.; Song, Y.S.; Peng, J.B. Capillary force induced air film for self-aligned short channel: Pushing the limits of inkjet printing. *Soft Matter* **2018**, *14*, 9402–9410. [CrossRef]

144. Han, L.; Liu, C.M.; Dong, S.L.; Du, C.X.; Zhang, X.Y.; Li, L.H.; Wei, Y. Enhanced conductivity of rGO/Ag NPs composites for electrochemical immunoassay of prostate-specific antigen. *Biosens. Bioelectron.* **2017**, *87*, 466–472. [CrossRef]

145. Rad, A.S.; Jahanshahi, M.; Ardjmand, M.; Safekordi, A.-A. Hydrogen Peroxide Biosensor Based on Enzymatic Modification of Electrode Using Deposited Silver Nano Layer. *Int. J. Electrochem. Sc.* **2012**, *7*, 2623–2632.

146. Wang, C.; Sun, Y.; Yu, X.; Ma, D.; Zheng, J.; Dou, P.; Cao, Z.; Xu, X. Ag–Pt hollow nanoparticles anchored reduced graphene oxide composites for non-enzymatic glucose biosensor. *J. Mater. Sci. Mater. Electron.* **2016**, *27*, 9370–9378. [CrossRef]

147. Misra, N.; Goel, N.K.; Varshney, L.; Kumar, V. Radiolytically Synthesized Noble Metal Nanoparticles: Sensor Applications. In *Reviews in Plasmonics 2015*; Geddes, C.D., Ed.; Springer International Publishing: Cham, Switzerland, 2016; pp. 51–67.

148. He, H.; Sydänheimo, L.; Virkki, J.; Ukkonen, L. Experimental Study on Inkjet-Printed Passive UHF RFID Tags on Versatile Paper-Based Substrates. *Int. J. Antenn. Propag.* **2016**, *2016*, 9265159. [CrossRef]

149. Ren, Y.; Sydänheimo, L.; Virkki, J.; Ukkonen, L. Optimisation of manufacturing parameters for inkjet-printed and photonically sintered metallic nanoparticle UHF RFID tags. *Electron. Lett.* **2014**, *50*, 1504–1505. [CrossRef]

150. Salmerón, J.F.; Molina-Lopez, F.; Briand, D.; Ruan, J.J.; Rivadeneyra, A.; Carvajal, M.A.; Capitán-Vallvey, L.F.; de Rooij, N.F.; Palma, A.J. Properties and Printability of Inkjet and Screen-Printed Silver Patterns for RFID Antennas. *J. Electron. Mater.* **2013**, *43*, 604–617. [CrossRef]

151. Jung, M.; Kim, J.; Noh, J.; Lim, N.; Lim, C.; Lee, G.; Kim, J.; Kang, H.; Jung, K.; Leonard, A.D.; et al. All-Printed and Roll-to-Roll-PrinTable 13.56-MHz-Operated 1-bit RF Tag on Plastic Foils. *IEEE Trans. Electron. Devices* **2010**, *57*, 571–580. [CrossRef]

152. Sanchez-Romaguera, V.; Wünscher, S.; Turki, B.M.; Abbel, R.; Barbosa, S.; Tate, D.J.; Oyeka, D.; Batchelor, J.C.; Parker, E.A.; Schubert, U.S.; et al. Inkjet printed paper based frequency selective surfaces and skin mounted RFID tags: The interrelation between silver nanoparticle ink, paper substrate and low temperature sintering technique. *J. Mater. Chem. C* **2015**, *3*, 2132–2140. [CrossRef]

153. Sanchez-Romaguera, V.; Ziai, M.A.; Oyeka, D.; Barbosa, S.; Wheeler, J.S.R.; Batchelor, J.C.; Parker, E.A.; Yeates, S.G. Towards inkjet-printed low cost passive UHF RFID skin mounted tattoo paper tags based on silver nanoparticle inks. *J. Mater. Chem. C* **2013**, *1*, 6395. [CrossRef]

154. Matsuhisa, N.; Inoue, D.; Zalar, P.; Jin, H.; Matsuba, Y.; Itoh, A.; Yokota, T.; Hashizume, D.; Someya, T. Printable elastic conductors by in situ formation of silver nanoparticles from silver flakes. *Nat. Mater.* **2017**, *16*, 834–840. [CrossRef]

155. Park, M.; Im, J.; Shin, M.; Min, Y.; Park, J.; Cho, H.; Park, S.; Shim, M.B.; Jeon, S.; Chung, D.Y.; et al. Highly stretchable electric circuits from a composite material of silver nanoparticles and elastomeric fibres. *Nat. Nanotechnol.* **2012**, *7*, 803–809. [CrossRef]

156. Chung, S.; Lee, J.; Song, H.; Kim, S.; Jeong, J.; Hong, Y. Inkjet-printed stretchable silver electrode on wave structured elastomeric substrate. *Appl. Phys. Lett.* **2011**, *98*, 153110. [CrossRef]

157. Albrecht, A.; Bobinger, M.; Salmerón, J.; Becherer, M.; Cheng, G.; Lugli, P.; Rivadeneyra, A. Over-Stretching Tolerant Conductors on Rubber Films by Inkjet-Printing Silver Nanoparticles for Wearables. *Polymers* **2018**, *10*, 1413. [CrossRef]

158. Chun, K.Y.; Oh, Y.; Rho, J.; Ahn, J.H.; Kim, Y.J.; Choi, H.R.; Baik, S. Highly conductive, printable and stretchable composite films of carbon nanotubes and silver. *Nat. Nanotechnol.* **2010**, *5*, 853–857. [CrossRef]

159. Kim, J.; Wang, Z.; Kim, W.S. Stretchable RFID for Wireless Strain Sensing With Silver Nano Ink. *IEEE Sens. J.* **2014**, *14*, 4395–4401. [CrossRef]

160. Matsuhisa, N.; Kaltenbrunner, M.; Yokota, T.; Jinno, H.; Kuribara, K.; Sekitani, T.; Someya, T. Printable elastic conductors with a high conductivity for electronic textile applications. *Nat. Commun.* **2015**, *6*, 7461. [CrossRef]

161. Mohammed, A.; Pecht, M. A stretchable and screen-printable conductive ink for stretchable electronics. *Appl. Phys. Lett.* **2016**, *109*, 184101. [CrossRef]

162. Park, J.Y.; Lee, W.J.; Kwon, B.-S.; Nam, S.-Y.; Choa, S.-H. Highly stretchable and conductive conductors based on Ag flakes and polyester composites. *Microelectron. Eng.* **2018**, *199*, 16–23. [CrossRef]

163. Araki, T.; Nogi, M.; Suganuma, K.; Kogure, M.; Kirihara, O. Printable and Stretchable Conductive Wirings Comprising Silver Flakes and Elastomers. *IEEE Electron. Device Lett.* **2011**, *32*, 1424–1426. [CrossRef]

164. Kumar, A.; Saghlatoon, H.; La, T.-G.; Mahdi Honari, M.; Charaya, H.; Abu Damis, H.; Mirzavand, R.; Mousavi, P.; Chung, H.-J. A highly deformable conducting traces for printed antennas and interconnects: Silver/fluoropolymer composite amalgamated by triethanolamine. *Flex. Print. Electron.* **2017**, *2*, 045001. [CrossRef]

165. Zhang, S.; Li, Y.; Tian, Q.; Liu, L.; Yao, W.; Chi, C.; Zeng, P.; Zhang, N.; Wu, W. Highly conductive, flexible and stretchable conductors based on fractal silver nanostructures. *J. Mater. Chem. C* **2018**, *6*, 3999–4006. [CrossRef]

166. Tybrandt, K.; Voros, J. Fast and Efficient Fabrication of Intrinsically Stretchable Multilayer Circuit Boards by Wax Pattern Assisted Filtration. *Small* **2016**, *12*, 180–184. [CrossRef]

167. Liang, J.; Tong, K.; Pei, Q. A Water-Based Silver-Nanowire Screen-Print Ink for the Fabrication of Stretchable Conductors and Wearable Thin-Film Transistors. *Adv. Mater.* **2016**, *28*, 5986–5996. [CrossRef]

168. Lee, S.; Shin, S.; Lee, S.; Seo, J.; Lee, J.; Son, S.; Cho, H.J.; Algadi, H.; Al-Sayari, S.; Kim, D.E.; et al. Ag Nanowire Reinforced Highly Stretchable Conductive Fibers for Wearable Electronics. *Adv. Funct. Mater.* **2015**, *25*, 3114–3121. [CrossRef]

169. Lee, I.; Lee, J.; Ko, S.H.; Kim, T.S. Reinforcing Ag nanoparticle thin films with very long Ag nanowires. *Nanotechnology* **2013**, *24*, 415704. [CrossRef]

170. Xu, F.; Zhu, Y. Highly conductive and stretchable silver nanowire conductors. *Adv. Mater.* **2012**, *24*, 5117–5122. [CrossRef]

MDPI

St. Alban-Anlage 66

4052 Basel

Switzerland

Tel. +41 61 683 77 34

Fax +41 61 302 89 18

www.mdpi.com

International Journal of Molecular Sciences Editorial Office

E-mail: ijms@mdpi.com

www.mdpi.com/journal/ijms

www.ingramcontent.com/pod-product-compliance
Lightning Source LLC
Chambersburg PA
CBHW051848210326
41597CB00033B/5824